T0305776

Bioelectrochemical Oxidation Processes for Wastewater Treatment

Toxic and hazardous pollutant treatment of wastewater is a longstanding challenge faced in every region across the globe. Growing urbanization, combined with the increased use of detergent soaps, cleansing agents with new formulations, chemical fertilizers, and pesticides, has greatly added to the global wastewater pollution burden. Conventional methods of wastewater treatment are somewhat successful in decontamination, but these current techniques require more time and energy than newer, novel techniques. Bioelectrochemical oxidation systems (BEOSs), for example, have greatly aided in wastewater treatment sustainability and efficiency, and offer promising solutions for different types of energy recovery options. *Bioelectrochemical Oxidation Processes for Wastewater Treatment* examines the latest hybrid technologies utilizing algae, bacteria, and various other chemical agents, and discusses the major challenges in large-scale operations, as well as forward-looking techniques to make treatment processes even more sustainable in the future. It:

- Discusses the fundamentals of biological wastewater treatment and bioelectrochemical oxidation systems, as well as their advantages and limitations.
- Presents the recent trends and developments in bioelectrochemical oxidation systems for achieving environmentally sustainable wastewater treatment.
- Describes carbon capture and resource recovery from wastewater using bioelectrochemical oxidation systems.
- Addresses the challenges of large-scale implementation of bioelectrochemical oxidation systems in existing and new wastewater treatment plants.

Dr. Maulin P. Shah has been an active researcher and scientific writer in his field for over 20 years. He received a B.Sc. degree in Microbiology from Gujarat University, Godhra (Gujarat), India (1999), and earned his Ph.D. degree in Environmental Microbiology from Sardar Patel University, Vallabh Vidyanagar (Gujarat) India (2005). His research interests include biological wastewater treatment, environmental microbiology, biodegradation, bioremediation, and phytoremediation of environmental pollutants from industrial wastewaters. He has published more than 350 research papers in national and international journals on various aspects of microbial biodegradation and bioremediation of environmental pollutants and is the editor of 150 books of international repute. He has edited 25 special issues specifically in the areas of industrial wastewater research, microbial remediation, and biorefinery of wastewater treatment. He is associated as an Editorial Board Member with 25 highly reputed journals.

Bioelectrochemical Oxidation Processes for Wastewater Treatment

Edited by
Maulin P. Shah

CRC Press is an imprint of the
Taylor & Francis Group, an **informa** business

Designed cover image: Shutterstock

First edition published 2025
by CRC Press
2385 NW Executive Center Drive, Suite 320, Boca Raton FL 33431

and by CRC Press
4 Park Square, Milton Park, Abingdon, Oxon, OX14 4RN

CRC Press is an imprint of Taylor & Francis Group, LLC

ISBN: 978-1-032-43695-1 (hbk)
ISBN: 978-1-032-43696-8 (pbk)
ISBN: 978-1-003-36847-2 (ebk)

DOI: 10.1201/9781003368472

Typeset in Times
by KnowledgeWorks Global Ltd.

Contents

List of Contributors vii

Chapter 1 Recovery and Removal of Heavy Metals from Industrial Effluents 1

Chapter 2 Microbial Electrochemical Systems-Anaerobic Digestion-A Hybrid System for Industrial Effluent Treatment and Energy Recovery 13

Chapter 3 Microbes and Wastewater Treatment 45

Chapter 4 Bioelectrochemical System-Based Removal of Hazardous Pollutants from Wastewater 63

Chapter 5 Microbial Carbon Capture Cell: Carbon Capture, Energy Recovery, and Wastewater Treatment 83

Chapter 6 Sewage Treatment and Energy Recovery by Bioelectrochemical Oxidation System 100

Chapter 7 Constructed Wetland-Bioelectrochemical Oxidation Systems: A Hybrid System for Wastewater Treatment 126

Chapter 8 Current Status of Wastewater in India/Other Countries/Regions 152

Chapter 9 Implementation of Bioelectrochemical Oxidation Systems in Existing Wastewater Treatment Plant: Challenges in Retrofitting 166

Chapter 10 Microbial Desalination Cell 189

Chapter 11 Treatment of Wastewater and Energy Recovery by Bioelectrochemical Oxidation Systems 218

Chapter 12 Bioelectrochemical Oxidation Systems: Basic Concepts and Types 239

Index 273

List of Contributors

Khan Muhammad Adil Nawaz
Quaid-i-Azam University,
Islamabad, Pakistan.

Chatterjee Ankita
REVA University, Bengaluru,
Karnataka, India.

Anurupa Anwesha
SRM Institute of Science and
Technology, Chengalpattu,
Tamil Nadu, India.

JohnRajan Jemina
SRM Institute of Science and
Technology, Chengalpattu,
Tamil Nadu, India.

Panwar Joginder Singh
Lovely Professional University,
Phagwara, Punjab, India.

Kavya L.
REVA University, Bangalore,
Karnataka, India.

Pathi Thulluru Lakshmi
Indian Institute of Technology,
Kharagpur, West Bengal, India.

Saini Lalit
Lovely Professional University,
Phagwara, Punjab, India.

Ramya T.R.
REVA University, Bangalore,
Karnataka, India.

Ghangrekar Makarand M.
Indian Institute of Technology,
Kharagpur, West Bengal, India.

Badshah Malik
Quaid-i-Azam University,
Islamabad, Pakistan.

Sharma Monika
Lovely Professional University,
Phagwara, Punjab, India.

Ali Naeem
Quaid-i-Azam University,
Islamabad, Pakistan.

Reddy Neema
REVA University, Bangalore,
Karnataka, India.

Padmanabhan Padmini
Birla Institute of Technology,
Mesra, Jharkhand, India.

Pandya Pranav
RK University, Rajkot, Gujarat, India.

Kumar Prasann
Lovely Professional University,
Phagwara, Punjab, India.

Devi Priyanka
Lovely Professional University,
Phagwara, Punjab, India.

Saravanathamizhan R.
Anna University, Chennai, India.

Dey Shipa Rani
Lovely Professional University,
Phagwara, Punjab, India.

Ahmed Safia
Saheed Benazir Bhutto Women
University, Peshawar, Pakistan.

Achuth M. Sai
Anna University, Chennai, India.

Khan Samiullah
Quaid-i-Azam University, Islamabad, Pakistan.

Chowdhury Shamik
Indian Institute of Technology, Kharagpur, West Bengal, India.

Sugumar Shobana
SRM Institute of Science and Technology, Chengalpattu, Tamil Nadu, India.

Anand Shreya
Birla Institute of Technology, Mesra, India.

Ghosh Sougata
Kasetsart University, Bangkok, Thailand.

Singh Joginder
Lovely Professional University, Phagwara, Punjab, India.

Sriraman Vidya
SRM Institute of Science and Technology, Chengalpattu, Tamil Nadu, India.

1 Recovery and Removal of Heavy Metals from Industrial Effluents

Ramya TR, Neema Reddy, Kavya L, and Ankita Chatterjee

1.1 INTRODUCTION

The environmental problems brought on by fast industrialization and globalization are increasing every year. Therefore, procedures that are efficient and effective are required, especially for remediation. The presence of heavy metals in industrial and wastewater effluent is a serious concern for environmental pollution. With the expansion of industry and human activity, there are more heavy metals in wastewater. Examples include the plating and electroplating industry, batteries, pesticides, mining industry, rayon industry, metal rinse processes, tanning industry, fluidized bed bioreactors, textile industry, metal smelting, petrochemicals, paper manufacturing, and electrolysis applications (Masindi and Muedi 2018). Wastewater that has been contaminated with heavy metals makes its way into the environment, endangering both human health and the ecosystem (see Table 1.1). As heavy metals cannot be biodegraded, they may cause health hazards, and thus their presence in water should be avoided (Alengebawy et al. 2021). The release of industrial waste, fertilizers, and pesticides; mining, smelting, metal plating/metal finishing operations; generation of waste from automobile batteries, vehicle emissions, and fly ash from combustion/incineration processes are various examples, which can contaminate subsurface soils and groundwater with heavy metals (Rahman et al. 2014).

The removal of heavy metal-contaminated aqueous streams is attempted using a number of physiochemical techniques. Traditionally, methods for removing heavy metals from wastewater have included chemical precipitation, membrane filtration, carbon adsorption, and ion exchange, among others (Peng and Guo 2020). However, they are pricey and unsuitable for environments with higher concentration of heavy metals. Thus, there is a need for alternatives, which are less expensive, easier to use, and require less upkeep. The adsorption approach has become more popular among all removal methods because of its low cost, high efficiency, availability, and ease of use (Ali et al. 2012). For the elimination of heavy metals, biosorbents have shown promise as alternatives. Biosorbents that may remove metalloids and heavy metals from aqueous solutions through biosorption include algae, fungi, and plants, which are biomass-derived.

DOI: 10.1201/9781003368472-1

TABLE 1.1
Permissible Limits on Toxic Heavy Metals and Their Toxic Effects on Human Health

Contaminants	Recommended Safe Limits by WHO in Drinking Water (mg/L)	Recommended Safe Limits by WHO in Wastewater (mg/L)	Toxic Effects
Copper	< 2	1	Gastrointestinal effects, carriers of the gene for Wilson disease, metabolic disorder of copper homeostasis
Zinc	< 3	2–5	Skin irritations, anaemia, nausea, and vomiting
Manganese	< 0.12	< 0.2	Neurological effects following inhalation exposure, psychological symptoms (irritability, emotional lability)
Arsenic	< 0.01	n/a	Chronic arsenicism including dermal lesions such as hyper- and hypopigmentation, peripheral neuropathy, skin cancer, bladder and lung cancers, and peripheral vascular disease
Cadmium	0.003–0.005	0.003	Carcinogenic by the inhalation route, kidney problem due to accumulation in the kidneys
Chromium	< 0.05	0.05	Chromium (VI) is carcinogenic to humans, lung cancer
Lead	< 0.01	0.01	Accumulates in the skeleton and creates bone problems, adverse effects on central and peripheral nervous systems
Selenium	< 0.01	n/a	Long-term exposure is manifested in nails, hair, and liver, effects on synthesis of a liver protein
Mercury	< 0.006	0.05	Oral ingestion creates problems in gastrointestinal tract, causes kidney damage, and increases the incidence of some benign tumours
Nickel	0.02–0.07	0.02	Irritability, nausea, vomiting, difficulty sleeping

1.2 REMOVAL OF HEAVY METALS FROM INDUSTRIAL EFFLUENTS BY TRADITIONAL METHODS

It is possible to use traditional treatment methods to remove heavy metals from inorganic wastewater. Chemical precipitation, coagulation, complexation, activated carbon adsorption, ion exchange, solvent extraction, foam flotation, electrodeposition, cementation, and membrane operations are just a few of the treatment strategies that can be used to remove heavy metals from industrial wastewater (Gunatilake 2015). The numerous heavy metal removal treatment procedures and tactics are described below.

1.3 PHYSICO-CHEMICAL TECHNIQUES

Researchers have explored and utilized the various physico-chemical methods to remove heavy metals. Physical separation techniques are mostly applicable to discrete particles, particulate forms of metals, or particles that contain metal (Cho et al. 2021). Physical separation involves mechanical screening, hydrodynamic classification, gravity concentration, flotation, magnetic separation, electrostatic separation, and attrition scrubbing. The effectiveness of physical separation depends on a variety of soil characteristics, including particle size distribution, particulate shape, clay content, moisture content, humic content, heterogeneity of soil matrix, density between soil matrix and metal contaminants, magnetic properties, and others. The traditional chemical methods for removing heavy metals from wastewater include a variety of techniques, including chemical precipitation, flotation, adsorption, ion exchange, and electrochemical deposition (Shrestha et al. 2021).

1.4 PRECIPITATION

Due to its ease of application, chemical precipitation is among the most frequently used methods in industry for removing heavy metals from inorganic effluent. These traditional chemical precipitation procedures result in the formation of insoluble heavy metal precipitates in the forms of hydroxide, sulphide, carbonate, and phosphate. The mechanism of this procedure is based on the reaction of precipitant and dissolved metals in the solution to create insoluble metal precipitation. Chemical precipitants, coagulants, and flocculation procedures are used to increase the particle size of the extremely small particles produced during the precipitation process so that they can be removed as sludge. Low metal concentrations can be discharged after metals precipitate and solidify, making them simple to remove. The removal percentage of metal ions in the solution might be maximized by changing chemical parameters like pH, ion charge, temperature, etc. (Qasem et al. 2021).

1.5 FLOCCULATION AND COAGULATION

In order to identify the electrostatic interaction between pollutants and coagulant-flocculant agents, the coagulation-flocculation mechanism is based on zeta potential measurement. The electrostatic repulsion technique is used to stabilize colloidal particles by stabilizing the net surface charge of the particles through coagulation. Through additional collisions and interactions with the inorganic polymers created by the organic polymers introduced, the flocculation process continuously raises the particle size to distinct particles. Discrete particles can be removed or separated by filtration, straining, or flocculation after they have been combined with bigger particles through flocculation. The process's main downsides include the creation of sludge, the use of chemicals, and the transfer of harmful substances into the solid phase (Castro-Riquelme et al. 2023).

1.6 ELECTRONIC CHEMICAL PROCEDURES

Electrolysis: One of the various methods for removing metals from wastewater streams is electrolytic recovery. An electrical current is sent through an aqueous metal-bearing solution including a cathode plate and an insoluble anode during this procedure. Moving electrons from one element to another can produce electricity. The heavy metals are precipitated as hydroxides in a weak acidic or neutralized catholyte as part of the electrochemical procedure used to treat wastewater containing heavy metals. Electrodeposition, electrocoagulation, electro flotation, and electro-oxidation are electrochemical processes used to treat wastewater (Nancharaiah et al. 2019).

Coagulation and precipitation by hydroxide formation to acceptable levels are terms used to describe the electrode stabilization of colloids. The process of electrolytic oxidation of coagulants by destabilizing impurities to generate floc is the most popular heavy metal precipitation method. A suitable anode material is electrolytically oxidized during the electrocoagulation process to produce the coagulant on-site. This method involves letting an anion in the effluent react with charged ionic metal species in the wastewater to eliminate them. Reduced sludge production, minimal need for chemical use, and simplicity of operation are the distinguishing features of this technique (Patil and Mane 2022). Chemical precipitation, on the other hand, needs a lot of chemicals to get metals down to a permissible level for disposal. Significant sludge generation, sluggish metal precipitation, inadequate settling, the accumulation of metal precipitates, and the metals.

1.7 ATOMIC EXCHANGE

Ion exchange, the most used technique in the water treatment business, can draw soluble ions from the liquid phase to the solid phase. Ion exchange is a cost-effective method that has been shown to be very effective at removing heavy metals from aqueous solutions, particularly when used to treat water with low concentrations of heavy metals. It typically uses inexpensive materials and simple procedures. To remove metal ions from the solution, this technique uses cations or anions that contain specialized ion exchangers. Synthetic organic ion exchange resins are ion exchangers that are often employed. This approach is extremely sensitive to the pH of the aqueous phase and can only be used with low-concentration metal solutions. Ion exchange resins are insoluble in water solids (Lebron et al. 2021).

1.8 APPLICATION OF MEMBRANE FILTER

For the treatment of inorganic wastewater, membrane filtration has drawn a lot of interest. It has the ability to eliminate suspended solids, organic substances, and inorganic pollutants like heavy metals. Different forms of membrane filtration, including ultrafiltration (UF), nanofiltration, and reverse osmosis (RO), can be used to remove heavy metals from wastewater, depending on the size of the particles that can be retained (Hube et al. 2020).

Using a permeable membrane, UF separates suspended particles, macromolecules, and heavy metals from inorganic solution based on the separating substances' molecular weights and pore sizes (5–20 nm) (1000–100,000 Da). With a metal content ranging from 10 to 112 mg/L, UF can achieve more than 90% of removal effectiveness depending on the membrane properties (Aloulou et al. 2020).

The polymer-supported ultrafiltration (PSU) approach creates a free targeted metal ion effluent while using water-soluble polymeric ligands to bind metal ions and form macromolecular complexes. Low energy needs for UF, extremely quick reaction kinetics, and greater selectivity of separation of selective bonding agents in aqueous solution are benefits of PSU technology (Gunatilake 2015).

A further comparable method called complexation-UF shows promise as a replacement for technologies based on precipitation and ion exchange. A hybrid method to selectively concentrate and recover heavy metals in the solution involves the use of water-soluble metal-binding polymers in conjunction with UF. To enhance their molecular weight with a higher molecular weight, the cationic forms of heavy metals are first complexed by a macro-ligand in the complexation-UF process (Saleh et al. 2022).

In RO, pressure is used to push a solution through a membrane, which keeps the solute on one side while letting the pure solvent flow to the other (Qasim et al. 2019; Shah 2021). The membrane in this case is semipermeable, which means that metals cannot travel through it but solvents can. In the polymer matrix of the RO membranes, where the majority of separation takes place, there is a dense barrier layer. Industrial procedures involve RO, which can remove a wide variety of chemicals and ions from solutions, including microorganisms. Due to the diffusive mechanism used in RO, the separation efficiency is influenced by the solute concentration, pressure, and water flux rate (Agenson and Urase 2007).

1.9 ELECTRODIALYSIS

By providing an electric potential to an ion exchange membrane, electrodialysis (ED) allows ionized species in the solution to flow through the membrane. Thin plastic sheets that have anionic or cationic properties make up the membranes. Anions and cations migrate across the anion-exchange and cation-exchange membranes when a solution containing ionic species moves through the cell compartments, moving in the direction of the anode and the cathode, respectively. Replacement of membranes and the corrosion process are also notable drawbacks. Better cell performance might be obtained by using membranes with larger ion exchange capacities (Moon and Yun 2014).

1.10 BIOLOGICAL APPROACHES

The biological removal of heavy metals from wastewater involves the removal of contaminants from wastewater using biological methods. Microorganisms aid in this process by helping the solids in the solution to settle. For the treatment of wastewater, trickling filters, stabilization ponds, and activated sludge are frequently utilized. The most popular alternative, activated sludge, uses microorganisms to break down organic material through aeration and agitation before allowing solids to settle out. To speed up organic decomposition, "activated sludge" containing bacteria is continuously returned to the aeration basin. The suspended growth-activated sludge technique has been the focus of the majority of research on heavy metal removal in biological systems (Chaemiso and Nefo 2019).

Another technique for removing heavy metals from wastewater is biosorption. The term "sorption process" really refers to a range of processes, including adsorption and precipitation reactions, where ions are transferred from the solution phase to the solid phase (Basu et al. 2022). Adsorption has emerged as one of the major wastewater treatment options. Adsorption, in its simplest form, is a mass transfer process where substances are bonded to solid surfaces by physical or chemical interactions. Recently, several low-cost adsorbents have been created and used to remove heavy metals from water that has been contaminated with metals. These materials can be derived from agricultural waste, industrial by-products, natural materials, or modified biopolymers. Organic removal has been the focus of activated carbon use in water and wastewater treatment. Research has focused on the removal of metallic ions from inorganics by activated carbon (Kumari et al. 2019; Cheng et al. 2021). Chemical modifications can be made to industrial by-products such as fly ash, waste iron, iron slags, and hydrous titanium oxide to improve their ability to remove metal from wastewater (Krishnan et al. 2021).

Recently, research has been focused on using agricultural by-products as adsorbents through the biosorption process in order to remove heavy metals from industrial effluent. After being chemically altered or heated into activated carbon or biochar, new resources, such as hazelnut shells, rice husks, pecan shells, jackfruit, rice straw, rice husk, coconut shell, etc., can be employed as adsorbents for the removal of heavy metals. They discovered that the presence of cellulose, lignin, carbohydrates, and silica in those biomasses' adsorbents caused the highest metal removal by those materials (Thakur et al. 2022).

Biopolymers have a variety of functional groups, including hydroxyls and amines, which improve the effectiveness of metal ion uptake. They are extensively used in industry because they can reduce transition metal ion concentrations to parts per billion levels. In order to remove heavy metals from wastewater, new polysaccharide-based compounds are referred to as biopolymer adsorbents (made from chitin, chitosan, and starch). Polysaccharide-based materials have complex sorption mechanisms that are pH-dependent. In addition, cross-linked hydrophilic polymers known as hydrogels are frequently utilized in the treatment of wastewater. Water diffusion into the hydrogel, especially in the absence of firmly binding sites, primarily controls the removal by bringing heavy metals within. With rising temperatures, maximum binding capacity rises (Hamza et al. 2019).

1.11 EVALUATION OF PROCESSES FOR REMOVING HEAVY METAL

An essential component of environmental research is the removal of heavy metals from wastewater. The researchers used a variety of techniques, including physical, chemical, and biological ones. Physical-chemical treatments have a number of benefits, including quick processing, simple operation and control, a wide range of input loads, etc. Chemical plants are adaptable whenever needed. These treatment systems need less room and are less expensive to install. Their disadvantages, which include high operating costs because of the chemicals they utilize, high energy consumption, and handling expenses for disposing of toxic sludge, outweigh their advantages. Physical-chemical treatments have been identified as one of the most effective treatments for inorganic materials if chemical costs can be reduced in any way.

Heavy metals can be removed from wastewater using a variety of methods. Precipitation and ion exchange are the main operating principles. Both processes result in wastes that are largely ineffective for recovering metals, such as lime sludge, exhausted ion exchange resins, and wastewater from deionizer regeneration that is heavily metal-loaded. Hazardous trash makes up the majority of these remnants, which signify the completion of pipe treatment. The electrolysis process is a well-known technique for recovering metal. The lower criterion for an efficient application for the majority of these technical electrowinning systems is, respectively, a minimum metal concentration of 30 mg/L and 50 mg/L. Metal recovery is increased utilizing special procedures (such as those that use fluidized, packed beds, or mesh cathode cells).

Wetland metal mineralization is a low-tech alternative to the depletion of metals in wastewater. In contrast to the approaches outlined above, the efficacy of heavy metal removal using this biogenic method is quite low. Additional methods that use microorganisms that accumulate metal, such as bacteria, algae, yeasts, and fungus, are intriguing alternatives but are only slightly used in the industrial treatment of effluents that contain metal.

1.12 EFFICACY OF MICROBIAL FUEL CELL (MFC) IN REMOVAL OF HEAVY METALS

MFCs are the bioelectrochemical combinations that use electrons generated by exo-electrogenic microorganisms during their metabolic activities for production of electricity. The first article regarding MFC was framed by researcher Potter in the year 1911 after observing production of electric current by the *Saccharomyces cerevisiae* (Gustave et al. 2021). MFCs are developed for heavy metals remediation via cathodic reduction reaction. MFCs are proven to be economically cheaper and environmentally safe technology in heavy metal remediation. Bioelectrochemical systems that are methodically framed for conversion of chemical energy in the organic matter to electrical energy with the help of microorganisms are considered as MFC. Typically, an MFC contains of an anode and a cathode. The biofilm formed by anaerobic microorganisms acts as catalysts for energy production from organic contents present in the wastewater. Substrate oxidation mechanism occurs as the protons and electrons are dispersed into aqueous solution. Studies related to MFCs still need to be explored more to frame a proper justification of their pros and cons since very limited facts and data are present at the current stage. The removal mechanism follows reduction and deposition at the surface of the cathode, as well as, in certain cases, involvement of chemical precipitation and electrochemical reduction of heavy metals are also noticed (Drendel et al. 2018). The efficiency of MFC is affected by various parameters, such as initial substrate, pH, temperature, electron transfer mechanism of the anodic chamber, electron acceptor, and the electrode material. The heavy metals uptake using MFC results to be effective when the reduction process occurs in thermodynamically favourable condition. The metals, which exhibit lower redox potential, such as cadmium, lead, nickel, and zinc, are unable to accept electrons from the cathode directly and would require certain external power supply for the transfer of electron to the cathode (Ezziat et al. 2019). (Figure 1.1.)

Sediment microbial fuel cell (SMFC) is an innovated MFC, which is used in heavy metal remediation. The prominent characteristics of SMFC include their lack of membrane of totally anoxic nature. The sediment here refers to the sink for the pollutants, majorly targeting heavy metals and organic contaminants. During the initial phase of using and researching about MFC, researchers used certain chemical mediators, such as thonin, natural red, potassium ferricyanide, and many more, for transferring the electrons to the electrodes from the bacteria. However, these chemical mediators were found to be unstable and hazardous in nature (Logan 2008). Thereafter, with various research studies, it was reported that the bacteria would be able to transfer the electrons without the incorporation of exogenous materials. Later, it was also studied that organic or inorganic materials from marine sources can act as substrate in the MFC for generation of electricity (Kim et al. 1999; Reimer et al. 2001; Bond et al. 2002). SMFC contains a cathode and anode made of graphite. The cathode and anode are placed in dissolved oxygen-contained water and anaerobic sediment, respectively. The functioning of SMFC is highly impacted by the gradient of sediment and water interface.

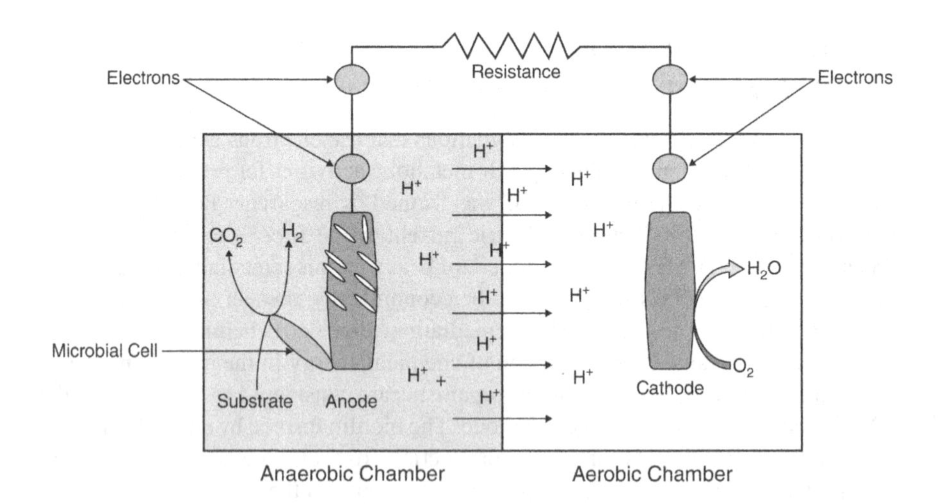

FIGURE 1.1 Microbial fuel cell (Fang and Achal 2019).

The microorganisms specialized in accepting electrons from the electrodes are termed as electrotrophs. The application of electrotrophs in remediation of heavy metals has introduced a new aspect to remediation strategies. Several microbial strains, such as *Pseudomonas* and *Proteobacter* as well as certain consortiums of microbes, have been used as effective electrotrophs (Abbas et al. 2017). *Geobacter* sp. is reported for eliminating uranium, fumarate, chlorinated solvents, and nitrates by using them as electron acceptors. The bacterium is known for its ability to accept electrons from the electrodes. Many bacterial species use electron shuttles for indirect electron transfer (Gregory et al. 2004; Thrash and Coates 2008). SMFC can be utilized in reduction of various heavy metals, such as cadmium, chromium, copper, zinc, and uranium. The electrotrophs existing in the sediments act on the heavy metals by reducing them. The reduced heavy metals are then discharged into the water or adsorbed effectively on the surface of the electrodes. Once absorbed on the electrode, the heavy metals can be successfully removed. Heavy metals, like mercury and arsenic, are reported to be difficult for removal by reduction process, and thus these heavy metals require oxidation mechanism. The heavy metals, oxidized at the anode, form ferric oxides as precipitation. In case the mercury is reduced, they are converted into even more toxic form, methyl mercury. However, to reduce the methyl mercury production, oxidized sediment environment is much required (Li and Yu 2015).

1.13 FUTURE PERSPECTIVES OF THE TECHNIQUES

MFCs are yet unable to treat every type of metal ion, and thus MFCs should be explored more with evolved mechanisms. It is noticed that in case of thermodynamically favourable reduction process, the metal ions act as electron acceptors in absence of external power supply. The metals with lower redox potentials, like cadmium, lead, and nickel, are unable to accept electrons spontaneously and thus require a support from external power supply (Wang and Ren 2014). The size of MFC greatly affects the performance. It has been reported that the performance quality of MFC decreases as the size of MFC expands. The probable reason for this might be the requirement for the transfer of mass in larger-sized electrodes (Ge et al. 2014). Further studies must be conducted in order to specify and analyze the importance of MFCs in large-scale wastewater treatment.

1.14 CONCLUSION

Traditional ways of wastewater treatment are dependent on the origin and source of water as well as the further use of the water. Various techniques, such as precipitation, electrochemical reaction, membrane filtration, ion exchange methods, as well as biological approaches, are explored across industries for removal of heavy metals from industrial wastewater. Biological approaches are environmental friendly, cost-effective, and sustainable when compared to the physico-chemical techniques. Applications of microbes, plant and agricultural wastes, and enzymes are promising biological sources used for heavy metal recovery from effluents. MFC is a newly developed approach for remediation of heavy metals from wastewater. MFCs are able to treat and recover the metals. However, certain challenges, like application of MFC in treating all types of metals and upscaling the application of MFCs, are required to be explored in future studies.

REFERENCES

Abbas SZ, Rafatullah M, Ismail N, Nastro RA. Enhanced bioremediation of toxic metals and harvesting electricity through sediment microbial fuel cell. International Journal of Energy Research. 2017 Nov;41(14):2345–55.

Agenson KO, Urase T. Change in membrane performance due to organic fouling in nanofiltration (NF)/reverse osmosis (RO) applications. Separation and Purification Technology. 2007 Jun 15;55(2):147–56.

Alengebawy A, Abdelkhalek ST, Qureshi SR, Wang MQ. Heavy metals and pesticides toxicity in agricultural soil and plants: Ecological risks and human health implications. Toxics. 2021 Feb 25;9(3):42.

Ali I, Asim M, Khan TA. Low cost adsorbents for the removal of organic pollutants from wastewater. Journal of Environmental Management. 2012 Dec 30;113:170–83.

Aloulou W, Aloulou H, Khemakhem M, Duplay J, Daramola MO, Amar RB. Synthesis and characterization of clay-based ultrafiltration membranes supported on natural zeolite for removal of heavy metals from wastewater. Environmental Technology & Innovation. 2020 May 1;18:100794.

Basu A, Ali SS, Hossain SS, Asif M. A review of the dynamic mathematical modeling of heavy metal removal with the biosorption process. Processes. 2022 Jun 8;10(6):1154.

Bond DR, Holmes DE, Tender LM, Lovley DR. Electrode-reducing microorganisms that harvest energy from marine sediments. Science. 2002 Jan 18;295(5554):483–5.

Castro-Riquelme CL, López-Maldonado EA, Ochoa-Terán A, Alcántar-Zavala E, Trujillo-Navarrete B, Pérez-Sicairos S, Miranda-Soto V, Zizumbo-López A. Chitosan-carbamoylcarboxylic acid grafted polymers for removal of metal ions in wastewater. Chemical Engineering Journal. 2023 Jan 15;456:141034.

Chaemiso TD, Nefo T. Removal methods of heavy metals from laboratory wastewater. Journal of Natural Sciences Research. 2019;9(2):36–42.

Cheng N, Wang B, Wu P, Lee X, Xing Y, Chen M, Gao B. Adsorption of emerging contaminants from water and wastewater by modified biochar: A review. Environmental Pollution. 2021 Mar 15;273:116448.

Cho K, Kim H, Purev O, Choi N, Lee J. Physical separation of contaminated soil using a washing ejector based on hydrodynamic cavitation. Sustainability. 2021 Dec 27;14(1):252.

Drendel G, Mathews ER, Semenec L, Franks AE. Microbial fuel cells, related technologies, and their applications. Applied Sciences. 2018 Nov 25;8(12):2384.

Ezziat L, Elabed A, Ibnsouda S, El Abed S. Challenges of microbial fuel cell architecture on heavy metal recovery and removal from wastewater. Frontiers in Energy Research. 2019 Jan 30;7:1.

Fang C, Achal V. The potential of microbial fuel cells for remediation of heavy metals from soil and water—Review of application. Microorganisms. 2019 Dec 13;7(12):697.

Ge Z, Li J, Xiao L, Tong Y, He Z. Recovery of electrical energy in microbial fuel cells: Brief review. Environmental Science & Technology Letters. 2014 Feb 11;1(2):137–41.

Gregory KB, Bond DR, Lovley DR. Graphite electrodes as electron donors for anaerobic respiration. Environmental Microbiology. 2004 Jun;6(6):596–604.

Gunatilake SK. Methods of removing heavy metals from industrial wastewater. Methods. 2015 Nov 30;1(1):14.

Gustave W, Yuan Z, Liu F, Chen Z. Mechanisms and challenges of microbial fuel cells for soil heavy metal (loid) s remediation. Science of the Total Environment. 2021 Feb 20;756:143865.

Hamza MF, Gamal A, Hussein G, Nagar MS, Abdel-Rahman AA, Wei Y, Guibal E. Uranium (VI) and zirconium (IV) sorption on magnetic chitosan derivatives–effect of different functional groups on separation properties. Journal of Chemical Technology & Biotechnology. 2019 Dec;94(12):3866–82.

Hube S, Eskafi M, Hrafnkelsdóttir KF, Bjarnadóttir B, Bjarnadóttir M, Axelsdóttir S, Wu B. Direct membrane filtration for wastewater treatment and resource recovery: A review. Science of the Total Environment. 2020 Mar 25;710:136375.

Kim BH, Ikeda T, Park HS, Kim HJ, Hyun MS, Kano K, Takagi K, Tatsumi H. Electrochemical activity of an Fe (III)-reducing bacterium, Shewanella putrefaciens IR-1, in the presence of alternative electron acceptors. Biotechnology Techniques. 1999 Jul;13:475–8.

Krishnan S, Zulkapli NS, Kamyab H, Taib SM, Din MF, Abd Majid Z, Chaiprapat S, Kenzo I, Ichikawa Y, Nasrullah M, Chelliapan S. Current technologies for recovery of metals from industrial wastes: An overview. Environmental Technology & Innovation. 2021 May 1;22:101525.

Kumari P, Alam M, Siddiqi WA. Usage of nanoparticles as adsorbents for waste water treatment: An emerging trend. Sustainable Materials and Technologies. 2019 Dec 1;22:e00128.

Lebron YA, Moreira VR, Amaral MC. Metallic ions recovery from membrane separation processes concentrate: A special look onto ion exchange resins. Chemical Engineering Journal. 2021 Dec 1;425:131812.

Li WW, Yu HQ. Stimulating sediment bioremediation with benthic microbial fuel cells. Biotechnology Advances. 2015 Jan 1;33(1):1–2.

Logan BE. Microbial Fuel Cells. 2008 Feb 8. John Wiley & Sons.

Masindi V, Muedi KL. Environmental contamination by heavy metals. Heavy Metals. 2018 Jun 27;10:115–32.

Moon SH, Yun SH. Process integration of electrodialysis for a cleaner environment. Current Opinion in Chemical Engineering. 2014 May 1;4:25–31.

Nancharaiah YV, Mohan SV, Lens PN. Removal and recovery of metals and nutrients from wastewater using bioelectrochemical systems. In Microbial Electrochemical Technology 2019 Jan 1 (pp. 693–720). Elsevier.

Patil MP, Mane S. An overview of electrochemical coagulation technology for treatment of industrial wastewater. International Journal of Research and Analytical Reviews (IJRAR). 2022 Apr;9(2):31–6.

Peng H, Guo J. Removal of chromium from wastewater by membrane filtration, chemical precipitation, ion exchange, adsorption electrocoagulation, electrochemical reduction, electrodialysis, electrodeionization, photocatalysis and nanotechnology: A review. Environmental Chemistry Letters. 2020 Nov;18:2055–68.

Qasem NA, Mohammed RH, Lawal DU. Removal of heavy metal ions from wastewater: A comprehensive and critical review. Npj Clean Water. 2021 Jul 8;4(1):36.

Qasim M, Badrelzaman M, Darwish NN, Darwish NA, Hilal N. Reverse osmosis desalination: A state-of-the-art review. Desalination. 2019 Jun 1;459:59–104.

Rahman MM, Salleh MA, Rashid U, Ahsan A, Hossain MM, Ra CS. Production of slow release crystal fertilizer from wastewaters through struvite crystallization—A review. Arabian Journal of Chemistry. 2014 Jan 1;7(1):139–55.

Reimers CE, Tender LM, Fertig S, Wang W. Harvesting energy from the marine sediment-water interface. Environmental Science & Technology. 2001 Jan 1;35(1):192–5.

Saleh TA, Mustaqeem M, Khaled M. Water treatment technologies in removing heavy metal ions from wastewater: A review. Environmental Nanotechnology, Monitoring & Management. 2022;17:100617.

Shah MP. Removal of Emerging Contaminants through Microbial Processes. 2021. Springer.
Shrestha R, Ban S, Devkota S, Sharma S, Joshi R, Tiwari AP, Kim HY, Joshi MK. Technological trends in heavy metals removal from industrial wastewater: A review. Journal of Environmental Chemical Engineering. 2021;9(4):105688.
Thakur AK, Singh R, Pullela RT, Pundir V. Green adsorbents for the removal of heavy metals from wastewater: A review. Materials Today: Proceedings. 2022 Jan 1;57:1468–72.
Thrash JC, Coates JD. Direct and indirect electrical stimulation of microbial metabolism. Environmental Science & Technology. 2008 Jun 1;42(11):3921–31.
Wang H, Ren ZJ. Bioelectrochemical metal recovery from wastewater: A review. Water Research. 2014 Dec 1;66:219–32.

2 Microbial Electrochemical Systems-Anaerobic Digestion-A Hybrid System for Industrial Effluent Treatment and Energy Recovery

Muhammad Adil Nawaz Khan, Sayed Muhammad Ata Ullah Shah Bukhari, Samiullah Khan, Safia Ahmed, Naeem Ali, Malik Badshah

2.1 INTRODUCTION

With an ever-increasing population in the world, meeting the energy requirement of such a vast population has posed to be the single most critical challenge to mankind (Shah et al., 2022; Vyas et al., 2022a). The requirement for energy has continued to go up exponentially as the population has grown over the years (Vyas et al., 2022b). However, our reliance on fossil fuels to generate the majority of our energy has not altered, and we continue to utilize fossil fuels such as coal, oil, and others to meet our needs (Devda et al., 2021), which eventually results in the release of pollutants into the environment, necessitating further remediation (Do et al., 2020; Varjani and Upasani, 2017). A lot of energy is used to run a number of industries around the globe, leaving behind by-products like wastewater (Varjani et al., 2020a). The different wastes generated in such processes require treatment before disposal, which in turn requires a lot of energy. Thus, the production and treatment energy nexus continues (Di Fraia et al., 2018). Wastewater is one of the most abundant by-products of practically all industries. Treatment of wastewater alone, before its disposal, requires a surplus amount of energy (Chattopadhyay et al., 2022). Apart from industries, a lot of municipal wastewater is generated on a daily basis, which also requires energy for its treatment. About 3% of the global electricity consumption is accounted for by the treatment of municipal wastewater (Grobelak et al., 2019). The environmental concerns arising from the use of fossil fuels as the source of energy have also led to the increased demand for alternative sources of energy and the refined processes to harness the energy from such unconventional sources (Asongu et al., 2020; Patel et al., 2021; Varjani, 2017). The

DOI: 10.1201/9781003368472-2

energy embedded in the wastewater is at least 2–4 times more than the amount of energy required for the treatment of the same wastewater. Therefore, attempts must be directed toward developing and refining such technologies that can produce energy from sources otherwise deemed waste (Thangamani et al., 2021). Additionally, other value-added products like chemicals, metals, clean water, etc. can be extracted from wastewater stream to maximize productivity and efficiency of the facility (Liu et al., 2021). The composition of wastewater greatly depends upon the source of generation (Varjani et al., 2021). Wastewater is generally a mixture of various organic and inorganic components, which can generate by-products like H_2 during treatment. The different sources of wastewater can be domestic wastewater, landfill wastewater, industrial wastewater, refinery wastewater, livestock, and dairy wastewater, to name a few (Kataki et al., 2021). The microbes and treatment technology used greatly depend on the substrate present in the wastewater. Similarly, end product derived from treatment of wastewater also depends upon the substrate. Ditzig et al. demonstrated the production of hydrogen using the domestic wastewater as a substrate using the MECs along with the treatment of wastewater, which resulted in the reduction of biological oxygen demand (BOD) and chemical oxygen demand (COD) (Ditzig et al., 2007). Among the most efficient technologies for the process of waste-to-product conversion is the Microbial Electrolysis Cell (MEC) (Janani et al., 2022; Mohan et al., 2019). The MEC setup uses the exoelectronic microbes to convert the biodegradable matter, like the organic matter present in the wastestream, into electric current and protons (Ahmad et al., 2022; Jain and He, 2018). The microbes are present at the anode and act upon the biodegradable waste to generate electricity and protons. The electrons are then transferred to the cathode, where they reduce the protons for H_2 production (Adesra et al., 2021; Huang et al., 2022). The type of microorganism used in the MEC depends on the substrate. For *Aeromonas hydrophila, Thermincola sp, Geothrix fermentans,* and *Gluconobacter oxydans* substrate are acetate; for *S. putrefaciens and Shewanella oneidensis* substrate are acetate lactate; and for *K. pneumoniae*, *Rhodoferax ferrireducens,* and *Enterobacter cloacae* substrate are acetate glucose (Kadier et al., 2020a). The microbial electrolysis is an endothermic process, and therefore, for the production of H_2 at the cathode, a small voltage is applied between the two electrodes to forcefully initiate the current generation. However, the applied voltage is only about 0.2–0.8 V and can be supplied by low-grade microbial fuel cells (MFCs) or small solar panels (Rousseau et al., 2020). The production of H_2 using MECs has not only been hailed for commercialization because of better H_2 production performance in comparison to other H_2 technologies like fermentation but also because of its higher performance efficiency than other microbial electrochemical systems like MFCs, etc. (Rousseau et al., 2020). This review aims to explore MEC applications in the following instances: an overview of MECs for energy generation and recycling, such as hydrogen, methane, and formic acid; contaminant removal, particularly complex organic and inorganic pollutants; and resource recovery. In this review, new MEC technology concepts are discussed, such as coupling with other technologies for value-added applications such as MEC-anaerobic digestion, MEC with anaerobic Membrane Bioreactor (MBR) and acidogenic, MEC-Thermoelectric microconverter, Dark fermentation, and MFC–MEC, and MEC with Microbial reverse-electrodialysis electrolysis cell systems. Finally, challenges, prospects, and the life cycle assessment of the process are discussed.

2.2 INDUSTRIAL WASTEWATER COMPOSITION

Wastewater from various industrial sectors contains many pollutants that are toxic and have hazardous effects on human and aquatic life as well as on agriculture. Such pollutants include heavy metals like chromium (Cr), zinc (Zn), lead (Pb), copper (Cu), iron (Fe), cadmium (Cd), nickel (Ni), arsenic (As), and mercury (Hg) (Q. Wang and Z. Yang, 2016). Most of these heavy metal pollutants are released by paint and dye manufacturing, textile, pharmaceutical, paper, and fine chemical industries. Phenol and phenolic compounds are also two of the major pollutants present in industrial wastewater. They are mostly released by oil refineries, phenol-formaldehyde resin, and bulk drug manufacturing industries. A number of poorly biodegradable refractory pollutants like petroleum hydrocarbons, sulfides, aniline, naphthalenic acid, organochlorines, olefins, nitrobenzene, alkanes, and chloroalkanes, generated by the petrochemical industries, are present in wastewater. The composition of petrochemical wastes is chemically very complex, and their treatment by biological methods is slow and not very effective. Even after the primary biological treatment, the organic pollutants are retained in the secondary effluents. They require chemical oxidants for the formation of inorganic end products and thus exhibit a low ratio of BOD to COD. Suspended solids and highly organic materials are the major water pollutants released by the paper and pulp industry. Depending on the quality of paper produced and pulp processing, the characteristics of the effluent change. The constituents of the effluents can be adsorbable organic halogens (AOX), phenolic compounds, biocides, colors, resin acids, non-biodegradable organic materials, tannins, sterols, lignin-derived compounds, *etc.* (Buyukkamaci and Koken, 2010; Lindholm-Lehto et al., 2015). Urea, ammonium nitrogen (NH_4–N), and other nitrogenous and phosphorus wastes come from various textile printing and dyeing industries, which use water in many steps while processing. Various textile industries generate compounds ranging from heavy metals like chromium to surfactants, bleaching agents including hydrogen peroxide and chlorine, AOX, sodium silicate, and alkaline bases (Z. Wang et al., 2011). Perfluoroalkyl acids (PFAAs) are used as surface protectors for their excellent high surface activity, stability, and oil–water repellence. However, two PFAAs with potential health risks are perfluorooctane sulfonate (PFOS) and perfluorooctanoic acid (PFOA). PFOS is released mainly by textile treatment, metal plating, and semiconductor industries, while PFOA is released by the fluoropolymer production and processing industries.

(Z. Liu et al., 2017). They are mainly circulated through the wastewater released from these industrial facilities. Besides all these pollutants, the high salinity of wastewater also has many adverse effects on life forms. Salt removal from wastewater has become as important as removal of organic matter and other pollutants in many countries. High salinity (mainly NaCl) wastewater is generated by petroleum, leather, food processing, and agro-based industries (Das et al., 2018).

2.3 ANAEROBIC DIGESTION (AD) PROCESS

AD is a natural process that occurs in the absence of oxygen, and usually takes place in watercourses, sediments, and wet soils, from raw materials like industrial and municipal wastewater, agro-industrial, municipal, food activities, and vegetal wastes, resulting in the production of a gas mixture known as biogas (Ward et al., 2008). AD is the only current biological treatment process able to produce biogas, which is a mixture of methane (CH_4) and carbon dioxide (CO_2). Instead of composting, AD allows biodegradation of organic matters in the absence of oxygen, resulting in a digested product that can be used as soil fertilizer. AD allows the treatment of high organic loading wastes in order to reduce their volume and load while recovering biogas, which can be used to produce heat, electricity, and biofuels for vehicles (Awe et al., 2017). The technology of AD can be classified according to the dry matter of the initial substrate as: wet digestion, where the substrate must have a dry matter content below 10%; dry digestion, in which the substrate should have dry matter content exceeding 20%; and dry matter digestion intermediate, also defined as semi-dry (Gkamarazi, 2015). Wet AD was largely employed in the first anaerobic reactors for treatment of sewage sludge in wastewater treatment plants, as well as for treatment of liquid waste and/or organic waste with high moisture content and low organic loading. Dry AD was specifically developed to approach treatment of non-separated municipal solid waste (MSW) or organic waste rich in dry fractions or badly diluted (Gikas et al., 2017). Usually, the biogas produced in these processes is collected in the gas tank, where it can be directly exported to the national gas system or sent for combustion in cogeneration system to generate electricity (with a yield in the range of 0.7–2.0 kWh/m^3 biogas) and heat. AD process is a complex dynamic system involving microbiological, biochemical, and physical–chemical processes. Among the possible methods for treating biowaste, AD has been identified as the most environmentally sustainable option since it makes possible not only to deviate biodegradable materials from landfills but also to produce bioenergy and potential by-products, such as soil biofertilizer (Monson et al., 2004). The project described by Gkamarazi (2015), for example, uses discontinuous reactors for dry AD of solid waste and was designed to work with a total throughput of 80,000 tons of MSW per year. According to the document on the best techniques available for treatment of industrial wastes (COM, 2006), the AD process leads to a theoretical methane production of 348 Nm^3/ton of COD. This methane production depends on the raw materials, but typically, 80–200 m^3 CH_4 per ton of MSW, has been reported in the literature (Mes et al., 2003). Several studies in the literature have shown that the AD process is a robust process with efficient treatment system and high rates of organic load removal (Berni et al., 2014; Jabeen et al., 2015; Mamimin et al., 2015).

2.3.1 Microbiology of AD

Generally, four trophic groups are considered relevant to AD processes, namely hydrolytic, acidogenic, acetogenic, and methanogenic bacteria (Kwietniewska and Tys, 2014). Carbohydrates, lipids, and proteins are broken by hydrolytic bacteria into sugars, long-chain fatty acids, and amino acids, respectively. Then, these molecules are converted into volatile fatty acids, alcohols, CO_2, and H_2 in the acidogenic phase. These molecules are further converted by acetogenic bacteria, mainly to acetic acid, H_2, and CO_2. Finally, all these intermediate products are transformed into CH_2, CO_2, and water in the last stage, where methanogenic bacteria are involved.

2.3.1.1 Hydrolytic Microorganisms

During the first phase of the AD process, the most complex particulate organic matter (polymers) is transformed into dissolved simple materials. This conversion process is given by the activity of hydrolytic microorganisms (*Clostridia, Micrococci, Bacteroides, Butyrivibrio, Fusobacterium, Selenomonas, and Streptococcus*) and requires the production of exoenzymes (cellulase, cellobiase, xylanase, amylase, protease, and lipase) excreted by the fermentative bacteria, which break down complex compounds such as proteins, amino acids, and carbohydrates into mono- and disaccharides and also enable the conversion of lipids into long-chain fatty acids and glycerin (Merlin Christy et al., 2014). This step is the most time-consuming of the AD process due to the formation of toxic or unwanted compounds (Ariunbaatar et al., 2014). Hydrolysis also limits the AD speed if the used substrate is in the form of particles. Then, an intensification of hydrolysis process leads to an increase in the performance of digestion (Yu et al., 2016). Biological, chemical, and mechanical pretreatments, or a combination thereof, can be used to accelerate hydrolysis because they can cause lysis or disintegration of the substrate and allow the release of intracellular matter allowing greater accessibility of anaerobic microorganisms, thus reducing the retention time in the digester (Ferrer et al., 2008).

2.3.1.1.1 Acidogenic Microorganisms

The second phase of AD occurs due to the action of acidogenic fermentative microorganisms (*Streptococcus, Lactobacillus, Bacillus, Escherichia coli, and Salmonella*), which convert soluble products of hydrolysis into compounds such as fatty acids, alcohols, lactic acid, CO_2, hydrogen, ammonia, and hydrogen sulfide (Merlin Christy et al., 2014). The conversion of organic material into organic acids causes the pH of the system to drop. This acidic condition favors the action of acidogenic and acetogenic microorganisms, which prefer a slightly acidic pH (4.5–5.5). The acetic and butyric acids produced in this step are important in the overall performance of the AD, since these acids are the preferred precursors for methane production, which occurs in the last step of the organic material degradation process (Hwang et al., 2001).

2.3.1.1.2 Acetogenic Microorganisms

In the third phase, acetogenic bacteria convert the compounds generated during the acidogenic phase, producing hydrogen, CO_2, and acetate. During acetic and propionic acids formation, a large amount of hydrogen ions is formed, causing a decrease in the pH of the aqueous medium (Gkamarazi, 2015; Mes et al., 2003). The optimum pH for the action of acetogenic microorganisms is around 6. These microorganisms are slow-growing and sensitive to fluctuations in organic loadings and environmental changes (Merlin Christy et al., 2014).

2.3.1.1.3 Methanogenic Microorganisms

The last stage of the AD process is called methanogenesis. In this phase, methanogenic archaea promote the breakdown of organic compounds derived from the acetogenic phase. Methanogenic archaea are divided into two main groups: acetoclastic, which degrades acetic acid or methanol to produce methane, and hydrogenotrophic, which uses hydrogen and CO_2 to produce methane (Gkamarazi, 2015; Mes et al., 2003). Hydrogenotrophic methanogens are more resistant to environmental changes than acetoclastic methanogens. Unlike other microorganisms, methanogens prefer slightly alkaline environment (Merlin Christy et al., 2014) around 6.5–8 (Kothari et al., 2014; Shah M.P., 2020). A well-balanced AD process occurs when all the products obtained in a metabolic stage are converted in the next stage without accumulation of intermediate products, resulting in a complete breakdown of organic matter into final products of interest, such as methane, CO_2, hydrogen sulfide, and ammonia (Bouallagui et al., 2005). However, if the objective is to ensure hydrogen production by means of fermentative processes, it is an essential blockage or inhibition of methanogenic bacteria by controlling the temperature, pH, or even by using additives, since such bacteria consume hydrogen for methane production (Wang and Zhao, 2009).

2.4 MICROBIAL ELECTROCHEMICAL SYSTEM

Everybody believes that people need to move away from using fossil fuels and toward carbon-free energy sources, but it is not clear how that transition can be done. However, all of these resources are rare, demanding appropriate maintenance techniques. Hydrogen generation via water electrolysis is one of the more efficient ways, with tremendous potential for the change of chemical energy from the electrical energy that may be stored, transmitted, and then used or converted back to electricity on command. Water electrolysis technologies of various types have been developed, but more progress is needed before they can be integrated into massive and expensive electrical infrastructure (Koffi and Okabe, 2021; Shah M.P., 2021). The kinetics of electrocatalysts, in particular, should be enhanced, along with their cost-effectiveness. Solar-powered electrolysis, for instance, is not yet cost-effective when contrasted with hydrogen synthesis from fossil fuels. The growing need for fossil fuels and the demand to prevent releasing dangerous elements into the environment have necessitated the development of alternative and sustainable energy sources. In this context, bioelectrochemical systems (BESs) have gained considerable interest from all around the world. For well over a decade, scientists have been fascinated by microorganisms that can produce current and microorganisms that can transmit electrons directly without the use of intermediaries or electron shuttles. These systems are in different stages of evolution into diverse technologies, such as microbiological fuel cells (MFCs) for the treatment of wastewater and power generation; MECs, which are devices that use electricity to generate methane or hydrogen gas; and other microbial electrochemical technologies (METs) for desalinating water and producing products, such as hydrogen peroxide (H_2O_2) (Peera et al., 2021; Shah M.P., 2021). An MEC is a biohydrogen-producing reactor that combines an MFC and electrolysis. An MEC comprises an anode compartment and cathode compartment, a power supply, and a separator. A separator, typically a cation/anion exchange membrane, separates the cathode electrodes. At the anode, microbial strains colonize the electrode surface to form an active biofilm by producing electrons, protons, and CO_2 via biological oxidation using organic materials, such as sewage sludge, wastewater, or sugar solutions, as their energy source. The electrons produced at the anode travel through an external circuit and the solution to the cathode, where they merge with protons to form hydrogen. The protons are transferred from the anode to the cathode through the proton exchange membrane (PEM). In the case of the anion exchange membrane (AEM), the OH ions produced from the cathodic oxygen reduction reaction are transferred to the anode via the AEM. An MEC is used to produce hydrogen peroxide, methane, and hydrogen/biohydrogen to remove pollutants (Afify et al., 2017).

2.4.1 MECs

MECs are similar to MFCs in that they have two chambers connected by an ion-exchange membrane. Several various combinations of MECs have been modified throughout the decades to increase efficiency and are discussed below. A fundamental H-type cell with gas collection pieces coupled to a cathode compartment was used in prior designs (Anwer et al., 2020). Subsequently, significant improvements were made to create dual-compartmental MECs that were simple to operate. Following a comprehensive evaluation of numerous configurations, a single-compartmental MEC exhibited larger fabrication/recovery rates and current densities than a dual-compartment MEC. As a result, a lot of time and effort has been put into fine-tuning this grouping for usage in scale-up investigations. Several sorts of reactor upgrades were put together based on the results; single-chambered, dual-chambered, combined, and many others (Rahimnejad et al., 2015).

2.4.1.1 Single-Chambered MECs

The initial configuration used a glass container with an overall capacity of 50 mL. The subsequent setup applied borosilicate glass vials with a total volume of 10 mL; the cells generally contained a mixed culture or pure culture. To keep the cathode and anode, which measured 4 Ã 5 cm and 3.5 Ã 4 cm, 2 cm apart, plastic screws were utilized. The anode was constructed of type A carbon, while the cathode was type B carbon with platinum. Single-chambered MECs lack a membrane. Since hydrogen is moderately insoluble in water, when production rates are higher, the microbial conversion of methane from hydrogen is slowed. In membrane-less MECs as shown in Figure 2.1, energy losses are reduced, and the energy recovery phase is effective (Park et al., 2017).

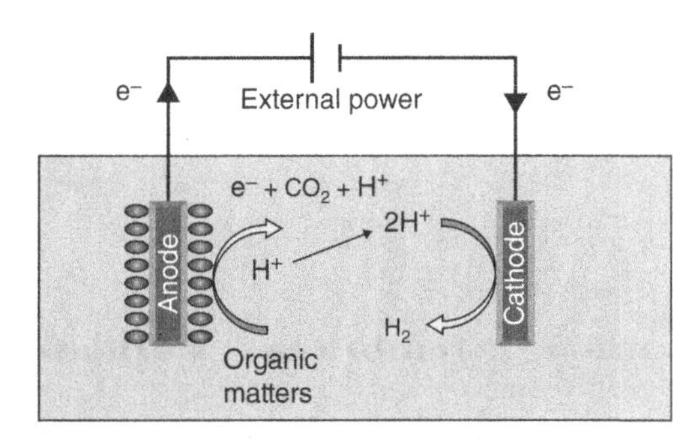

FIGURE 2.1 Diagrammatic representation of single chamber MEC without membrane.

2.4.1.2 Dual-Chambered MECs

The anodic and cathodic chambers of dual-chambered MECs are divided by a membrane. Due to their complex structures and high volumes with greater internal resistance, dual-compartmental MECs are difficult to scale-up. The application of a membrane serves two purposes. It shortens the transition from the anode compartment to the cathode compartment, preventing short circuits and preserving the quality of the cathode-side product. The PEM as shown in Figure 2.2 is the most commonly used one because it is designed to allow only freely available protons to pass through while using $-SO_3$-type functional groups. Secondary membranes, including anion-exchange membranes such as bipolar membranes, AMI7001, and charge-mosaic membranes, have been studied in MECs, along with regular membranes (Munoz-Cupa et al., 2021).

2.4.1.3 Proton Exchange Membranes

The primary role of a PEM in an MEC-based technique is to separate reactants and transfer protons from the anode to the cathode. A PEM is a semipermeable membrane formed of ionomers that are developed to transfer protons while being impermeable to gases, such as oxygen and hydrogen. Polymeric membranes or mixed membranes, wherein additional materials are embedded in a polymer matrix, can be used to construct PEMs. Nafion is the most popular PEM material, with a hydrophobic Teflon-like backbone (-CF2-CF2-) and hydrophilic side chains terminating with ion-conducting sulfonic acid groups ($-SO_3-H$). However, the Nafion membrane is costly, prone to fuel and gas crossovers, and has limited proton selectivity. By combining with the gases produced in the anode compartment, an MEC decreases hydrogen purity in the cathode compartment. As a result, a variety of new membrane

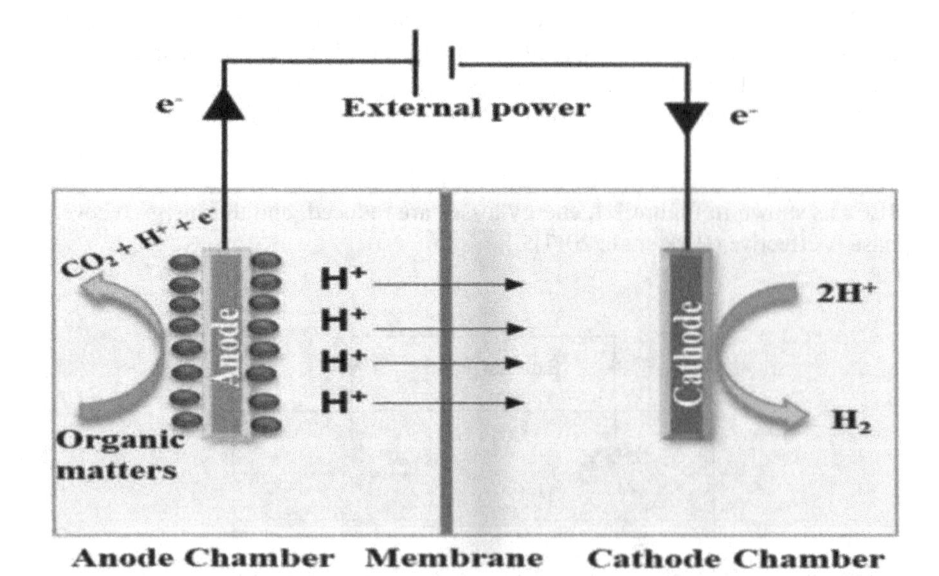

FIGURE 2.2 Diagrammatic representation of double chamber MEC having a proton exchange membrane.

types have been developed that use different proton (or ion) conductors. Therefore, for commercial applications of those technologies, it is important to develop alternative membranes to the expensive Nafion. The alternative choices to Nafion are considered based on a few previous works reported by researchers (Park et al., 2017). A nanofiber-reinforced composite proton exchange membrane (NFR–PEM) based on sulfonated polyether ether ketone (SPEEK) as a proton conductor was prepared and studied for MECs. A sulfonated poly(arylene ether sulfone) (SPAES)/polyimide nanofiber (PIN) composite PEM was developed for use in MECs, where diverse cations that compete with protons coexist in high concentrations.

2.4.1.4 A Cathode-On-Top Single-Chamber MEC

The reactor was made up of two parts: a top cover and the main chamber, both of which were constructed of glass and had a capacity of 0.4 L (Kadier et al., 2016b). The substrate and electrolyte were pumped in via the bottom intake, and the produced gas was collected from the cathode with the use of a gasbag. Hydrogen production rate (HPR) increased from 0.03 L·L −1 day−1 to 1.58 L·L −1 ·day−1 in a 24-hr batch test when the applied voltage was expanded from 0.2 V to 1.0 V, and total hydrogen recoveries rose from 26.03% to 87.73% when the applied voltage was increased from 0.2 V to 1.0 V. The greatest total energy recovery was 86.78% at an applied voltage of 0.6 V (Guo et al., 2013).

2.5 FUNDAMENTALS OF MECs

The first study on MECs was published by Liu et al. (2005). Since then, exponentially growing numbers of papers have been published (Kadier et al., 2016b; Lu and Ren, 2016). Basically, the exoelectrogens (i.e., the bacteria that can transfer electrons through extracellular mechanism) in MECs use anode as an electron sink to oxidize organics and produce protons and electrons. The protons diffuse via a PEM to cathode and are reduced by the electrons transferred via the external electric circuit to produce H_2. The MEC system cannot work spontaneously due to the thermodynamic barrier, and additional energy is required to drive the reduction reaction (Kadier et al., 2015; Kumar et al., 2017). In theory, any carbon compounds that can be digested by the exoelectrogens can be used in anode of MECs. This has led to a large variety of organic sources studied in MECs, ranging from defined pure chemicals to mixture of real wastewaters (Escapa et al., 2016; Kadier et al., 2014). With different carbon sources, the anode potential can range from 0.2 to 0.5 V. The half-reaction at the cathode is the reduction of protons to H_2 gas. Based on the Nernst equations, the cathode potential will be affected by the concentration of free protons (i.e., pH of electrolyte), temperature, and partial pressure of H_2. The cathode potential decreases with increases in temperature, pH value, and partial pressure of H_2. With respect to influence on cathode potential, pH is the most critical parameter (Rozendal et al., 2006). The energy input is essential for MECs because of their thermodynamic limitations (Call and Logan, 2008; Liu et al., 2005). Under standard conditions (pH of 7, H_2 partial pressure (pH_2) of 1 bar, and temperature of 25°C), the cathode potential for H_2 evolution reaction is about -0.41 V (Rozendal et al., 2006), which is more negative than the anode potentials using most of the carbon sources. Theoretically, it should be possible to make the reaction happen spontaneously with glucose as substrate, but glucose in that case has to be fully oxidized to CO_2, which does not happen in anaerobic fermentation (Wünschiers and Lindblad, 2002). Another approach to reduce or remove the energy requirement of MECs is to increase the cathode potential by changing the pH or pH_2. However, maintaining a low pH in cathode would block the proton migration through the membrane (Rozendal et al., 2006), subsequently resulting in decreased pH in anode, which is harmful to the microorganisms. On the other hand, maintaining an extremely low pH_2 is impractical during the operations.

2.6 MEC FOR WASTEWATER TREATMENT

2.6.1 Working Principles

MEC is a recently developed BES that interacts with bacterial metabolism with electrochemistry (Liu et al., 2005). MEC is related to MFCs; however, MEC reverses the process of generation of bio-H_2 from organic matter present in biomass by applying an external electrical output in the form of direct current (or voltage). Anode-respiring bacteria (ARB) or exoelectrogens present in anodic compartment serve as biocatalysts to transfer electrons generated from organic compounds to the anode electrode during the oxidation process. The generated electrons during anodic reactions further passed through the external circuit and were reduced with protons to form hydrogen gas in the cathodic chamber as shown in Figure 2.3. External power supply boosts the applied voltage (Eap) of the electrons that move to the cathode so that they have enough energy to reduce the protons to H_2. The mechanism of the hydrogen evolution reaction (HER) comprises multistep reactions in which the first step is the electrochemical Volmer reaction, which is followed by a chemical Tafel reaction or an electrochemical Heyrovsky reaction (Jeremiasse, 2011; Varanasi et al., 2019). Kadier et al., 2020 states that the rate of hydrogen production depends on inoculum properties, type of substrate used, applied potential, anodic as well as cathodic parameters, operating conditions, etc.

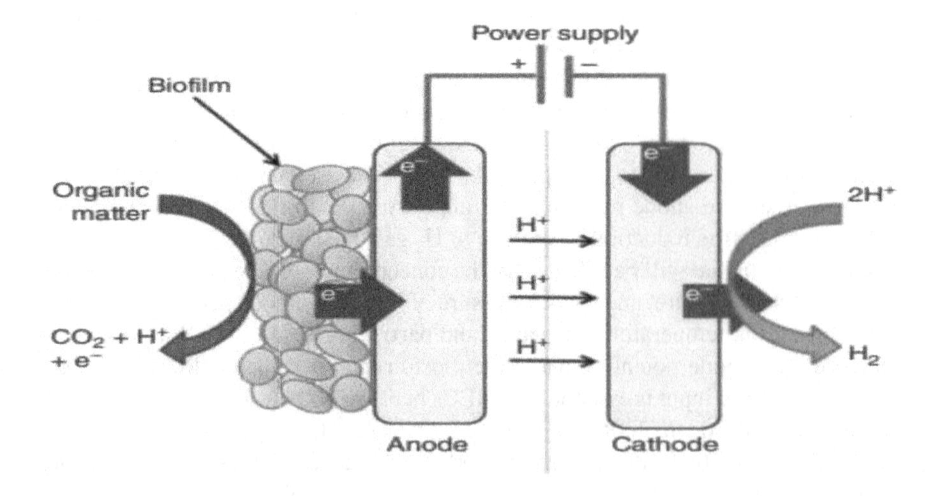

FIGURE 2.3 Schematic representation of a typical MEC with a proton exchange membrane and its operation.

2.7 CURRENT WASTEWATER TREATMENT METHODS AND LIMITATIONS

Conventional wastewater treatment relies on sewerage to transport wastewater to a centralized wastewater treatment plant and uses end-of-pipe technologies to treat the waste in a linear fashion. Transition to a closed-loop model is required (Werner et al., 2016), in which water is treated while simultaneously recovering energy and nutrients. Compared to centralized solutions, decentralized wastewater treatment plants can demonstrate reduction in operation and capital expenditure (OPEX). Even though decentralized solutions are in their infancy in terms of deployment and optimization compared to centralized solutions (Garrido-Baserba et al., 2018), the benefits of decentralization are showing, and trends in wastewater treatment plants are making a transition toward decentralization with resource recovery. Centralized water systems are not applicable in many parts of the world where there is rapid population growth and a swing toward urbanization (United Nations, 2018). The shift from rural to urban living has put wastewater infrastructure under pressure and, in many cases, meant that well-designed and effective wastewater treatment systems have not been implemented. Creating centralized wastewater treatment systems retrospectively is difficult and expensive. Shifting toward distributed and non-networked technologies creates a decentralized wastewater treatment infrastructure that can provide greater resilience and a lower economic and environmental cost compared to conventional systems (Starkl et al., 2015).

2.8 MECHANISM OF MECs FOR WASTEWATER TREATMENT

MECs offer an alternative to centralized wastewater treatment methods. The embodied chemical energy potential of wastewater, generated from its organic components, is approximately 9.3 times greater than the energy needed to treat it (Kim et al., 2018). Harnessing energy from biomass using biological anaerobic wastewater treatment methods is already popular due to their relative simplicity and robustness (Escapa et al., 2015). Today, AD is widely used to produce biogas from sludge. AD generates energy via the digestion of acetate to form CH_4 and CO_2 (Van Eerten-Jansen, 2011). However, the process is relatively slow and produces biogas with high levels of CO_2, reducing its energy density. The high CO_2 content makes the biogas difficult to store and requires extensive chemical treatment, including cryogenic separation, to remove CO_2 before use (Awe et al., 2017). MECs offer an alternative approach to produce both CH_4 and H_2 MECs contain electrodes poised at a set potential and utilize microorganisms as biocatalysts to reduce activation overpotential of a given redox event, increasing the MECs' voltage efficiency and product yield (Kitching et al., 2017). Microorganisms form a biofilm on the surface of the anode that can convert chemical energy, present in organic substances, into electrical energy, which is then used to create other valuable products such as H_2 and CH_4 at the cathode (Pant et al., 2011). These microorganisms are electrochemically active and, therefore, can exchange electrons with the electrode in order to maintain cellular function and growth (Sleutels et al., 2012). MECs can be configured for many different outputs, including H_2 and CH_4 production, through different reactor architecture designs, which may include the use of multiple chambers and membranes. MECs designed for

H_2 production often require a separator between the bioanode and cathode to stop H_2 evolution to CH_4, with microorganisms only adhering to the anode. MECs for CH_4 production have a biocathode where methanogenic microorganisms are used to generate CH_4 from H_2 and CO_2 at the cathode (Nelabhotla et al., 2019). These systems are called MEC-ADs (Yu et al., 2018) or methanogenic microbial electrolysis cells (MMECs) (Siegert et al., 2015) and use the process of electromethanogenesis to produce CH_4. The anode and cathode are connected to a power source, under anaerobic conditions. Wastewater enters the reactor, and electricity is supplied to the electrodes at precisely controlled potential differences. Biogas is produced, and wastewater is treated, meaning effluent water has significantly lower levels of organic contaminants.

At the anode, electroactive Bacteria (EAB) (s) grow to form a biofilm, which uses electrical energy to oxidize organic matter to CO_2 (Kadier et al., 2016a).

$$\text{Anode}: C_2H_4O_2 + 2H_2O \rightarrow 2CO_2 + 8H^+ + 8e^-$$

At the cathode, CH_4 production takes place. There are several different pathways to CH_4 production through direct or indirect electron transfer.

Investigations have been conducted to determine the most favorable route of methanogenesis. Theoretically, CH_4 can be produced by direct electron transfer at relatively low voltages. However, in practice, this is very energy intensive as an electrocatalyst is required to lower the overpotentials. Therefore, a lack of an appropriate catalyst to reduce overpotentials means that high electrode potentials are needed to drive the reaction. CH_4 generated in ADs mainly originates from acetate (~70%) via acetotrophic methanogenesis, where acetate is produced as an intermediate. However, analysis of the microbial consortium present at the cathode in an MEC-AD indicated that intermediate hydrogen production was favored (Zeng et al., 2015; Rozendal et al., 2008). Villano et al. demonstrated that only a fraction of CH_4 produced was via direct electron transfer, with the majority from H_2-mediated methanogenesis via hydrogenotrophic methanogens. This is an important factor to consider when evaluating cathode materials (Villano et al., 2010). If H_2-mediated methanogenesis is favored, then the hydrogen evolution ability of the cathode material must be considered. Cathode materials, therefore, can be seen as acting as biocatalysts by enhancing electrode-microbe electron transfer that improves the rate of formation of products (Zhang et al., 2012). The review aims to analyze the critical aspects of the treatment process, namely the electrode materials, feedstock, and inoculum, to determine the most efficient parameters for the treatment of wastewater using MEC, then to situate this with economic analysis to evaluate each parameter in the context of what is economically viable to facilitate industrial adoption of this technology. This study seeks to act as a bridge between research and industry, analyzing the future steps toward commercialization of the technology.

2.9 PARAMETERS AFFECTING THE DESIGN OF MEC SYSTEMS FOR INDUSTRIAL WASTEWATER TREATMENT

Accelerating the commercialization of this technology will increase its impact within the sector while increasing funding and research for continuous improvements. There are multiple variables to consider when analyzing and comparing MECs for wastewater treatment and energy production. This review identifies the strength and type of feedstock, the anode and cathode materials, the size and electrode surface area to reactor volume, the system architecture, the outputs, and the costs as the most significant parameters affecting the system design. These parameters affect the system performance and the economic viability of commercializing the technology. To make MECs commercially viable, a balance must be struck between system-optimizing performance and economic efficiency of all components. Future MEC design needs to incorporate anode cost reduction, improvements in organic loading rates (OLRs), understanding of maintenance requirements, and electrode lifetime expectancy (Aiken et al., 2019). The study will evaluate the parameters in the context of industrial use, where cost and the ease of manufacturing at all stages are considered. In order to identify the next steps, researchers should look toward accelerating the adoption of this technology by industry.

2.9.1 Feedstock

The identity of the feedstock has a large impact on the performance of MECs. The strength of the wastewater refers to the level of contamination of the water and dictates the treatment time, size of the reactor, and the energy requirement and production. High strength wastewater has a higher organic load; therefore, more energy can be recovered by treatment methods. However, high strength wastewater usually takes longer to treat, leading to greater hydraulic retention times (HRT). Research should look at optimizing the HRT to help reduce the reactor size and affect energy production/consumption. The organic loading drastically changes depending on the wastewater sources, which in turn affects the OLR. With improvements in the cost of MECs, a viable OLR for a system ranges between 800–1400 mg/L. Urban wastewater is low strength with a COD value of 300–500 mg/L, therefore, requiring a shorter HRT of 5–9 hrs, whereas swine waste, with a higher COD of 18,300 mg/L, requires a very longer HRT of 314 hrs. Longer HRTs act as a significant barrier to industrial adoption, reducing the amount of waste that can be treated per day as well as increasing the size of reactors needed. However, high strength wastewaters produce more energy and, therefore, can be more economically viable in terms of energy production, as higher organic loads contain higher energy potential. A normalized net energy production of 76.2 $kWh/m^3/day$ was achieved for pig slurry, a high strength wastewater, at a volume of 16 L and an anodic and cathodic surface area of 11.25 m^2/m^3 (San-Martín et al., 2019). However, as for low strength wastewater (2000 mg COD/L/day), the theoretical HRT for the different waste streams has been calculated, as shown in Table 2.1. Table 2.1 also indicates that it is possible to increase the HRT and still be competitive against activated sludge treatment. MECs also have a significantly lower HRT than AD due to more efficient treatment. The HRT of AD systems is typically between 20–30 days. The long treatment time of AD suggests that using MEC-ADs instead would allow for reduced facility size, as a higher OLR can be used. MEC-ADs provide multiple further benefits over AD including increased biological stability, higher energy production, and an increase in substrate removal, especially of complex compounds that are difficult to remove (Yu et al., 2018). The improvement in performance stems from both the presence of electrodes,

TABLE 2.1
Integration of MEC with Other Technologies for Energy Production

S/No	Type of Reactor	Type of Integration	Applied Energy (Voltage/ Current)	Hydrogen	Type of Electrode	Current or Power Density
1	Single chamber	Dark fermentation and Microbial Fuel Cell–Microbial Electrolysis Cell coupled system	0.33–0.47 V	0.48 m^3 $H_2/m^3/d$	Anode: Carbon brush Cathode: Platinum-coated carbon cloth	52 A/m^3 (MEC)
2	Double chamber	Thermoelectric microconverter–Microbial Electrolysis Cell coupled system	0.17–0.83 V	0.16 m^3 $H_2/m^3/d$	Anode: Plain CF Cathode: Carbon paper with Platinum	0.28 to 1.10 A/m^2
3	Single chamber	Microbial Electrolysis Cell with Hydrogen Bioreactor (HBR)	0.6 V	0.53 mmol/ hour	Anode: Graphite plate Cathode: Graphite plate	N/A
4	Single chamber	Microbial Electrolysis Cell Anaerobic Digestion coupled	0.0–0.8 V	CH_4 at 0.4 V	Anode: Carbon brush Cathode: Ti/RuO_2	Steady increase in current with applied voltage

(*Source*: Adapted from Dange et al., 2021; Koul et al., 2022)

greater surface area for bacterial attachment, and the applied voltage. Arvin et al. found that with electrodes present but with no voltage applied, COD removal was 49% compared to 58.4%, 64.7%, and 72.6% for 0.6 V, 0.8 V, and 1.0 V, respectively (Arvin et al., 2019). The energy demands for traditional wastewater treatment are high, especially for activated sludge treatment, equating to 60% of the total energy needed. Using MEC-ADs for wastewater treatment has resulted in a 1.7 times higher energy production and an accelerated substrate removal compared to AD, indicating the potential of highly efficient treatment for any commercial waste streams already used for AD. Furthermore, reducing the need for post-treatment would significantly reduce the energy requirements within the treatment process. Using MECs to reduce the energy consumption of wastewater treatment would have significant impacts throughout the world. Research into the performance of MECs in treating different waste streams is increasing and is helping to generate an understanding of how different microbial communities perform with the substrates. If MECs and MEC-ADs are to become commercially viable, they will compete with other technologies in the field: AD systems that focus on energy generation or activated sludge systems that focus on nutrient and COD removal. Going forward, researchers should assess the economic viability of treatment of particular wastes and compare MEC and MEC-AD performance to the state-of-the-art technology used by industry.

2.9.2 Electrode Material

2.9.2.1 Anode Material

The anode's performance is critical for BESs that rely on bioelectrochemical reactions, which take place at the anode, and which can be replaced by MFCs, MECs, and MEC-ADs. It was established that anode activity is a restrictive component in the overall performance of the system (Lim et al., 2017). As an electroactive bacterium (EAB) attaches itself to the bioanode surface, producing a biofilm, it provides the bioanode with energy. As a result of this oxidation by EAB, organic molecules are converted to CO_2. Because of their remarkable ability to adhere to EAB, their huge surface area, and their abundance, carbon-based materials have become the most extensively used electrode material (Savla et al., 2020). It has been shown that carbon compounds help to increase interfacial microbial colonization and, thus, the production of biofilms. Electrically conductive current collectors of metals are employed as electron acceptors, to overwhelm poor conductivity. Titanium wire is often utilized because of its corrosion resistance. Additionally, the capability to simulate interfacial microbial colonies allows for improved current density by developing a beneficial microenvironment for electron transport that compensates for the decreased conductivity (Li et al., 2017). Graphite is affordable, plentiful, and conductive, and, because of this, it has become one of the most extensively used electrode materials (Liu et al., 2018). Graphite electrodes have been implemented in many ways, including brush, granular, rods, felts, and foams. Nevertheless, graphite's molecular structure and morphology are both planar in comparison to other carbon materials due to its low surface porosity required for bacterial adhesion. Surface area-increasing porous 3D carbon materials, such as carbon brushes, felts, meshes, and foams, have been the focus of recent research (Guo et al., 2015; Saheb-Alam., 2018). Carbon fiber (CF) electrodes have previously proven to be effective and are currently being used to achieve good outcomes (Carlotta-Jones et al., 2020). Carbon nanotubes (CNTs) have incredible electrical, mechanical, biological, and thermal properties, making them ideal for real-time applications. Despite extensive study and application possibilities of CNTs, many issues, such as biodegradability, biotoxicity, and biosafety, remain difficult to address and should be addressed with caution prior to design and manufacturing (Rasheed et al., 2020). Anodes of a mesh-like design (i.e., porous, woven, or multilobed) tend to generate greater current densities (more current flows) than flat or plate-shaped anodes because of improved mass transfer, surface area, and biofilm growth. CF brushes give excellent test results; however, because of their very expensive cost, they are seldom employed in large-scale BES. Based on an independent study, which used recycled CF anodes and found that, in comparison to graphite felt anodes, the use of recycled CF electrodes produced better results while also being cheaper, Carlotta-Janes et al. found that it was possible to improve performance while cutting costs if recycled CF anodes were used. As there is a considerable portion of the anode in the current model without a biofilm, increasing the surface area of the anode is more likely to result in a greater increase in biofilm density and adherence to the anode. Reduction in anode size will be advantageous for commercial viability since the anode material constitutes about 70% of the whole system, which will need a 90% drop in cost to make it profitable. Additionally, molybdenum anodes showed excellent overall durability, neither corroding nor lowering in current production for over 350 days. Another important consideration is the endurance of the electrode materials. Unfortunately, there are no data on electrode materials' long-term durability, and most experiments last about 1 year. Material dissipation in electrodes is often underestimated, which may lead to significant issues when determining which materials to utilize commercially. Stainless steel is also good since it has several characteristics that can be utilized (Pocaznoi et al., 2012). In terms of conductivity and scale-up potential, stainless steel outperforms carbon anodes owing to lower capital expenditures, despite a relatively flat surface, which reduces its biocompatibility. Stainless steel has a high nickel concentration and may efficiently catalyze the HER. Stainless-steel brush cathodes, for example, produced hydrogen at a rate of 1.7 m^3 m^{-3} day^{-1} and had a cathodic efficiency of 84%, comparable to Pt cathodes in single-chamber MECs. The high Ni content (8–11%) and the large specific surface area were also implicated for the rapid hydrogen generation (810 m^2 m^{-3}). Flame spray oxidation improves the biocompatibility of stainless steel by producing an iron oxide coating on the surface, which facilitates the adherence of iron-reducing bacteria and increases surface roughness without sacrificing corrosion resistance (Guo et al., 2014). Because stainless steel has yet to be tested on pilot systems, more research into its durability is needed. Cotterill et al. compared a 30 L tank to a 175 L tank to examine how tank capacity influences H_2 production. Hydrogen generation was fourfold greater in the small MEC in comparison to the bigger MEC, when the anode surface area was reduced from 1 m^2 to 0.06 m^2. The larger MEC had a lesser performance, demonstrating a negative relationship between scale and gas output, implying that efficiency decreases as size increases (Cotterill et al., 2017). As part of the commercialization process, a cost–benefit analysis of anode materials must be completed, which considers the material's availability, corrosion resistance, and capacity to scale-up.

2.9.2.2 Cathode Material

The necessity to create either CH_4 or H_2 dictates reactions at the cathode, and this is dependent on the need and potential to manufacture and utilize H_2 or CH_4 on-site. The rate at which H_2 is consumed is determined by the amount of methanogenic activity present. The hydrogen will very probably be consumed throughout the reaction if the device is operated as a single chamber without a membrane, and the biogas generated will be in the form of CH_4 (Bo et al., 2014). Temperature has a direct influence on methanogen activity, with temperatures exceeding 35°C considerably boosting methanogenic activity (Ahn et al., 2017). Hydrogenotrophic methanogens are more prevalent in MECs where CH_4 is produced, according to the study. Hydrogenotrophic methanogens produce CH_4 through the intermediate synthesis of H_2. The CH_4 synthesis route reveals that the cathode material's hydrogen evolution capacity is a critical design factor. As a result, the cathode serves as both a biocatalyst and an electrocatalyst, enhancing HERs by increasing electrode–microbe electron transfer (Zhang et al., 2012). The presence of hydrogen-scavenging bacteria in waste streams necessitates the use of membranes if pure hydrogen is required. Multiple investigations have shown that membrane systems can achieve hydrogen purity above 98% (Mohammed and Ismail,

2018). Corrosion resistance, good conductivity, high specific surface area, biocompatibility, and outstanding mechanical qualities are all required of successful cathode materials (Wang et al., 2019). Furthermore, cathodic materials must minimize significant hydrogen evolution overpotentials. Cathode fabrication for industrial application must be low-cost, utilizing easily accessible materials and conventional production procedures, for the large-scale deployment to be practicable. Metals have been investigated because they conduct electricity more efficiently than carbon-based materials (Santoro et al., 2017), and they have greater biocompatibility, as well as cathode potential, which prevents corrosion. Platinum has the strongest HER activity, which leads to improved H_2 evolution (Sangeetha et al., 2016). Platinum, on the other hand, has some disadvantages, including being expensive and having substantial mining environmental effects; as a result, the invention of new metallic electrode materials is required (Siegert et al., 2015). Stainless steel and nickel have performed well as nonprecious metals (San-Martín et al., 2019). Stainless steel is a typical material for electrode construction because it is a relatively inexpensive metal. When it comes to hydrogen production, stainless steel with a large specific surface area can be as effective as a platinum catalytic electrode containing carbon. Because of its high conductivity as a transient metal, stainless-steel mesh is thought to have outstanding ohmic resistance and electron transport resistance. Meshes and brushes made of stainless steel have a low cost and excellent performance, making them an ideal cathode made of a nonprecious metal for further evaluation and scale-up operations. The findings are consistent with the use of meshes and wool in pilot-scale systems to produce a low-cost, high-surface-area cathode with a low cost and large surface area. Nickel, like other non-platinum metals, has high corrosion resistance as well as high hermetic electron transfer activity. Nickel is also more corrosion-resistant than stainless steel, which is important for an electrode because it must be long-lasting in order to be commercially viable. Hydrogenotrophic methanogens play a role in the enhanced performance, implying that nickel's high HER activity relative to other materials helps it perform better (Siegert et al., 2014). HER activity must be a key focus of study to maximize the efficiency of cathodes. Stainless steel is now commonly used. Due to its availability, a pilot study has shown it to be the ideal cathode material for scale-up. Cost and machinability are two factors to consider. On a pilot scale, a comparison was made between nickel and stainless steel. It would be wise to experiment with nickel cathodes to determine if the improved performance justifies the additional expense.

2.9.3 Effect of Electrolyte pH

Because the HER at the cathode depends on electrolyte pH and has the most crucial impact on overall performance of MECs. High overpotentials can occur owing to a difference in redox potentials between anode and cathode chambers; it was observed that more cation instead of proton percolates through the cation-exchange membrane. Consequently, the cathode becomes alkaline while the anode becomes acidic. Theoretically, 59 mV of voltage loss is incurred due to a difference in pH level of 1 between the anode and cathode. Microbial activities are pH-dependent; microbes are highly sensitive to surrounding pH, and its variability may cause modifications in microbial respiration and, consequently, extracellular electron transfer.

In fact, because microbes are mostly active at neutral pH, most MEC studies are conducted at pH 7. Moreover, many other parameters (ion transfer, conductivity, substrate oxidation, etc.) are directly or indirectly associated with pH. Researchers have reported that low cathode and high anode pH improved hydrogen production (Ki et al., 2017; Liu et al., 2012; Zhang and Angelidaki, 2014). Protons accumulate under high pH, thereby increasing the electrogen proliferation due to conducive environment. Research suggests that periodic polarity reversal can be used to stabilize pH in two-chambered MECs (Borole and Mielenz, 2011). An electrolyte, including a weak acid, operates as an electric charge at high pH, increasing MEC characteristics, and the deprotonation process may increase the conductivity of the electrolyte while lowering the impedance between the anode and cathode. However, we must evaluate the possible impacts of weak acid catalysis and solution resistance for a lower pH electrolyte to determine whether the reactor can function more effectively. However, certain experimental findings revealed that the presence of phosphate species and some weak electrolyte acids, as a charge carrier for improving conductivity, had a beneficial impact on a stainless-steel brush cathode and also reduced the Pt/C cathode's overpotential. Merrill and Logan (2009) found that lowering the pH improves MEC performance by lowering solution resistance and cathode overpotential. Munoz et al. (2010) found that using phosphate as an electrolyte may increase the rate of hydrogen generation and current density in MECs. Yossan et al. (2013) investigated five kinds of catholytes in MECs, namely, deionized water, tap water, NaCl solution, acidified water, and a phosphate buffer. Due to its greater buffer capacity, a 100-mM phosphate catholyte in an MEC exhibited the best rate of hydrogen generation. As a result, phosphate is the most often utilized electrolyte in MECs.

2.9.4 Temperature

Temperature is a key element in MECs that affects their function because it improves exoelectrogen selectivity and production. Most microbes prefer an optimum temperature range of 35–40°C for growth, enzyme activity, and the development of a durable biofilm, which increases substrate degradation, mass transfer, and power generation. According to the COD removal efficiency and microbe loading at the anode, Omidi et al. reported that 31°C is the most efficient operating temperature for MECs (Omidi and Sathasivan, 2013). As a result, the test temperature of an MEC is typically kept at about 30°C. Lu et al., on the other hand, demonstrated that utilizing a single-chambered MEC produces hydrogen at low temperatures, such as 4°C, while simultaneously reducing the generation of methane (Lu et al., 2011). Additionally, the anode biofilm and hydrodynamic force both impact hydrogen generation in MECs (Ajayi et al., 2010), with the hydrodynamic force having a larger effect on hydrogen production than the anode biofilm (Wang et al., 2010).

2.9.5 Applied Potentials

As explained in the above sections, a minimum of 0.2 V is required to break the thermodynamic barrier for feasibly producing hydrogen at the MEC cathode. Large cathodic overpotentials reduce the efficiency of the overall process. Even though hydrogen evolution increases with increasing applied potential (Sun et al., 2008), optimum potential ranging from 0.2 V to 0.8 V must be applied for achieving process scalability (Nam et al., 2011). Researchers have reported that varying applied potential can decrease cell metabolism and increase cell lysis (Chae et al., 2009).

2.10 COUPLING MEC WITH OTHER TECHNOLOGIES FOR VALUE-ADDED APPLICATIONS

2.10.1 MEC-AD Coupled System

AD occurs as a result of anaerobic microorganism metabolisms that degrade biodegradable materials and create biogas or CH_4 gas (Waqas et al., 2018). This is a technique that has been utilized commercially for concurrent CH_4 production and waste/effluent remediation (Zhang and Angelidaki, 2015). Electrochemical techniques and AD can be merged for improved productivity, with stillage from a certain process utilized as feedstock in the earlier and energy retrieved from the earlier utilized in the latter (Sadhukhan et al., 2016). In another study, an upflow anaerobic sludge blanket reactor was compared to an upflow anaerobic bioelectrochemical reactor for CH_4 production from acidic distillery effluent. The upflow anaerobic bioelectrochemical reactor was set to 300 mVolt, and both the upflow anaerobic sludge blanket reactor and the upflow anaerobic bioelectrochemical reactor were run in a constant state. The combination, that is, upflow anaerobic bioelectrochemical reactor, resulted in a considerably greater methane yield of 407 milliliters per gram CODr at 4.0-gram COD/L day than upflow anaerobic sludge blanket reactor (282 milliliters per gram CODr). By integrating an MEC technology into the reactor, De Vrieze et al. attempted to assess the rational structure underlying the enhanced operation of AD. They found that adding electrodes and balancing them at 0.75 and 1.205 V enhanced the stability of the AD treating treacle. Methane generation in the control reactors decreased to 50% of the starting rate (on day 91), whereas it stayed constant in the MEC-AD reactors, implying a stabilizing impact. Surprisingly, when the electrodes from these reactors were placed in the control reactors, the CH_4 output jumped threefold to fourfold. This demonstrated that the electrochemically active biofilm generated on the surface of the electrode, instead of the electric current, should have improved the constancy of AD. Integrating MECs and AD is a great illustration of how MECs may be utilized in a modular fashion to improve the productivity of current technology (Aryal et al., 2018). The underlying premise for organic substrate breakdown is identical in the AD and MEC procedures. The ultimate end products differ because the ultimate electron acceptor is varied. In MECs, CH_4 can be created electrochemically as well as through fermentation processes. It has been proposed that combining these two procedures can help overcome the limits of separate methods. Cerrillo et al. observed that the MEC system retrieved limiting products of the AD, such as NH_3, hence increasing the production of CH_4 (Cerrillo et al., 2017; Zhi et al., 2019). Since few traditional AD techniques have been marketed, the development of integrated AD–MEC systems appears to be feasible in the coming years.

2.10.2 MEC with Anaerobic Membrane Bioreactor (MBR) (MBR and Acidogenic)

Membrane bioreactors are effluent remediation systems that combine a semipermeable membrane with a suspended growth bioreactor. The following are some examples of how membrane filtering can be included in BESs: (1) Between the electrodes, as a separator; (2) In the anode/cathode section, there is an inner filtration element; or (3) Prior to or after the BES, there is an outer remediation technique. More effective treatment has the benefits of integrated systems, including increased energy efficiency, lower investment, less fouling, and/or long-term desalination (Yuan and He, 2015). Katuri et al. created the anaerobic electrochemical membrane bioreactor. A unique anaerobic remediation technology; a mix of MEC and membrane filtration using nickel-based hollow-fiber membranes that are electrically conductive and porous (Katuri et al., 2014). The anaerobic electrochemical membrane bioreactor was used to treat low-organic strength effluents and solutions, as well as to retrieve biogas. The nickel-based hollow-fiber membrane fulfilled two purposes: as a cathode electrode for producing hydrogen and as a membrane for filtering the purified water. At a 700-megavolt applied voltage, greater than 95% of the COD (starting COD: 320 mg/L) was removed, and up to 71% of the substrate energy (methane-rich biogas, 83%) was retrieved. In addition, the anaerobic electrochemical membrane bioreactor had little membrane fouling than the control reactor, which was related to the generation of hydrogen bubbles, a lower cathode voltage, and a localized higher pH at the cathode surface. Although there are numerous advantages to integrating a membrane bioreactor with an MEC, more information regarding energy production and consumption in the connected system is required.

2.10.2.1 Acidogenic Bioreactors

Acidogenic bioreactors can be paired with MECs in the same way as AD and MECs may be mixed. Babu et al. tested biohydrogen generation in a single chamber MEC using CH_3COO^- (acetate), $C_3H_7COO^-$ (butyrate), and $C_3H_5O_2^-$ (propionate) as substrates at varied voltages (Babu et al., 2013b). At 600 megavolts, the highest HPR of 2.42 millimole/hour was reported, with around 53% of synthetic acids removed (Babu et al., 2013a). The same researcher's group also paired an acidogenic bioreactor with MECs to boost product retrieval and hydrogen production. MEC was run at 3000 milligram/L volatile fatty acid concentrations under various poised potentials, with the highest HPR of 0.53 millimole/hour and 49.8% volatile fatty acid consumption observed at 600 megavolts. A unique bio-electrohydrolysis system based on self-inducing electrogenic activity was developed as a pretreatment tool to boost hydrogen generation efficiency via the remediation of food waste in another research. Hydrolysis (1st stage) was preceded by acidogenic fermentation for hydrogen generation (2nd stage) in a two-stage coupled or hybrid system. Bio-electrohydrolysis produced more hydrogen (29.12 milliliter/hour) as a result of pretreatment than the control (26.75 milliliter/hour). Furthermore, substrate breakdown was increased with the bio-electrohydrolysis-pretreated substrate (a 52.42% reduction in COD) as compared to the control (COD removal of 43.68%) (Modestra et al., 2015), additionally developed a one-chambered MEC with an acid-pretreated biocatalyst for electro-fermentation of wastewaters for further hydrogen generation while simultaneously treating the wastewater. At 200 mVolt and 600 mVolt applied potentials, the effect of volatile fatty acid concentration (4000 milligrams/L and 8000 milligrams/L) on biohydrogen generation with simultaneous treatment was investigated. At 600 mVolt, the highest HPR was 0.057 mVolt/hour, with a volatile fatty acid utilization of 68% (Modestra et al., 2015). As a result, by coupling MECs with acidogenic reactors, wastewater from these reactors might be transformed into usable chemicals like hydrogen.

2.10.3 Thermoelectric Microconverter-MEC Coupled System

Industrial procedures, such as those in the automobiles and steel sectors, produce waste heat as a by-product, which thermoelectric converters can use as a source of energy, and this energy is known as thermoelectricity. Thermoelectricity changers work by using a temperature gradient in the medium, which can be liquid or other solid phases. Industrial processes produce a lot of heat, which can generate a temperature gradient that can help with thermoelectricity production. Cooler temperatures, on which there is little research, can also be used to capture thermoelectricity. It is more eco-friendly to retrieve waste heat from cold temperatures. The operation of MECs for the creation of hydrogen necessitates only a little quantity of electrical energy. Thermoelectricity is a natural nonconventional resource (Chen et al., 2016). As a result, combining MECs with thermoelectricity can improve the long-term viability of the hydrogen generation procedure. Chen et al. investigated the impacts of various temperature ranges on hydrogen synthesis from CH_3COO^- as a carbon source in a thermoelectric microconverter-MEC linked system. The thermoelectric microconverters produced electric potential and were discovered to affect the systems' HPR. The voltage ranged between 170 mVolt and 830 mVolt depending on the temperature (between 35–55°C). At 55°C on the hot side, the highest hydrogen generation of 0.16 m^3/day and output of 2.7 mol/mol CH_3COO^- were reported, with an average voltage of 700 mVolt and current density ranging from 0.28 to 1.10 ampere/meter square (Kadier et al., 2016a).

2.11 APPLICATIONS OF THE MEC

2.11.1 Obtaining Value-Added Products

2.11.1.1 Hydrogen Production

A viable treatment solution for improved hydrogen generation from food waste was determined to be a one-chamber MEC with negative pressure control (Harirchi et al., 2022). To obtain efficient H_2 retrieve, food waste was used as a substrate in a combined reactor that included one-chamber MEC remediation and AD (Feng et al., 2018). Throughout continuous AD-MECs operation, some investigations used a combined reactor to integrate single-chamber H_2 production (511.02-milliliter hydrogen gram 1 VS), which was greater than that attained by AD (49.39-milliliter hydrogen gram 1 VS). In AD-MECs, H_2 recovery was 96% and electrical energy recovery was 238.7%, accordingly (Huang et al., 2020a). Key components of food waste (lipids,

VFAs, carbohydrates, and protein) were studied to determine use of organic material to find the mechanism of hydrogen generation rise. The clearance efficiencies of proteins and carbohydrates in the dissolved phase in AD MECs were raised by four times and 2.3 times, significantly, as compared to AD treatment. Volatile fatty acid elimination by AD-MEC was raised by 4.7 times, indicating that the AD reactor in combination with MEC technologies increased the usage of the primary organic substances and hence improved hydrogen generation. As a result, study illustrates feasibility of minimizing food waste amounts while still producing biohydrogen. MEC technique is regarded as a resource-efficient solution for food waste treatment with a variety of engineering applications (Al Afif and Amon, 2019; Velásquez et al., 2019; Yun et al., 2018).

2.11.1.2 Methane Production

AD in combination with an MEC is anticipated to speed up CH_4 synthesis from biomass hydrolysate in MEC-AD feeding with raw waste-activated sludge and heat-pretreated waste-activated sludge, accordingly, the methanogenesis efficiency and responsiveness of operational microbes (Shah et al., 2022). The study described a low-cost adaptive technique for increasing CH_4 productivity and output by using waste-activated sludge as a substrate for MEC-AD. The CH_4 production and productivity were increased by 9.5 and 7.8 times, respectively, when the applied voltage was 0.8 V greater than the standard open circuit (stage I), and by 6.3 and 6.2 times, respectively, when the voltage was returned to 0 V (Wang et al., 2021a). When raw waste-activated sludge was used in MEC-AD, the hydrolysis fermentation and acidogenesis were significantly boosted, leading to a greater synergy of acetogenic bacteria and hydrogenotrophic methanogens, as compared to the methods for enhancing methanogenesis with heat-pretreated waste-activated sludge as substrate. As a result, MECs could be employed as a viable option to boost operational microorganisms in order to produce high and steady CH_4 generation in traditional AD while also lowering the greater operating costs associated with pretreatment and voltage (Liu et al., 2019; Yu et al., 2018).

2.11.1.3 Hydrogen Peroxide Production

MECs can also create hydrogen peroxide (H_2O_2), an essential industrial chemical. The viability of producing hydrogen peroxide by combining microbial oxidation of organic materials in the anode with oxygen reduction in the cathode of MECs has recently been established. This system was able to create hydrogen peroxide at a rate of 1.17 millimole/L/hour in the aerated cathode with an applied voltage of 0.5 V, leading to a massive efficiency of 83% based on CH_3COO^- oxidation. Hydrogen peroxide synthesis in MECs uses far less energy than in typical electrochemical methods, with 0.93 kWh/kg hydrogen peroxide reported in the research. More work should be placed toward improving hydrogen peroxide concentration for this innovation to mature. The highest hydrogen peroxide concentration currently achievable in MECs is 0.13 wt.%, which seems to be an order of magnitude below the expected level for actual industrial ramifications (Kadier et al., 2016a; Ki et al., 2017).

2.12 ADVANTAGES AND DISADVANTAGES OF MEC TECHNOLOGY

MEC is a cutting-edge technique for the treatment of wastewater because (i) it produces the energy output in the form of hydrogen gas, (ii) it can reduce solid production and thus lower sludge handling costs, (iii) it can be useful to recover value-added products from waste, and (iv) it possibly decreases the release of odors. The MEC system has several advantages over other bio-H_2 processes as microbial population present in inoculum can oxidize a wide variety of substrates ranging from simple to complex wastewater as well as industrial and lignocellulosic waste. Compared with DF process, the HPR was about fivefold higher (Call and Logan, 2008). Considering the amount of energy required for input voltage supply, there is a need to make more cost-effective and economical process to make MEC comparable with the existing conventional wastewater treatment technologies (Logan et al., 2008). The energy required to maintain the external applied potential in MEC is approximately equivalent to the energy consumption for operating the aerator in an activated sludge process (Tchobanoglous et al., 2003). To make MEC system economical and cost-effective, the efficiency should be improved in terms of hydrogen production and other value-added products. In this regard, process optimization and use of more efficient processes would help to improve the environmental performance as well as the economic balance. Additionally, MEC offers advantages of nutrient removal and recovery over AD system (Haddadi et al., 2014). Also, the gas produced in MEC is more valuable than AD. On the basis of COD loading rates, MECs could be comparable with other wastewater treatment technologies such as activated sludge process and AD (Liu et al., 2018). Also, compared to conventional water electrolysis, the applied voltage energy investment is significantly low and thus reduces overall cost. Recently developed MET, that is, MEC, offers effective wastewater treatment along with simultaneous hydrogen and other value-added product recovery (Rozendal et al., 2007; Yu et al., 2018).

However, there are some disadvantages of MEC technology like (a) Over a period of time, the yield of H_2 decreases due to the several undesired electrons sinks in various metabolisms (San-Martín et al., 2019); (b) The setting up of an MEC reactor is determined by the configuration of the reactor, the materials used in its making, and the type of substrate that will be used in the reactor system. These configurations are set theoretically, but the actual setting may vary from the theoretically set conditions and can alter the results (Miller et al., 2019); (c) For an efficient MEC, understanding the microbes and their relationship amongst themselves is essential to ensure the competition amongst the microbial species does not affect substrate utilization and product formation. Therefore, a thorough understanding of the microbes and their associated behavior is critical (Kadier et al., 2018); and (d) The use of MECs as a treatment for industrial waste can significantly reduce the organic constituents of the waste stream; however, it faces difficulty in meeting the standards set for the effluent discharge. Therefore, it is essential to refine this process to maximize its efficiency (Leicester et al., 2020).

2.13 CONCLUSION

MECs show great potential in not only reducing the pollution quotient in the wastewater stream of the industry but also helping in aiding in the resource recovery from the industrial wastewater. This system not only helps in reducing biodegradable matter in the stream but also helps in generating electricity and hydrogen, which is used as a clean fuel due to its non-polluting properties. Due to its increased demand globally, the use of MEC has proved to be an efficient source of hydrogen generation. This system is energy-positive and carbon-negative, and its integration with traditional technologies leads to higher output and refined processes. However, there are certain shortcomings in this process as well, like the competition amongst the microbes, structural shortcomings, etc., and an extensive study and research needs to be directed in this regard to increase the efficiency of the process and maximize the yield produced in the process. Efforts to enhance reactor design and substrate selection can be made to maximize the efficiency of the setup. The MEC technology can prove to be a great asset for energy generation and resource recovery and can significantly help in reducing the dependence on conventional sources of energy like fossil fuels, which are also polluting in nature. This technology can therefore pave a new way for energy generation and help meet the global energy requirement.

REFERENCES

Adesra A, Srivastava VK, Varjani S. Valorization of dairy wastes: Integrative approaches for value added products. Indian J Microbiol. 2021;61:270–278. https://doi.org/10.1007/s12088-021-00943-5

Afify AH, Gwad AAE, Rahman NAE. Effect of power supply and bacteria on bio-hydrogen production using microbial electrolysis cells (MECs). J Agric Chem Biotechn. 2017;8:221–224. https://www.researchgate.net/publication/352933973

Ahmad A, Chowdhary P, Khan N, et al. Effect of sewage sludge biochar on the soil nutrient, microbial abundance, and plant biomass: A sustainable approach towards mitigation of solid waste. Chemosphere. 2022;287:132112.

Ahn Y, Im S, Chung J-W. Optimizing the operating temperature for microbial electrolysis cell treating sewage sludge. Int J Hydrog Energy. 2017;42:27784–27791.

Aiken DC, Curtis TP, Heidrich ES. Avenues to the financial viability of microbial electrolysis cells [MEC] for domestic wastewater treatment and hydrogen production. Int J Hydrogen Energy. 2019;44:2426–2434.

Ajayi FF, Kim K-Y, Chae K-J, Choi M-J, Kim IS. Effect of hydrodymamic force and prolonged oxygen exposure on the performance of anodic biofilm in microbial electrolysis cells. Int J Hydrog Energy. 2010;35:3206–3213.

Al Afif R, Amon T. Mesophilic anaerobic co-digestion of cow manure with three-phase olive mill solid waste. Energy sources, part a recover. Util Environ Eff. 2019;41:1800–1808.

Anwer AH, Khan MD, Joshi R. Microbial electrochemical cell: An emerging technology for waste water treatment and carbon sequestration. In Modern Age Waste Water Problems. Springer: Cham, Switzerland, 2020; pp. 339–360.

Ariunbaatar J, Panico A, Esposito G, Pirozzi F, Lens PNL. Pretreatment methods to enhance anaerobic digestion of organic solid waste. Appl Energy. 2014;123:143–156. https://doi.org/10.1016/j.apenergy.2014.02.035

Arvin A, Hosseini M, Amin M-M, Najafpour-Darzi G, Ghasemi Y. Efficient methane production from petrochemical wastewater in a single membrane-less microbial electrolysis cell: The effect of the operational parameters in batch and continuous mode on bioenergy recovery. J Environ Heal Sci Eng. 2019;17:305–317.

Aryal N, Kvist T, Ammam F, et al. An overview of microbial biogas enrichment. Bioresour Technol. 2018;264:359–369.

Asongu SA, Agboola MO, Alola AA, et al. The criticality of growth, urbanization, electricity and fossil fuel consumption to environment sustainability in Africa. Sci Total Environ. 2020;712:136376. [CrossRef][10.1016/j.scitotenv.2019.136376]

Awe OW, Zhao Y, Nzihou A, Minh DP, Lyczko N. A review of biogas utilisation, purification and upgrading technologies. Waste Biomass-Valoriz. 2017;8:267–283. [CrossRef]

Babu ML, Sarma PN, Mohan SV. Microbial electrolysis of synthetic acids for biohydrogen production: Influence of biocatalyst pretreatment and pH with the function of applied potential. J Microb Biochem Technol S. 2013a;6:2.

Babu ML, Subhash GV, Sarma PN, et al. Bioelectrolytic conversion of acidogenic effluents to biohydrogen: An integration strategy for higher substrate conversion and product recovery. Bioresour Technol. 2013b;133:322–331.

Berni M, Dorileo I, Nathia G, Forster-Carneiro T, Lachos D, Santos BG. Anaerobic digestion and biogas production: combine effluent treatment with energy generation in UASB reactor as biorefinery annex. Int J Chem Eng. 2014;1: 543529.

Bo T, Zhu X, Zhang L, Tao Y, He X, Li D, Yan Z. A new upgraded biogas production process: Coupling microbial electrolysis cell and anaerobic digestion in single-chamber, barrel-shape stainless steel reactor. Electrochem Commun. 2014;45:67–70.

Borole AP, Mielenz JR. Estimating hydrogen production potential in biorefineries using microbial electrolysis cell technology. Int J Hydrog Energy. 2011;36:14787–14795.

Buyukkamaci N, Koken E. Economic evaluation of alternative wastewater treatment plant options for pulp and paper industry. Sci Total Environ. 2010;408:6070–6078.

Call D, Logan BE. Hydrogen production in a single chamber microbial electrolysis cell (MEC) lacking a membrane. Environ Sci Technol. 2008;42 (9):3401–3406.

Carlotta-Jones DI, Purdy K, Kirwan K, Stratford J, Coles SR. Improved hydrogen gas production in microbial electrolysis cells using inexpensive recycled carbon fibre fabrics. Bioresour Technol. 2020;304:122983.

Cerrillo M, Viñas M, Bonmatí A. Unravelling the active microbial community in a thermophilic anaerobic digester-microbial electrolysis cell coupled system under different conditions. Water Res. 2017;110:192–201.

Chae K-J, Choi M-J, Kim K-Y, Ajayi FF, Chang I-S, Kim IS. A solar-powered microbial electrolysis cell with a platinum catalyst-free cathode to produce hydrogen. Environ Sci Technol. 2009;43:9525–9530.

Chattopadhyay I, Banu RJ, Usman TMM, et al. Exploring the role of microbial biofilm for industrial effluents treatment. Bioengineered. 2022;2022250. https://doi.org/10.1080/21655979.2022

Chen Y, Chen M, Shen N, et al. H_2 production by the thermoelectric microconverter coupled with microbial electrolysis cell. Int J Hydrogen Energy. 2016;41:22760–22768.

Cheng S, Li Z, Mang HP, Huba EM. A review of prefabricated biogas digesters in China. Renew Sust Energ Rev. 2013;28:738–748.

COM. Emissions from Storage (EFS BREF), European Commission, JRC IPTS EIPPCB, 2006.

Cotterill SE, Dolfing J, Jones C, Curtis TP, Heidrich ES. Low temperature domestic wastewater treatment in a microbial electrolysis cell with 1 M2 anodes: Towards system scale-up. Fuel Cells. 2017;17:584–592.

Dange P, Pandit S, Jadhav D, Shanmugam P, Gupta PK, Kumar S, Kumar M, Yang YH, Bhatia SK. Recent developments in microbial electrolysis cell-based biohydrogen production utilizing wastewater as a feedstock. Sustainability. 2021;13(16):8796.

Das GC, Mondal S, Singh AK, Singh B, Prasad KK, Singh. Effluent treatment technologies in the iron and steel industry – A state of the art review. Water Environ Res. 2018;90:395–408.

Devda V, Chaudhary K, Varjani S, et al. Recovery of resources from industrial wastewater employing electrochemical technologies: Status, advancements and perspectives. Bioengineered. 2021;12:4697–4718.

Di Fraia S, Massarotti N, Vanoli L. A novel energy assessment of urban wastewater treatment plants. Energy Convers Manag. 2018;163:304–313.

Ditzig J, Liu H, Logan BE. Production of hydrogen from domestic wastewater using a bioelectrochemically assisted microbial reactor (BEAMR). Int J Hydrogen Energy. 2007;32:2296–2304.

Do MH, Ngo HH, Guo W, et al. Microbial fuel cell-based biosensor for online monitoring wastewater quality: A critical review. Sci Total Environ. 2020;712:135612.

Escapa A, Mateos R, Martínez EJ, Blanes J. Microbial electrolysis cells: An emerging technology for wastewater treatment and energy recovery. From laboratory to pilot plantand beyond. Renew Sustain Energy Rev. 2015;55:942–956.

Feng H, Huang L, Wang M, et al. An effective method for hydrogen production in a single-chamber microbial electrolysis by negative pressure control. Int J Hydrogen Energy. 2018;43:17556–17561.

Ferrer I, Ponsá S, Vázquez F, Font X. Increasing biogas production by thermal (70 c) sludge pre-treatment prior to thermophilic anaerobic digestion. Biochem Eng J. 2008;42:186–192. https://doi.org/10.1016/j.bej.2008.06.020

Garrido-Baserba M, Vinardell S, Molinos-Senante M, Rosso D, Poch M. The economics of wastewater treatment decentralization: A techno-economic evaluation. Environ Sci Technol. 2018;52:8965–8976.

Gikas P. Towards energy positive wastewater treatment plants. J Environ Manag. 2017; 203:621–629.

Gkamarazi N. Implementing anaerobic digestion for municipal solid waste treatment: Challenges and prospects. Paper presented at the International Conference on Environmental Science and Technology, CEST, Rhodes, Greece, 3–5 September 2015, p 6.

Grobelak A, Grosser A, Kacprzak M, et al. Sewage sludge processing and management in small and medium-sized municipal wastewater treatment plant-new technical solution. J Environ Manage. 2019;234:90–96.

Guo K, Donose BC, Soeriyadi AH, Prévoteau A, Patil SA, Freguia S, Gooding JJ, Rabaey K. Flame oxidation of stainless steel felt enhances anodic biofilm formation and current output in bioelectrochemical systems. Environ Sci Technol. 2014;48:7151–7156.

Guo K, Prévoteau A, Patil SA, Rabaey K. Engineering electrodes for microbial electrocatalysis. Curr Opin Biotechnol. 2015;33:149–156.

Guo X, Liu J, Xiao B. Bioelectrochemical enhancement of hydrogen and methane production from the anaerobic digestion of sewage sludge in single-chamber membrane-free microbial electrolysis cells. Int J Hydrog Energy. 2013;38:1342–1347.

Haddadi S, Nabi-Bidhendi G, Mehrdadi N. Nitrogen removal from wastewater through microbial electrolysis cells and cation exchange membrane. J Environ Health Sci Eng. 2014;12 (1):48. https://doi.org/10.1186/2052-336X-12-48

Harirchi S, Wainaina S, Sar T, et al. Microbiological insights into anaerobic digestion for biogas, hydrogen or volatile fatty acids (vfas): A review. Bioengineered. 2022;13(3):6521–6557. https://doi.org/10.1080/21655979.2022.2035986

Huang J, Feng H, Huang L, et al. Continuous hydrogen production from food waste by anaerobic digestion (AD) coupled single-chamber microbial electrolysis cell (MEC) under negative pressure. Waste Manag. 2020a;103:61–66.

Huang Q, Liu Y, Dhar BR. A critical review of microbial electrolysis cells coupled with anaerobic digester for enhanced biomethane recovery from high-strength feedstocks. Crit Rev Environ Sci Technol. 2020c;52(1):1–40.

Hwang S, Lee Y, Yang K. Maximization of acetic acid production in partial acidogenesis of swine wastewater. Biotechnol Bioeng. 2001;75:521–529.

Jabeen M, Yousaf S, Haider MR, Malik RN. High-solids anaerobic co-digestion of food waste and rice husk at different organic loading rates. Int biodeterior biodegrad. 2015;102:149–153.

Jain A, He Z. "NEW" resource recovery from wastewater using bioelectrochemical systems: Moving forward with functions. Front Environ Sci Eng. 2018;12:1–13.

Janani R, Baskar G, Sivakumar K, et al. Advancements in heavy metals removal from effluents employing nano-adsorbents: Way towards cleaner production. Environ Res. 2022;203:111815.

Jeremiasse AW. Cathode innovations for enhanced H_2 production through microbial electrolysis. Academic thesis. Wageningen University, Wageningen, 2011.

Kadier A, Al-Shorgani NK, Jadhav DA, Sonawane JM, Mathuriya AS, Kalil MS, Hasan HA, Alabbosh KFS. Microbial electrolysis cell (MEC) an innovative waste to bioenergy and value-added by-product technology. In Bioelectrosynthesis: Principles and Technologies for Value-Added Products. Wiley-VCH Verlag GmbH & Co., 2020; pp. 95–128.

Kadier A, Jain P, Lai B, et al. Biorefinery perspectives of microbial electrolysis cells (MECs) for hydrogen and valuable chemicals production through wastewater treatment. Biofuel Res J. 2020a;7:1128–1142.

Kadier A, Kalil MS, Abdeshahian P, et al. Recent advances and emerging challenges in microbial electrolysis cells (MECs) for microbial production of hydrogen and value-added chemicals. Renewable Sustainable Energy Rev. 2016a;61:501–525.

Kadier A, Kalil MS, Chandrasekhar K, et al. Surpassing the current limitations of high purity H_2 production in microbial electrolysis cell (MECs): Strategies for inhibiting growth of methanogens. Bioelectrochemistry. 2018;119:211–219.

Kadier A, Simayi Y, Abdeshahian P, et al. A comprehensive review of microbial electrolysis cells (MEC) reactor designs and configurations for sustainable hydrogen gas production. Alexandria Eng J. 2016b;55(1):427–443.

Kadier A, Simayi Y, Chandrasekhar K, Ismail M, Kalil MS. Hydrogen gas production with an electroformed Ni mesh cathode catalysts in a single-chamber microbial electrolysis cell (MEC). Int J Hydrogen Energy. 2015;40(41), 14095–14103.

Kadier A, Simayi Y, Kalil MS, Abdeshahian P, Hamid AA. Review of the substrates used in microbial electrolysis cells (MECs) for producing sustainable and clean hydrogen gas. Renew Energy. 2014;71:466–472.

Kataki S, Chatterjee S, Vairale MG, et al. Constructed wetland, an eco-technology for wastewater treatment: A review on types of wastewater treated and components of the technology (macrophyte, biolfilm and substrate). J Environ Manage. 2021;283:111986.

Katuri KP, Werner CM, Jimenez-Sandoval RJ, et al. A novel anaerobic electrochemical membrane bioreactor (AnEMBR) with conductive hollow-fiber membrane for treatment of low-organic strength solutions. Environ Sci Technol. 2014;48:12833–12841.

Ki D, Parameswaran P, Popat SC, Rittmann BE, Torres CI. Maximizing Coulombic recovery and solids reduction from primary sludge by controlling retention time and PH in a flat-plate microbial electrolysis cell. Environ Sci Water Res Technol. 2017;3:333–339. [CrossRef]

Ki D, Popat SC, Rittmann BE, et al. H_2O_2 production in microbial electrochemical cells fed with primary sludge. Environ Sci Technol. 2017;51:6139–6145.

Kim KN, Lee SH, Kim H, Park YH, In S-I. Improved microbial electrolysis cell hydrogen production by hybridization with a TiO_2 nanotube array photoanode. Energies. 2018;11:3184.

Kitching M, Butler R, Marsili E. Microbial bioelectrosynthesis of hydrogen: Current challenges and scale-up. Enzym Microb Technol. 2017;96:1–13.

Koffi NJ, Okabe S. Bioelectrochemical anoxic ammonium nitrogen removal by an MFC driven single chamber microbial electrolysis cell. Chemosphere. 2021;274:129715.

Kothari R, Pandey AK, Kumar S, Tyagi VV, Tyagi SK. Different aspects of dry anaerobic digestion for bio-energy: An overview. Renew Sustain Energy Rev. 2014;39:174–195. https://doi.org/10.1016/j.rser.2014.07.011

Koul Y, Devda V, Varjani S, Guo W, Ngo HH, Taherzadeh MJ, Chang JS, Wong JW, Bilal M, Kim SH, Bui XT. Microbial electrolysis: A promising approach for treatment and resource recovery from industrial wastewater. Bioengineered. 2022;13(4):8115–8134.

Kumar G, Saratale RG, Kadier A, Sivagurunathan P, Zhen G, Kim SH, Saratale GD. A review on bio-electrochemical systems (BESs) for the syngas and value added biochemicals production. Chemosphere. 2017;177, 84–92.

Kwietniewska E, Tys J. Process characteristics, inhibition factors and methane yields of anaerobic digestion process, with particular focus on microalgal biomass fermentation. Renew Sustain Energy Rev. 2014;34:491–500. https://doi.org/10.1016/j.rser.2014.03.041

Leicester D, Amezaga J, Heidrich E. Is bioelectrochemical energy production from wastewater a reality? Identifying and standardising the progress made in scaling up microbial electrolysis cells. Renewable Sustainable Energy Rev. 2020;133:110279.

Li S, Cheng C, Thomas A. Carbon-based microbial-fuel-cell electrodes: From conductive supports to active catalysts. Adv Mater. 2017;29:1602547.

Lim SS, Yu EH, Daud WRW, Kim BH, Scott K. Bioanode as a limiting factor to biocathode performance in microbial electrolysis cells. Bioresour Technol. 2017;238:313–324.

Lindholm-Lehto PC, Knuutinen JS, Ahkola HS, Herve SH. Refractory organic pollutants and toxicity in pulp and paper mill wastewaters. Environ Sci Pollut Res. 2015;22:6473–6499.

Liu C, Sun D, Zhao Z, et al. Methanothrix enhances biogas upgrading in microbial electrolysis cell via direct electron transfer. Bioresour Technol. 2019;291:121877.

Liu D, Roca-Puigros M, Geppert F, Caizán-Juanarena L, Na Ayudthaya SP, Buisman C, ter Heijne A. Granular carbon based electrodes as cathodes in methane-producing bioelectrochemical systems. Front Bioeng Biotechnol. 2018;6:78. [CrossRef]

Liu H, Grot S, Logan BE. Electrochemically assisted microbial production of hydrogen from acetate. Environ Sci Technol. 2005;39 (11):4317–4320.

Liu S, Ma Q, Wang B, Wang J, Zhang Y. Advanced treatment of refractory organic pollutants in petrochemical industrial wastewater by bioactive enhanced ponds and wetland system. Ecotoxicology. 2014;23:689–698.

Liu W, Huang S, Zhou A, Zhou G, Ren N, Wang A, Zhuang G. Hydrogen generation in microbial electrolysis cell feeding with fermentation liquid of waste activated sludge. Int J Hydrog Energy. 2012;37:13859–13864.

Liu Y, Deng YY, Zhang Q, et al. Overview of recent developments of resource recovery from wastewater via electrochemistry-based technologies. Sci Total Environ. 2021;757:143901.

Liu Z, Lu Y, Wang P, Wang T, Liu S, Johnson AC, Sweetman AJ, Baninla Y. Pollution pathways and release estimation of perfluorooctane sulfonate (PFOS) and perfluorooctanoic acid (PFOA) in central and eastern China. Sci Total Environ. 2017;580:1247–1256.

Liu Z, Zhou A, Zhang J, et al. Hydrogen recovery from waste activated sludge: Role of free nitrous acid in a prefermentation–microbial electrolysis cells system. ACS Sustainable Chem Eng. 2018;6 (3):3870–3878. https://doi.org/10.1021/acssuschemeng.7b04201

Logan BE, Call D, Cheng S, et al. Microbial electrolysis cells for high yield hydrogen gas production from organic matter. Environ Sci Technol. 2008;42 (23):8630–8640.

Lu L, Ren N, Zhao X, Wang H, Wu D, Xing D. Hydrogen production, methanogen inhibition and microbial community structures in psychrophilic single-chamber microbial electrolysis cells. Energy Environ Sci. 2011;4:1329–1336.

Lu L, Ren ZJ. Microbial electrolysis cells for waste biorefinery: A state of the art review. Bioresour Technol. 2016;215:254–264.

Mamimin C, Singkhala A, Kongjan P, Suraraksa B, Prasertsan P, Imai T, Sompong O. Two-stage thermophilic fermentation and mesophilic methanogen process for biohythane production from palm oil mill effluent. Int J Hydrog Energy. 2015;40(19): 6319–6328.

Merlin Christy P, Gopinath LR, Divya D. A review on anaerobic decomposition and enhancement of biogas production through enzymes and microorganisms. Renew Sustain Energy Rev. 2014;34:167–173. https://doi.org/10.1016/j.rser.2014.03.010

Merrill MD, Logan BE. Electrolyte effects on hydrogen evolution and solution resistance in microbial electrolysis cells. J Power Sources. 2009;191:203–208.

Mes T, Stams A, Reith J, Zeeman G. Methane production by anaerobic digestion of wastewater and solid wastes. Bio-Methane and Bio-Hydrogen. 2003;58–102.

Miller A, Singh L, Wang L, et al. Linking internal resistance with design and operation decisions in microbial electrolysis cells. Environ Int. 2019;126:611–618.

Modestra JA, Babu ML, Mohan SV. Electrofermentation of real-field acidogenic spent wash effluents for additional biohydrogen production with simultaneous treatment in a microbial electrolysis cell. Sep Purif Technol. 2015;150:308–315.

Mohammed AJ, Ismail ZZ. Slaughterhouse wastewater biotreatment associated with bioelectricity generation and nitrogen recovery in hybrid system of microbial fuel cell with aerobic and anoxic bioreactors. Ecol Eng. 2018;125:119–130.

Mohan SV, Sravan JS, Butti SK, et al. Microbial electrochemical technology. In: Microbial Electrochemical Technology: Emerging and Sustainable Platform. Elsevier, 2019; pp. 3–18.

Monson KD. The Effect of Various Control Actions on Anaerobic Digester Performance. University of South Wales, 2004.

Munoz LD, Erable B, Etcheverry L, Riess J, Basséguy R, Bergel A. Combining phosphate species and stainless steel cathode to enhance hydrogen evolution in microbial electrolysis cell (MEC). Electrochem Commun. 2010;12:183–186.

Munoz-Cupa C, Hu Y, Xu C, Bassi A. An overview of microbial fuel cell usage in wastewater treatment, resource recovery and energy production. Sci Total Environ. 2021;754:142429.

Nam J-Y, Tokash JC, Logan BE. Comparison of microbial electrolysis cells operated with added voltage or by setting the anode potential. Int J Hydrog Energy. 2011;36:10550–10556.

Nelabhotla ABT, Dinamarca C. Bioelectrochemical CO_2 reduction to methane: MES integration in biogas production processes. Appl Sci. 2019;9:1056.

Omidi H, Sathasivan A. Optimal temperature for microbes in an acetate fed microbial electrolysis cell (MEC). Int Biodeterior Biodegrad. 2013;85:688–692.

Pant D, Singh A, Van Bogaert G, Olsen SI, Nigam PS, Diels L, Vanbroekhoven K. Bioelectrochemical systems (BES) for sustainable energy production and product recovery from organic wastes and industrial wastewaters. R Soc Chem Adv. 2011;2:1248–1263.

Park S-G, Chae K-J, Lee M. A sulfonated poly(arylene ether sulfone)/polyimide nanofiber composite proton exchange membrane for microbial electrolysis cell application under the coexistence of diverse competitive cations and protons. J Membr Sci. 2017;540:165–173.

Patel GB, Rakholiya P, Shindhal T, et al. Lipolytic nocardiopsis for reduction of pollution load in textile industry effluent and Swiss model for structural study of lipase. Bioresour Technol. 2021;341:125673.

Peera SG, Maiyalagan T, Liu C, Ashmath S, Lee TG, Jiang Z, Mao S. A review on carbon and non-precious metal-based cathode catalysts in microbial fuel cells. Int J Hydrog Energy. 2021;46:3056–3089.

Pocaznoi D, Calmet A, Etcheverry L, Erable B, Bergel A. Stainless steel is a promising electrode material for anodes of microbial fuel cells. Energy Environ Sci. 2012;5:9645–9652.

Rahimnejad M, Adhami A, Darvari S, Zirehpour A, Oh S-E. Microbial fuel cell as new technology for bioelectricity generation: A review. Alex Eng J. 2015;54:745–756.

Rasheed T, Hassan AA, Kausar F, Sher F, Bilal M, Iqbal HMN. Carbon nanotubes assisted analytical detection—Sensing/Delivery cues for environmental and biomedical monitoring. Trends Anal Chem. 2020;132:116066.

Rozendal RA, Hamelers HVM, Euverink GJW, Metz SJ, Buisman CJN. Principle and perspectives of hydrogen production through biocatalyzed electrolysis. Int J Hydrogen Energy. 2006;31(12):1632–1640.

Rozendal RA, Hamelers HVM, Molenkamp RJ, Buisman CJN. Performance of single chamber biocatalyzed electrolysis with different types of ion exchange membranes. Water Res. 2007;41:1984–1994.

Rozendal RA, Jeremiasse AW, Hamelers HV, Buisman CJ. Hydrogen production with a microbial biocathode. Environ Sci Technol. 2008;42(2):629–634.

Sadhukhan J, Lloyd JR, Scott K, et al. A critical review of integration analysis of microbial electrosynthesis (MES) systems with waste biorefineries for the production of biofuel and chemical from reuse of CO_2. Renewable Sustainable Energy Rev. 2016;56:116–132.

Saheb-Alam S, Singh A, Hermansson M, Persson F, Schnürer A, Wilén BM, Modin O. Effect of start-up strategies and electrode materials on carbon dioxide reduction on biocathodes. Am. Soc Microbiol. 2018;84(4):e02242–17.

Sangeetha T, Guo Z, Liu W, Cui M, Yang C, Wang L, Wang A. Cathode material as an influencing factor on beer wastewater treatment and methane production in a novel integrated upflow microbial electrolysis cell (Upflow-MEC). Int J Hydrog Energy. 2016;41:2189–2196.

San-Martín MI, Sotres A, Alonso RM, Díaz-Marcos J, Morán A, Escapa A. Assessing anodic microbial populations and membrane ageing in a pilot microbial electrolysis cell. Int J Hydrog Energy. 2019;44:17304–17315.

Santoro C, Arbizzani C, Erable B, Ieropoulos I. Microbial fuel cells: From fundamentals to applications. A review. J Power Sources. 2017;356:225–244.

Savla N, Anand R, Pandit S, Prasad R. Utilization of nanomaterials as anode Modifiers for improving microbial fuel cells performance. Journal of Renewable Materials. 2020;8:1581–1605.

Shah AV, Singh A, Mohanty SS, et al. Organic solid waste: Biorefinery approach as a sustainable strategy in circular bioeconomy. Bioresour Technol. 2022;349:126835.

Shah MP. Microbial Bioremediation & Biodegradation. Springer, 2020.

Shah MP. Removal of Emerging Contaminants through Microbial Processes. Springer, 2021.

Shah MP. Removal of Refractory Pollutants from Wastewater Treatment Plants. CRC Press, 2021.

Siegert M, Yates MD, Call DF, Zhu X, Spormann A, Logan BE. Comparison of nonprecious metal cathode materials for methane production by electromethanogenesis. ACS Sustain Chem Eng. 2014;2:910–917.

Siegert M, Yates MD, Spormann AM, Logan BE. Methanobacterium dominates biocathodic archaeal communities in methanogenic microbial electrolysis cells. ACS Sustain Chem Eng. 2015;3:1668–1676.

Sleutels TH, Ter Heijne A, Buisman CJ, Hamelers HV. Bioelectrochemical systems: An outlook for practical applications. Chem Sus Chem. 2012;6:1012–1019.

Starkl M, Brunner N, Feil M, Hauser A. Ensuring sustainability of non-networked sanitation technologies: An approach to standardization. Environ Sci Technol. 2015;49:6411–6418.

Sun M, Sheng G-P, Zhang L, et al. An MEC-MFCCoupled system for biohydrogen production from acetate. Environ Sci Technol. 2008;42:8095–8100.

Tchobanoglous G, Burton FL, Stensel HD. Wastewater Engineering, Treatment and Reuse, Metcalf & Eddy, Inc., 4e. Boston, IL: McGraw-Hill, 2003.

Thangamani R, Vidhya L, Varjani S. Electrochemical technologies for wastewater treatment and resource reclamation. In: Microbe Mediated Remediation of Environmental Contaminants. Elsevier, 2021; pp. 381–389.

United Nations. The World's Cities in 2018; World's Cities 2018—Data Booklet; United Nations: New York, NY, USA, 2018; p. 34.

Van Eerten-Jansen MCAA, Ter Heijne A, Buisman CJN, Hamelers HVM. Microbial electrolysis cells for production of methane from CO_2: Long-term performance and perspectives. Int J Energy Res. 2011;36:809–819.

Varanasi JL, Veerubhotla R, Pandit S, Das D. Biohydrogen production using microbial electrolysis cell. Microb Electrochem Technol. 2019;843–869. https://doi.org/10.1016/b978-0-444-64052-9.00035-2

Varjani S, Rakholiya P, Ng HY, et al. Microbial degradation of dyes: An overview. Bioresour Technol. 2020a;314:123728.

Varjani S, Rakholiya P, Shindhal T, et al. Trends in dye industry effluent treatment and recovery of value added products. J Water Process Eng. 2021;39:101734.

Varjani SJ. Microbial degradation of petroleum hydrocarbons. Bioresour Technol. 2017;223:277–286.

Varjani SJ, Upasani VN. A new look on factors affecting microbial degradation of petroleum hydrocarbon pollutants. Int Biodeterior Biodegrad. 2017;120:71–83.

Velásquez F, Espitia J, Mendieta O, et al. Non-centrifugal cane sugar processing: A review on recent advances and the influence of process variables on qualities attributes of final products. J Food Eng. 2019;255:32–40.

Villano M, Aulenta F, Ciucci C, Ferri T, Giuliano A, Majone M. Bioelectrochemical reduction of CO_2 to CH_4 via direct and indirect extracellular electron transfer by a hydrogenophilic methanogenic culture. Bioresour Technol. 2010;101:3085–3090.

Vyas S, Prajapati P, Shah AV, et al. Municipal solid waste management: Dynamics, risk assessment, ecological influence, advancements, constraints and perspectives. Sci Total Environ. 2022a;814:152802.

Vyas S, Prajapati P, Shah AV, et al. Opportunities and knowledge gaps in biochemical interventions for mining of resources from solid waste: A special focus on anaerobic digestion. Fuel. 2022b;311:122625.

Wang A, Liu W, Ren N, Cheng H, Lee D-J. Reduced internal resistance of microbial electrolysis cell (MEC) as factors of configuration and stuffing with granular activated carbon. Int J Hydrog Energy. 2010;35:13488–13492.

Wang L, He Z, Guo Z, Sangeetha T, Yang C, Gao L, Wang A, Liu W. Microbial community development on different cathode metals in a bioelectrolysis enhanced methane production system. J Power Sources. 2019;444:227306.

Wang Q, Yang Z. Industrial water pollution, water environment treatment, and health risks in China. Environ Pollut. 2016;218,358–365.

Wang X, Zhao Y-C. A bench scale study of fermentative hydrogen and methane production from food waste in integrated two-stage process. Int J Hydrog Energy. 2009;34:245–254. https://doi.org/10.1016/j.ijhydene.2008.09.100

Wang XT, Zhao L, Chen C, et al. Microbial electrolysis cells (MEC) accelerated methane production from the enhanced hydrolysis and acidogenesis of raw waste activated sludge. Chem Eng J. 2021a;413:127472.

Wang Z, Xue M, Huang K, Liu Z. Textile dyeing waste water treatment. Advances in Treating Textile Effluent. 2011;5:91–116.

Waqas M, Rehan M, Aburiazaiza AS, et al. Wastewater biorefinery based on the microbial electrolysis cell: Opportunities and challenges. Elsevier B.V. 2018; 10.1016/B978-0-444-64017-8.00017-8

Ward AJ, Hobbs PJ, Holliman PJ, Jones DL. Optimisation of the anaerobic digestion of agricultural resources. Bioresource Technol. 2008;99(17):7928–7940.

Werner CM, Katuri KP, Hari AR, Chen W, Lai Z, Logan BE, Amy GL, Saikaly PE. Graphene-coated hollow fiber membrane as the cathode in anaerobic electrochemical membrane bioreactors—Effect of configuration and applied voltage on performance and membrane fouling. Environ Sci Technol. 2016;50:4439–4447.

Wünschiers R, Lindblad P. Hydrogen in education—A biological approach. Int J Hydrogen Energy. 2002;27(11–12):1131–1140.

Yossan S, Xiao L, Prasertsan P, He Z. Hydrogen production in microbial electrolysis cells: Choice of catholyte. Int J Hydrog Energy. 2013;38:9619–9624.

Yu H, Wang Z, Wu Z, Zhu C. Enhanced waste activated sludge digestion using a submerged anaerobic dynamic membrane bioreactor: Performance, sludge characteristics and microbial community. Sci Rep. 2016;6:20111.

Yu Z, Leng X, Zhao S, Ji J, Zhou T, Khan A, Kakde A, Liu P, Li X. A review on the applications of microbial electrolysis cells in anaerobic digestion. Bioresour Technol. 2018;255:340–348. https://doi.org/10.1016/j.biortech.2018.02.003

Yuan H, He Z. Integrating membrane filtration into bioelectrochemical systems as next generation energy-efficient wastewater treatment technologies for water reclamation: A review. Bioresour Technol. 2015;195:202–209.

Yun Y-M, Lee M-K, Im S-W, et al. Biohydrogen production from food waste: Current status, limitations, and future perspectives. Bioresour Technol. 2018;248:79–87.

Zeng M, Li Y. Recent advances in heterogeneous electrocatalysts for the hydrogen evolution reaction. J Mater Chem A. 2015;3:14942–14962. [CrossRef]

Zhang T, Nie H, Bain TS, Lu H, Cui M, Snoeyenbos-West OL, Franks AE, Nevin KP, Russell TP, Lovley DR. Improved cathode materials for microbial electrosynthesis. Energy Environ Sci. 2012;6:217–224.

Zhang Y, Angelidaki I. Counteracting ammonia inhibition during anaerobic digestion by recovery using submersible microbial desalination cell. Biotechnol Bioeng. 2015;112:1478–1482.

Zhang Y, Angelidaki I. Microbial electrolysis cells turning to be versatile technology: Recent advances and future challenges. Water Res. 2014;56:11–25.

Zhi Z, Pan Y, Lu X, et al. Electrically regulating co-fermentation of sewage sludge and food waste towards promoting biomethane production and mass reduction. Bioresour Technol. 2019;279:218–227.

3 Microbes and Wastewater Treatment

Shreya Anand and Padmini Padmanabhan

3.1 INTRODUCTION

More than 1 billion people throughout the world are unable to breathe clean air, and 3 million people die each year as a result of the rising pollution (WHO, 2006). The first quarter of this century has been the warmest in recorded history. Every year, pollution kills approximately 1 million seabirds and thousands of sea animals around the world. Every year, almost 3 million children under the age of five die as a result of numerous environmental pollutents. These numbers not only provide a vivid picture of the current precarious situation but also serve as a warning to think about a better tomorrow today. Population growth, human activities, and other factors have combined to create a situation that has never been seen before. We utilize about 60,000 chemicals in our daily lives. However, industrial development is necessary for emerging countries to achieve faster growth, and the pollution that comes with it must be managed in a sustainable manner. As a result, all researchers and decision-making organizations should prioritize effective management policy, appropriate remedial techniques, and sustainable resource exploitation without affecting the natural ecosystem.

Microorganisms, in this perspective, play a critical part in the ecosystem sustainability since they are extra proficient in quickly adapting to ecological variations. Microorganisms are the foremost evolved form of life; they are flexible to a diverse environmental condition (Abatenh et al., 2017). Microorganisms are ubiquitously present, have huge environmental impact, and are significant controllers that control the biogeochemical cycles (Seigle-Murandi et al., 1996; Shah M., 2020). Microorganisms have control over carbon, nitrogen fixation, and methane, sulfur metabolism, which in turn affect global biogeochemical cycles (Das et al., 2006). They yield a variety of enzymes that are used to safely remove contaminants, both by destroying the chemical or by transforming the toxins (Dash and Das, 2012). Microorganisms have given a helpful platform for an expanded model of bioremediation due to their adaptability (Doney et al., 2012).

In-depth research has been done on bioelectrochemical systems (BESs), a novel wastewater treatment method. The BES anodes' treatment capacity is constrained, and the anode wastewater is typically unable to be discharged or recycled directly. BES cathodes could be used for further treatment of pollutants to improve the treatment. Numerous methods—which can be divided into cathode-stimulated treatment and cathode-supported treatment—have been used to study this. In the former, impurities like nitrates or dye compounds are reduced immediately through electron transfer, whereas in the latter, contaminants can be removed through aerobic

DOI: 10.1201/9781003368472-3

oxidation, algae growth, the creation of powerful oxidants for further oxidation, and/or membrane treatment (Jain and He, 2018).

Environmental management faces a significant problem in removing pollutants from contaminated sites. Toxic compounds have been removed using a variety of techniques based on ion exchange resins or biosorbents; however, they are vulnerable to environmental factors (Wang and Chen, 2009; Shah M., 2021). Some of the conservative methods of treating wastewater are conventional coagulation, precipitation, adsorption, ion exchange, and reverse osmosis (U.S. EPA, 2007). Coagulants, including aluminum sulfate, iron salt, and lime, are used in the traditional coagulation method to remove hazardous components. When contaminants are present at high concentrations in waste materials, this procedure can be used. The fundamental flaw in this approach is the achievement of the preferred precipitation stages (Henneberry et al., 2011). The activated carbon adsorption method is extremely effective, with water reuse post-regeneration. The downside of this approach is sluggish response, suitable only for low contaminant concentrations, and higher cost of regeneration (Mohan et al., 2001). Use of naturally occurring waste as adsorbents is eco-friendly, cost-effective, and yields fewer secondary products (Ramadevi and Srinivasan, 2005; Shah M., 2021). Various ion exchange resins are used in the ion exchange procedures. These techniques have several advantages, including being insensitive to fluctuation, the capacity to achieve zero impurities, and the possibility to reuse the resins after regeneration (Chiarle et al., 2000).

The reverse osmosis method uses cellulose acetate or aromatic polyamide membranes under high pressure, resulting in the solvent removal from the membrane (Zhu and Elimelech, 1997). On the other hand, metal-contaminated locations have been remedied using low-cost biological methods such as phytoremediation (Narang et al., 2011). It's also a potential technology to use microbes to remove metal from contaminated places. However, the disadvantages of bacterial processes include the generation of enormous capacities of contaminant-laden biomass and difficult in disposing. Biological approaches, such as bacterial-mediated bioremediation, include sustainability, economic efficiency, and ease of in-situ application.

Bioremediation is a biological technique for recycling wastes into a form that other species may utilize and reuse. The globe is currently dealing with various forms of environmental contamination. Microorganisms are a critical component of a key alternative strategy for overcoming obstacles. Microorganisms can survive in every location on the biosphere due to their incredible metabolic activity, and they can thrive in a wide range of environmental conditions. Microorganisms have a wide range of nutritional capacities, which makes them ideal for bioremediation of environmental contaminants. Through the all-inclusive and action of microorganisms, bioremediation is heavily involved in the degradation, eradication, immobilization, or detoxification of various chemical wastes and physically dangerous chemicals from the surroundings.

Degrading and changing pollutants such as hydrocarbons, oil, heavy metals, pesticides, dyes, and other chemicals is the primary principle (Abatenh et al., 2017; Burghal et al., 2016; Maliji et al., 2013). Because it is carried out in an enzymatic manner by metabolizing, it has a significant contribution to the solution of many environmental problems. The pace of deterioration is determined by two types of factors: biotic and abiotic environments. Various tactics and strategies are currently being used in the domain in various parts of the world. Biostimulation, bioaugmentation, bioventing, biopiles, and bioattenuation are only a few examples. Because each bioremediation technology has its own unique application, each has its own set of benefits and drawbacks.

Microorganisms are widely distributed throughout the biosphere due to their outstanding metabolic abilities and their capacity to proliferate in a wide range of environmental circumstances. Microorganisms' nutritional adaptability can also be used for pollution biodegradation. Bioremediation is the word for this type of procedure. It is sustained by the ability of specific microbes to convert, adapt, and consume harmful contaminants in order to obtain energy and biomass production (Tang et al., 2007). Bioremediation is a microbiologically well-organized procedural activity that is used to break down or change contaminants into less harmful or harmless elemental and compound forms rather than merely collecting and storing them.

Bioremediators are biological substances that are applied during bioremediation to decontaminate contaminated locations. Fungi, bacteria, and archaea are frequently the primary bioremediators (Fazli et al., 2015; Strong and Burgess, 2008). The use of bioremediation, a biotechnological procedure employing microorganisms, to address and eliminate risks associated with various contaminants through environmental biodegradation (Singh and Majumdar, 2017). The terms "biodegradation" and "bioremediation" can be used interchangeably more often. In soil, water, and sediments, microorganisms play a key role in the removal of pollutants; this is largely because they outperform conventional remedial procedural techniques. Microorganisms are cleaning up the environment and halting more contamination (Demnerová et al., 2005). In this chapter, an overview of bioelectrochemical processes will be discussed, along with the role of microbes in bioelectrochemical processes for the treatment of industrial wastewater followed involvement of microbes in the efficient treatment of wastewater, and their mode of action.

3.2 BIOELECTROCHEMICAL PROCESS

Dynamic systems called "bioelectrochemical systems" (BESs) make use of the relationship between solid electron acceptors and microorganism donors to accomplish the removal of pollutants from wastewater while simultaneously producing electricity (such as an electrode) (Pant et al., 2012). A microbial fuel cell (MFCs) consists of microbes that oxidize organic materials in the anode and transmit electrons to an anode electrode (Gude, 2016). The cathode electrode subsequently transfers the electrons to a terminal electron acceptor, such as oxygen, for reduction (Zhang and Angelidaki, 2014). When a load is introduced to the circuit, this electrical current or power is generated by the flow of electrons. In order to produce hydrogen in microbial electrolysis cell, promote purification in microbial desalination cells (Sevda et al., 2015), or create value-added molecules in microbial electrosynthesis cells, researchers have harnessed the electron flow (MES).

BESs have made strides in recent years, especially when it comes to scaling up the system, and numerous large-scale set-ups are researched for treatment of diverse wastewaters. Additionally, there is a lot of interest in using BESs to recover important resources like nutrients, energy, and water from wastewater. Wastewater treatment is regarded as a fundamental task along the BES development and is crucial to the potential application of this technology. The major objective of wastewater treatment is organic removal, and in BESs, the efficacy of this removal is very varied and dependent on a number of variables, including the configuration and power of the wastewater, the hydraulic retaining period, apparatus designs, and temperature. When creating electrons, the BES anode removes the most material.

In general, BESs can effectively remove organic matter from simple organic substrates or low-strength wastewater; however, the removal effectiveness would drastically decrease the high-strength wastewater. However, since their removal would necessitate aerobic conditions, nutrients like nitrogen and phosphorus are difficult to extract from the anode of a BES. Consequently, the BES anode's effluent is a stream that has only been partially cleaned and still contains nutrients and organic residues. The elimination of both organic and ammonium chemicals was greatly improved by cathodic treatment, as shown in various prior investigations of MFCs handling actual domestic wastewater.

Consider using a BES's cathode for additional treatment in order to improve wastewater treatment in BESs without adding more reactors. A cathode can provide biological processes with greater room to run for prolonged therapy, and cathodic reduction reactions can be employed to get rid of a variety of contaminants that the anode cannot get rid of. A cathode with a membrane integrated will be able to deliver high-quality effluent while rejecting a variety of pollutants, including pathogens. Although cathodes have been researched for the treatment of wastewater, there hasn't been a lot of focus/attention on them in previous BES studies.

3.2.1 Treatment through Cathode

Through electron absorption, the cathode-stimulated treatment is accomplished. Theoretically, some substance that can be condensed can serve as an electron acceptor for a BES cathode, but, in practice, this lessening can be possible through a catalyst having lower activation energy, which consequently enables the transfer of electrons from a cathode electrode to acceptors of electron. Various chemicals, including refractory substances, metals, nitrogenous compounds, and carbon dioxide, can be eliminated via cathodic reduction processes (Cecconet et al., 2018).

A few resistant chemicals can be removed via reduction, while many need to be removed by an oxidation process. For instance, nitrobenzene, a substance used extensively in industry, can be degraded through cathode. When compared to traditional anaerobic degradation, this method could dramatically lower organic dosage. Numerous dye chemicals have been investigated for reduction through decolorization and removal in BESs (Hasan et al., 2013). A developing water pollutant called perchlorate has the ability to serve as an MFC's only electron acceptor in the cathode.

In the cathode of an MFC, perchlorate, a newly discovered drinking water pollutant, can serve as the only electron acceptor. Despite the potential advantages of reduced energy usage, there is a significant problem with BES removal of certain chemicals. As previously mentioned, nitrobenzene is transformed into aniline. It was discovered that the main byproduct of dye reduction in the BES cathode was sulfanilic acid. Therefore, post-treatment measures like aerobic therapy may still be required. Additionally, co-substrates were used in several investigations to accelerate the decolorization of dye molecules by incorporating quickly degradable organic chemicals into the cathode (Shah, 2013). This strategy raises concerns since the organic compounds that have been introduced are providers of electron and it contend by means of cathode electrode, decreasing the efficacy of the electron transfer.

Heavy metals are significant industrial pollutants because they are poisonous, nonbiodegradable, and widespread. Immobilization and/or filtering are typically used to remove them (Stelting et al., 2012). After reduction processes, numerous metals transit to an insoluble state, and the cathode is able to carry out this process. Chromium is a good example because it can be eliminated through precipitation after undergoing a cathodic reaction that is mediated by bacteria or other abiotic catalysts. Hexavalent chromium (Cr(VI)) is very soluble in water and is both carcinogenic and mutagenic (Sinha et al., 2011).

3.2.2 Treatment Supported by Cathode

Although treatment of wastewater through the support of cathode takes place in the cathode part, the cathode electrode is not used to directly accept electrons. In other words, this type of treatment makes use of the cathode compartment's physical area and may benefit or harm the cathode electrode or its reactions. These therapies include loop operation, algae bioreactor, enhanced oxidation, and membrane integration, to name a few. By feeding an anode effluent into cathode, loop action was developed in order to employ the section of cathode as just a secondary treatment unit. This contributes to the solution of the anode's inadequate organic elimination problem. It was claimed that cathode treatment of anode effluent might increase COD removal between 75% and 92% (Lefebvre et al., 2011).

The transfer of enhanced proton from anode to cathode counteracts the electrolyte alkalization and reduces the difference in the pH in the electrodes, which is a more substantial advantage that eliminates the need for pH correction. According to one study, compared to a negative control, a loop operation enhanced maximal generation of power, by 180% in batch mode and by 380% in continuous mode. But by adding organic substances to the anode effluent, electron donors are also introduced into the cathode, competing with both the electrode surface for terminal acceptors of electrons (e.g., oxygen). The cathodic reaction may be inhibited by this competition.

Furthermore, the cathode electrode's surface will develop a biofilm from the rapid development of heterotrophic bacteria on organic materials, thus reducing the cathode's efficiency. Therefore, it may not be the best strategy for BES to improve organic removal by employing the cathode. Nitrogen compounds can be eliminated in the cathode via previously discussed mixotrophic nitrification and bioelectrochemical denitrification if the concentration of carbon-based compounds can be brought down to a little level that has no impact on the cathodic reaction. As a result, a loop operation might be beneficial for removing organics from the anode and nitrogen from the cathode while also lowering the pH of the catholyte (Clauwaert et al., 2009).

3.3 WASTEWATER TREATMENT THROUGH BIOREMEDIATION

The process of biologically reducing organic wastes to a harmless state or concentrations below permissible regulatory limits is known as bioremediation. Microorganisms are used for degradation of pollutants due to the presence of enzymes. By giving them the correct nutrients and other chemicals, they require to break down and detoxify pollutants that are damaging to the environment and living things, bioremediation aims to put them to work. Enzymes are involved in every metabolic reaction. These are oxidoreductases, hydrolases, lyases, transferases, isomerases, and ligases, among others. Because of their nonspecific and specific substrate affinity, many enzymes have a very wide degradation capacity. Effective bioremediation depends on microorganisms attacking pollutants enzymatically and converting them to harmless molecules. Because bioremediation only works when the environment supports microbial activity and growth, environmental parameters are regularly changed to promote microbial growth and degradation (Kumar et al., 2011).

Biodegradation is the foundation of bioremediation technique. It refers to the transformation of organic, harmful pollutants into innocuous or naturally occurring substances (Jain and Bajpai, 2012). For the biodegradation of a wide range of organic molecules, numerous mechanisms and pathways have been discovered; for example, it occurs in the presence and absence of oxygen.

3.3.1 Factors Involved in the Process of Bioremediation

The optimization and control of bioremediation are complex systems with many variables. Toxicants must be available to the microbial population, the microbial population must be capable of decomposing the pollutants, and environmental issues must be taken into account. It is possible to isolate microorganisms from almost any environment. Microbes may adapt and grow in a variety of habitats, including those with severe temperatures (cold, hot, dry, desert, water, and anaerobic), a lot of oxygen, dangerous compounds, or any waste stream. A source of energy and a carbon source are the two primary requirements for bacteria to degrade the contaminants.

Carbon is the utmost important component in the living beings. Ninety-five percent of the weight of living beings consists of oxygen, nitrogen, and hydrogen. The bioremediation category is determined by the amount of pollutants in the soil as well as the presence or lack of other nutrients like phosphorus and sulfur. To separate pollutants with high concentrations from the soil, the soil is mixed with an interface active agent in water (Wolski, et al., 2012). After that, bioremediation is used to effectively clean the polluted soil. As a result of this environmentally friendly process, soil may be recycled and reused without any effort.

Bioremediation is the process of bacteria, fungi, and plants degrading, eliminating, modifying, immobilizing, or detoxifying various chemicals and physical pollutants from the environment (Jasińska et al., 2012). Microorganisms are involved in the process due to enzymatic pathways working as biocatalysts, accelerating biochemical reactions, and destroying the intended pollutant. Microorganisms only take action against pollution when they are given access to a wide range of materials and nutrients that help them produce energy and create new cells.

The effectiveness of bioremediation is influenced by the chemical make-up and concentration of pollutants, as well as by the physicochemical characteristics of the surroundings and their accessibility to microorganisms (El Fantroussi and Agathos, 2005). The rate of degradation is affected because bacteria and pollutants need not come into contact with one another. Furthermore, the distribution of pollutants and microorganisms in the environment is not uniform. Regulating and maximizing bioremediation activities are the complex system due to a number of factors. Environmental factors, the availability of contaminants to the microbial population, and the existence of a microbial population capable of degrading pollutants are all taken into account.

3.4 ROLE OF MICROORGANISMS

The metabolism involved in the boosting of hydrocarbons effectiveness helps in the usage of microbial consortia rather than pure cultures has expanded considerably. The biodegradation process becomes more efficient and faster when appropriate trapping methods can be used for efficient degradation.

The employment of biological agents to alter, degrade, or remove harmful or unwanted chemical pollutants from the environment is known as bioremediation. To return contaminated soils to their original non-contaminated form, biological mechanisms are applied. When manufacturing foods and beverages through controlled fermentation, humans have harnessed natural bioprocesses to develop biotechnological products. Many biotechnology discoveries and applications nowadays are often based on natural microbial activities. For bioremediation objectives, the prospect of isolating, cultivating, and studying wild microorganisms with unique metabolic pathways is of tremendous interest. It's worth noting that, unlike manufactured organisms, nonpathogenic "wild" microbes are not subject to stringent legal constraints (Plan and Van den Eede, 2010).

The most prevalent bioremediation concepts and procedures are generally utilized both singly and in combination. Biostimulation is the most often used bioremediation approach due to the easy acceptance of electrons in conditions of environmental pollution, as well as the safety of these nutrients and electron acceptors once they have been exposed to contamination. The chemical structure of a pollutant is one of the factors evaluated when determining its bioremediation capacity.

Hydrophobic structures have a higher propensity to block biological functions and worsen the effects of possible bioremediation. The bioavailability of certain drugs is an important parameter. A chemical's suitability for bioprocessing is not always implied by the fact that it is present in a specific environment. For instance, a substance may stop being capable of biodegradation or biotransformation if it is absorbed by the environment after a spill (e.g., if there are clays present). In other circumstances, a compound may be dissolved and recirculated by river flow until it is deposited in a specific area where it is more readily available for bioremediation. In other circumstances, a compound may be dissolved and recirculated by river flow until it is deposited in a specific area where it is more readily available for bioremediation. As a result, the presence of microorganisms capable of biodegrading chemicals in the environment, as well as the accessibility of appropriate ecological conditions, are critical for successful bioremediation. Environmental conditions can also be controlled to boost the natural environment's bioremediation capacity.

Similarly, it's worth considering that some tactics that wouldn't work in broad fields (in situ) might operate in isolated or limited settings (ex situ). It is possible to process the contamination with minimum environmental impact by first mobilizing a fraction of the polluted environment into a safe treatment location. Nonetheless, it is frequently impossible to mobilize even a small portion of the contaminated area.

- **Biosparging**: By pumping pressured air beneath the table, contaminants will be biologically broken down by naturally occurring microorganisms more quickly, and groundwater oxygen levels will rise. By enhancing blending in the saturated zone, it increases soil's contact with groundwater. The system's construction and design can be extremely flexible due to how easy and inexpensive it is to install small-diameter air injection locations.
- **Bioventing**: A prospective new method called bioventing encourages the in situ microbial degradation of almost any chemical that can be broken down aerobically by giving the already existing soil bacteria oxygen. Only the necessary amount of oxygen is provided by modest air flow rates to maintain microbial activity. The most common way to add oxygen is through high-speed injection into soil residue contamination through wells. The leftover fuel that has been adsorbed as well as volatile compounds will be biodegraded as vapors slowly move through biologically active soil.
- **Bioaugmentation**: Bioaugmentation is the process of introducing a variety of naturally occurring microbes or a genetically engineered strain to remediate contaminated soil or water. It is most frequently used to reactivate sludge bioreactors in the treatment of municipal wastewater. Tetra- and trichloroethylene are two examples of chlorinated ethanes that can contaminate soil and groundwater. Bioaugmentation is used at these locations to enable in situ microbes to completely break down these toxins into ethylene and chloride.
- **Biopiling**: Excavated soils are mixed with soil additives, placed on a treatment area, and bioremediated utilizing forced aeration in this full-scale method. Carbon dioxide and water are decreased from the pollutants. A treatment bed, an aeration system, an irrigation/nutrient system, and a leachate collection system are all part of a basic biopile system. Biodegradation is aided by controlling moisture, heat, nutrients, oxygen, and pH. The irrigation/nutrient system is buried beneath the soil and uses vacuum or positive pressure to transport air and nutrients.

The use of modified *E. coli* enclosed in silica to extract solidified mercurium from methylmercury has recently been discovered as an alternative. Another option is to biovolatilize the mercurium; this is why many of these characteristics must be considered in design techniques. It's also crucial to remember that some methods that work in one setting may not work in another, and that strategies that work in the lab may not repeat findings in the field. Natural microorganisms that can combat a specific pollutant are commonly used in these tactics.

However, the nature of such pollutants may occasionally interfere with microorganisms' natural ability to biodegrade or biotransform a specific type of contamination. This is the case with very hydrophobic chemicals, which can stymie microbial activity by collecting in cellular membranes and disturbing them. In this light, bacteria that can propagate in organic solvents have been discovered (Yadav et al., 2011). However, genome-based investigations revealed that the solvent-tolerant *Pseudomonas putida* S12, which comprises several controlled mobile components, can play a key role in stress management and adaptation (Hosseini et al., 2017). The protein named Phasin Pha-P obtained from the soil bacterium *Azotobacter* sp. strain FA8 has so far been discovered to be a polyhydroxyalkanoate granule-related protein that can be used to safeguard recombinant strains from ethanol accumulation (Mezzina et al., 2017). The ethanol resistance protein found in *Azotobacter* sp. and SrpABC from *P. putida* S12 are just two examples of possible natural defenses against contaminants. It's also worth noting that finding both resistance and biodegradation/biotransformation pathways in the same microbe would be a disadvantage.

But these occurrences also occur naturally, and it's possible that a community of microorganisms rather than a single one will exhibit the two characteristics more frequently. As a result, it is crucial to look into microbial communities at both the physiological and meta-genomic levels, as this is a useful method for figuring out the genetic basis for both toxin resistance and biodegradation performance. Engineering microorganisms with both of the above-mentioned functions is an intriguing approach. However, restrictions governing genetically modified organisms (GMOs) stymie this type of study.

3.4.1 Process of Degradation of Contaminants by Microbes

Although microorganisms can biodegrade practically all organic contaminants and many inorganic contaminants, bioremediation is now utilized commercially to clean up a limited spectrum of contaminants—mostly hydrocarbons found in gasoline. Bioremediation's utility could be extended well beyond its current use by utilizing a wide range of microbiological processes. The similar values are employed to encourage the proper microbial activity either the use is conventional or innovative.

The organisms can utilize the contaminants for their own growth and reproduction; hence, microbial transformation of organic contaminants is common. Organic pollutants have two objectives for organisms: they give a carbon source as fundamental building blocks of fresh cell components and arrange for electrons to be available for energy. Microorganisms obtain energy by accelerating chemical processes that produce energy by breaking chemical bonds and moving electrons away from the contamination. An oxidation-reduction reaction takes place when an organic pollutant is oxidized and the molecule that gets the electrons is reduced. Both the contaminant and the electron acceptor are referred to as electron donors and acceptors, respectively. Along with some of the electrons and carbon from the contaminant, the energy produced by these electron exchanges is then "invested" in the creation of more cells. These two materials, the electron donor and acceptor, are known as the primary substrates because they are required for cell growth. The electron acceptor of choice for many microorganisms, including humans, is molecular oxygen (O_2). Aerobic respiration is the process of eliminating organic compounds with the help of oxygen. Microbes employ O_2 to oxidize some of the carbon in the contaminant to carbon dioxide (CO_2), with the remaining carbon utilized to generate new cell mass in aerobic respiration.

Numerous microorganisms are devoid of oxygen and survive through anaerobic respiration, in which nitrate, sulfate, and metals such as iron and manganese can play the role of oxygen, accepting electrons from the degraded contaminant. As a result, inorganic compounds are used as electron acceptors in anaerobic respiration. Depending on the electron acceptor, anaerobic respiration can produce nitrogen gas, hydrogen sulfide, reduced forms of metals, and methane as byproducts in addition to new cell matter. Some of the metals used as electron acceptors by anaerobic microbes are considered pollutants. A recent study has shown that some microbes may reduce soluble uranium to insoluble uranium by using it as an electron acceptor. The organisms cause uranium to precipitate in this situation, lowering its concentration and mobility in the groundwater. Other microbes can employ inorganic molecules as electron donors in addition to those that use inorganic compounds as electron acceptors for anaerobic respiration. Ammonium and other inorganic electron donors deliver electrons to an electron acceptor to create energy for cell production. In most cases, bacteria with an inorganic molecule as their principal electron donor must acquire their carbon from atmospheric CO_2.

Fermentation is a kind of metabolism that can be useful in oxygen-deficient settings. Because the organic contaminant functions as both an electron donor and an electron acceptor, fermentation does not require any external electron acceptors. The organic pollutant is transformed into harmless molecules known as fermentation products through a sequence of internal electron transfers mediated by the bacteria. Acetate, propionate, ethanol, hydrogen, and carbon dioxide are examples of fermentation products. Other bacteria can biodegrade fermentation products, eventually turning them into carbon dioxide, methane, and water (Phulpoto et al., 2016).

Microorganisms can change pollutants in some situations, even if the transformation has little or no benefit to the cell. Secondary utilization is a broad word for such non-beneficial biotransformations, and cometabolism is a key example. The change of the contaminant is an unintended consequence of cometabolism, and it is catalyzed by enzymes involved in normal cell metabolism or particular detoxification reactions. When microorganisms oxidize methane, they develop enzymes that destroy the chlorinated solvent by accident, despite the fact that the solvent cannot support microbial growth.

3.4.2 Characteristics of Suitable Microbe for Wastewater Treatment

Bacteria have a number of characteristics that make them suitable for use in bioremediation procedures. The distinctive properties of bacteria have been presented in this chapter, with sea bacteria serving as a model microbe. More than 90% of the volume of the biosphere is made up of the ocean, which is also the world's largest habitat. Microorganisms in the ocean are also accountable for even more than 50% of the world's resource extraction and biogeochemical cycles (Lauro et al., 2009). Such marine bacteria are found in the sea, sediments, mangroves, normal marine animal flora, deep-sea hydrothermal vents, and sediments of marine habitats (McKew et al., 2007). They normally need sodium and potassium ions to grow and keep the osmotic balance in their cytoplasm in check. This sodium ion need is unique to marine bacteria, and it is ascribed to the formation of indole from tryptophan, the oxidation of L-arabinose, mannitol, and lactose, and the transfer of substrates into the cell (Häse et al., 2001).

Other physical characteristics that separate marine bacteria from terrestrial bacteria include facultative psychrophilicity, higher pressure tolerance than terrestrial bacteria, the ability to survive in seawater, primarily Gram-negative rods, and motile spore formers (Buerger et al., 2012). -Aminoglutaric acid, also known as -glutamate, is found in higher concentrations in marine sediments and is used as an osmolyte by marine bacteria. Some of the thermophilic marine microorganisms discovered in deep-sea hydrothermal vents can also produce nitrogen (Wu et al., 2015). The presence of rhodopsin, which has 2197 genes, is the greatest distinguishing trait of a photosynthetic marine bacterial genome. Furthermore, the gene contents of marine cyanobacteria follow a similar pattern that is linked to their isolation origins.

3.4.3 Adaptation of Microbes in the Adverse Environmental Condition

Microorganisms come in a wide variety of shapes and sizes, which is important for the functional role they perform in their natural environment. They are attractive for possible bioremediation and biological indicator purposes because they respond fast to changing environmental patterns. Variations in temperature and pH of the environment, sea level rise, shifting light patterns, terrestrial inputs, and tropical storms are all examples of periodic changes in the environment. Microorganisms in marine habitats are constantly exposed to fluctuations in oceanic temperature; however, the amount of exposure varies depending on the microbial niche. Some bacteria get over this challenge by moving their physical locations beneath sediments or forming symbiotic relationships with other species, which is primarily seen in pathogenic germs.

Additionally, pollutants and xenobiotics are introduced into seawater by rainfall and river flooding, which may change the microbial community's structure and function. On the other hand, bacteria change their rates of growth, gene expression, physiological or enzymatic activity, and their close or symbiotic connections with other species in response to such situations. It has been noted that bacteria can acquire a number of unusual mechanisms, such as the synthesis of bioactive chemicals, the creation of bacteria in the atmosphere, and the production of biosurfactants, when they are subjected to extremities in pressure, temperature, and salinity or when micronutrients are depleted (Mangwani et al., 2012). Any microbe utilized in bioremediation procedures needs to have the genotype that is resistant to the specific contaminant.

3.4.4 Use of Bacteria for Wastewater Treatment

The employment of bacteria for the biodegradation of numerous natural and man-made compounds, and hence the reduction of dangers, is gaining popularity. Bacteria have a wide range of bioremediation capabilities that are useful to both the environment and the economy. Microorganisms' naturally occurring metabolic ability to breakdown, modify, or accumulate harmful materials such as hydrocarbons, heterocyclic compounds, medicinal drugs, radionuclides, and toxic metals has been used in bioremediation and biotransformation approaches (Karigar and Rao, 2011).

The purpose of bioremediation is to provide nutrients and other compounds to microorganisms so that they can eliminate pollutants. Today's bioremediation technologies rely on microorganisms that are native to the contaminated locations and encourage them to work by supplying them with optimum levels of nutrients and other chemicals essential for their metabolism. Microorganisms obtain energy by catalyzing energy-producing chemical reactions involving the breakdown of chemical reactions, hence shifting electrons away from the contamination. The energy gained from these electron transfers is then used to produce additional cells, along with some electrons and carbon from the contaminant. When considering microbial bioremediation for site cleanup, there are at least five key aspects to consider.

These criteria include:

i. Magnitude of contaminant, its toxicity, and its mobility. It can be thoroughly explored to govern:
   - Extent of contamination;
   - Types of pollutants; and
   - Likelihood of future contamination mobility, which is dependent on the physical characteristics.

ii. Human and environmental receptor proximity: The rate and extent of contaminant degradation determine whether bioremediation is an appropriate clean-up treatment for any location.

iii. Contaminant degradability: The degradability of a compound is determined by the compound's natural occurrence. High molecular weight chemicals, particularly those with complex ring structures and halogen substituents, breakdown more slowly than simpler straight chain hydrocarbons or low molecular weight molecules in some cases. Thus, the availability of electron acceptors and other nutrients determines the rate and extent to which the molecule is metabolized in the environment.

iv. Intended site use: The rate and extent of pollutant degradation are the most important factors in determining whether or not bioremediation is appropriate.

v. Monitoring ability: Ability to effectively monitor. Because of the structural, physiological, and microbiological variabilities of the contaminated matrix, there seems to be intrinsic uncertainty regarding the use of bioremediation for polluted soil and aquifers.

a. **Heavy metal removal**

One of the most serious environmental concerns is heavy metal contamination, which is generated by a variety of natural and man-made processes (Peña-Montenegro et al., 2015). Though numerous physical and chemical ways to remove such toxic metals from the environment have been proposed, they have shown to be ineffective in terms of cost, restrictions, and the formation of harmful chemicals (Wuana and Okieimen, 2011). Because microorganisms produce no byproducts and are very efficient even at low metal concentrations, they overcome these concerns. It was also demonstrated that a consortium of marine bacteria may simply eliminate mercury in a bioreactor utilizing a disturbance-independent technique. Additionally, a novel genetic system in microorganisms for the potential phenol and metal degradation was described (Singh et al., 2013).

b. **Degradation of hydrocarbons**

Because of their persistence, toxicity, mutagenicity, and carcinogenicity in nature, polyaromatic hydrocarbons (PAHs) are a major environmental problem. Many marine bacteria, on the other hand, have been reported to have the ability to bioremediate the same in the metabolic process to produce CO_2 and metabolic intermediates, acquiring energy and carbon for cell growth. The bioremediation capability of these bacteria can be boosted, as seen in an experiment in which a *Pseudomonas putida* catabolic plasmid harboring hydrocarbon exhibited degrading genotype in a marine bacterium, increasing its effectiveness.

c. **Plastic degradation**

Polyethylene, polypropylene, polystyrene, polyethylene terephthalate, and polyvinyl chloride are among the many types of plastic used in fishing, packing, and other applications that contaminate the environment. Microorganisms, on the other hand, can develop a way to breakdown the plastic into harmless forms. *Rhodococcus ruber* destroys 8% of dry weight of plastic in 30 days in concentrated liquid culture in vitro, according to new research. Similarly, bacterial isolates from the genera *Shewanella, Moritella, Psychrobacter*, and *Pseudomonas* identified in Japan's deep seas have the ability to efficiently degrade -caprolactone (Paranthaman and Karthikeyan, 2015). *Micrococcus, Moraxella, Pseudomonas, Streptococcus, and Staphylococcus*, which are present in mangroves, have also been found to breakdown 20% of plastic.

3.5 RADICAL PROCESSES INVOLVED IN BIOREMEDIATION

Molecular biology has played a significant influence in the development of cutting-edge bioremediation methods. In the subject of bioremediation, innovative approaches include the use of genetically engineered microbes. Due to the elasticity of the metabolic pathways, *Pseudomonas* sp. is a good bacteria for bioremediation. Nonetheless, the bigger genome sequence of *Pseudomonas* sp. is due to this extended and varied metabolome, which can be called a genetic and metabolic "burden."

In the first scenario, the outcome is quite encouraging, and it is expected that it will reduce extraneous routes that may be obstructing the activity, allowing *Pseudomonas* sp. to develop and perform better. However, it's worth noting that some of *Pseudomonas* sp. extended pathways aren't always tied to the biodegradation of a given molecule but are essential for the microbes' survival in severe environments. The second option is creation of low level bacteria through genetic engineering and metabolic pathway regulation from preexisting microbes or higher organisms (e.g., plants or fungi). *E. coli* is typically utilized as this designed host since it has a B25 minute doubling time at 37°C and there are many genetic tools accessible for their changes. Nevertheless, these features make it a superb option for lab settings and allow for the construction of the ideal framework for evaluating imported metabolic pathways and their operation in the breakdown of interesting substances.

3.5.1 Treatment of Wastewater Ahead of Genomic Period

The decreasing expense of DNA sequencing and the anticipated advances in the process of sequencing and gene expression analysis have led to creation of genome-based metabolic simulations that help biological systems can be retrieved, applied, and tested in a systematic manner (Pereira et al., 2014). Bioremediation strategies may also be intended by choosing, nourishing, and straightforwardly utilizing them in the domain via biostimulation and/or bioaugmentation techniques.

The only way to find acceptable metabolites for bioremediation is to isolate microbial environmental samples from a contaminated environment. However, there is a chance that enriching, isolating, and amplifying a specific microorganism with a specific metabolism will be impossible because that metabolism is dependent on the association of complex microbial community. Sequencing of the meta-genome and identifying genes accountable for the bioprocessing of a given substance is a useful strategy by which other "non-cultivable" metabolisms capable of performing bioremediation processes can be discovered. Finding these genes allows scientists to create microbes with regulatory systems that boost the effectiveness of putative biodegradation and biotransformations. Furthermore, because most of these biodegradation pathways are the product of a cometabolism, they may not be entirely optimized or thermodynamically suitable in nature. As a result, promiscuity is a feature of microbial activity that can be used for biocatalysis as well as bioremediation.

Currently, the UM-PPS has 249 biotransformation rules generated from UM-BBD and scientific literature reactions. PubChem and ChemSpide benefit from UM-BBD chemical data. The enormous volumes of genomic, proteomic, and metabolic data, as well as advances in synthetic biology, genetic engineering, and metabolic engineering, have made it possible to alter biological systems (Kumar and Prasad, 2011; Wanichthanarak et al., 2015). All of this data may be used to for the genetic sequences liable for specific enzyme activity, or to decide which microbe can be used to execute a biotransformation. By connecting heterologous metabolic pathways to compatible host microorganisms, not only facilitates the creation of cocktails from natural bacteria but also provides a framework for designing and engineering microbes for bioremediation.

3.6 CONCLUSION

The rapid growth of the human population, combined with rising demand for industrial products, results in a waste and pollution problem that has a significant influence on the environment. Finding cost-effective, environmentally friendly, and life-compatible waste treatment alternatives is a must. Because of its compatibility with life and cost-effectiveness, the employment of bioremediation technologies has proven to be an intriguing strategy. However, there are some issues to be resolved in terms of scaling it up from the bench to the field, as well as improving its time-effectiveness.

We must be aware of the difficulties involved with this strategy in addition to the possible benefits of employing cathodes to enhance wastewater treatment. First, pollutant eradication and energy recovery are always a trade-off in BESs. Since numerous contaminants must be removed using electrons, the better the contaminant removal will be, the more electrons will flow. BESs will therefore have to be run in a mode for high current generation, which doesn't generate a lot of power or energy. The advantage of improved pollutant removal may make up for reduced energy output. The second issue is how to support high current generation and pollutant removal while overcoming the cathode restriction. Despite the correlation between the two jobs, more research has to be done on the specific interactions between them. By coating the cathode surface (e.g., through metal deposition) or poisoning the catalysts, some intermediates or end products of contaminant removal, for instance, may restrict further current generation. Third, the performance of the BES may suffer from the introduction of unwanted electron acceptors or donors. By adding electron donors (such as organics and ammonia) to the cathode, such as an anolyte, terminal electron acceptors may face competition (e.g., oxygen).

REFERENCES

Abatenh, E., Gizaw, B., Tsegaye, Z., & Wassie, M. (2017). The role of microorganisms in bioremediation-a review. *Open Journal of Environmental Biology*, *2*(1), 038–046.

Adebajo, S. O., Balogun, S. A., & Akintokun, A. K. (2017). Decolourization of vat dyes by bacterial isolates recovered from local textile mills in southwest, Nigeria. *Microbiology Research Journal International*, 18(1), 1–8.

Buerger, S., Spoering, A., Gavrish, E., Leslin, C., Ling, L., & Epstein, S. S. (2012). Microbial scout hypothesis, stochastic exit from dormancy, and the nature of slow growers. *Applied and Environmental Microbiology*, *78*(9), 3221–3228.

Burghal, A. A., Abu-Mejdad, N. M. J. A., & Al-Tamimi, W. H. (2016). Mycodegradation of crude oil by fungal species isolated from petroleum contaminated soil. *International Journal of Innovative Research in Science, Engineering and Technology*, *5*(2), 1517–1524.

Cecconet, D., Zou, S., Capodaglio, A. G., & He, Z. (2018). Evaluation of energy consumption of treating nitrate-contaminated groundwater by bioelectrochemical systems. *Science of the Total Environment*, *636*, 881–890.

Chiarle, S., Ratto, M., & Rovatti, M. (2000). Mercury removal from water by ion exchange resins adsorption. *Water Research*, *34*(11), 2971–2978.

Clauwaert, P., Mulenga, S., Aelterman, P., & Verstraete, W. (2009). Litre-scale microbial fuel cells operated in a complete loop. *Applied Microbiology and Biotechnology*, *83*(2), 241–247.

Das, S., Lyla, P. S., & Khan, S. A. (2006). Marine microbial diversity and ecology: Importance and future perspectives. *Current Science*, 90(10), 1325–1335.

Dash, H. R., & Das, S. (2012). Bioremediation of mercury and the importance of bacterial mer genes. *International Biodeterioration & Biodegradation*, *75*, 207–213.

Demnerová, K., Mackova, M., Speváková, V., Beranova, K., Kochánková, L., Lovecká, P., & Macek, T., et al. (2005). Two approaches to biological decontamination of groundwater and soil polluted by aromatics—Characterization of microbial populations. *International Microbiology*, *8*(3), 205–211.

Doney, S. C., Ruckelshaus, M., Emmett Duffy, J., Barry, J. P., Chan, F., English, C. A., & Talley, L. D., et al. (2012). Climate change impacts on marine ecosystems. *Annual Review of Marine Science*, *4*, 11–37.

El Fantroussi, S., & Agathos, S. N. (2005). Is bioaugmentation a feasible strategy for pollutant removal and site remediation? *Current Opinion in Microbiology*, *8*(3), 268–275.

Fazli, M. M., Soleimani, N., Mehrasbi, M., Darabian, S., Mohammadi, J., & Ramazani, A. (2015). Highly cadmium tolerant fungi: Their tolerance and removal potential. *Journal of Environmental Health Science and Engineering*, *13*(1), 1–9.

Gude, V. G. (2016). Wastewater treatment in microbial fuel cells—An overview. *Journal of Cleaner Production, 122*, 287–307.

Häse, C. C., Fedorova, N. D., Galperin, M. Y., & Dibrov, P. A. (2001). Sodium ion cycle in bacterial pathogens: Evidence from cross-genome comparisons. *Microbiology and Molecular Biology Reviews, 65*(3), 353–370.

Hassan, M. M., Alam, M. Z., & Anwar, M. N. (2013). Biodegradation of textile azo dyes by bacteria isolated from dyeing industry effluent. *International Research Journal of Biological Sciences, 2*(8), 27–31.

Henneberry, Y. K., Kraus, T. E., Fleck, J. A., Krabbenhoft, D. P., Bachand, P. M., & Horwath, W. R. (2011). Removal of inorganic mercury and methylmercury from surface waters following coagulation of dissolved organic matter with metal-based salts. *Science of the Total Environment, 409*(3), 631–637.

Hosseini, R., Kuepper, J., Koebbing, S., Blank, L. M., Wierckx, N., & de Winde, J. H. (2017). Regulation of solvent tolerance in P seudomonas putida S12 mediated by mobile elements. *Microbial Biotechnology, 10*(6), 1558–1568.

Jain, A., & He, Z. (2018). Cathode-enhanced wastewater treatment in bioelectrochemical systems. *Npj Clean Water, 1*(1), 1–5.

Jain, P. K., & Bajpai, V. (2012). Biotechnology of bioremediation-a review. *International Journal of Environmental Sciences, 3*(1), 535–549.

Jasińska, A., Różalska, S., Bernat, P., Paraszkiewicz, K., & Długoński, J. (2012). Malachite green decolorization by non-basidiomycete filamentous fungi of penicillium pinophilum and myrothecium roridum. *International Biodeterioration & Biodegradation, 73*, 33–40.

Karigar, C. S., & Rao, S. S. (2011). Role of microbial enzymes in the bioremediation of pollutants: A review. *Enzyme Research, 2011*, 1.

Kumar, A., Bisht, B. S., Joshi, V. D., & Dhewa, T. (2011). Review on bioremediation of polluted environment: A management tool. *International Journal of Environmental Sciences, 1*(6), 1079–1093.

Kumar, R. R., & Prasad, S. (2011). Metabolic engineering of bacteria. *Indian Journal of Microbiology, 51*(3), 403–409.

Lauro, F. M., McDougald, D., Thomas, T., Williams, T. J., Egan, S., Rice, S., & Cavicchioli, R. (2009). The genomic basis of trophic strategy in marine bacteria. *Proceedings of the National Academy of Sciences, 106*(37), 15527–15533.

Lefebvre, O., Shen, Y., Tan, Z., Uzabiaga, A., Chang, I. S., & Ng, H. Y. (2011). Full-loop operation and cathodic acidification of a microbial fuel cell operated on domestic wastewater. *Bioresource Technology, 102*(10), 5841–5848.

Maliji, D., Olama, Z., & Holail, H. (2013). Environmental studies on the microbial degradation of oil hydrocarbons and its application in Lebanese oil polluted coastal and marine ecosystem. *International Journal of Current Microbiology and Applied Sciences, 2*(6), 1–18.

Mangwani, N., Dash, H. R., Chauhan, A., & Das, S. (2012). Bacterial quorum sensing: Functional features and potential applications in biotechnology. *Journal of Molecular Microbiology and Biotechnology, 22*(4), 215–227.

McKew, B. A., Coulon, F., Osborn, A. M., Timmis, K. N., & McGenity, T. J. (2007). Determining the identity and roles of oil-metabolizing marine bacteria from the Thames estuary, UK. *Environmental Microbiology, 9*(1), 165–176.

Mezzina, M. P., Álvarez, D. S., Egoburo, D. E., Diaz Pena, R., Nikel, P. I., & Pettinari, M. J. (2017). A new player in the biorefineries field: Phasin PhaP enhances tolerance to solvents and boosts ethanol and 1, 3-propanediol synthesis in Escherichia coli. *Applied and Environmental Microbiology, 83*(14), e00662–17.

Mohan, D., Gupta, V. K., Srivastava, S. K., & Chander, S. (2001). Kinetics of mercury adsorption from wastewater using activated carbon derived from fertilizer waste. *Colloids and Surfaces A: Physicochemical and Engineering Aspects, 177*(2–3), 169–181.

Narang, U., Bhardwaj, R., Garg, S. K., & Thukral, A. K. (2011). Phytoremediation of mercury using Eichhornia crassipes (mart.) Solms. *International Journal of Environment and Waste Management*, *8*(1–2), 92–105.

Pant, D., Singh, A., Van Bogaert, G., Olsen, S. I., Nigam, P. S., Diels, L., & Vanbroekhoven, K. (2012). Bioelectrochemical systems (BES) for sustainable energy production and product recovery from organic wastes and industrial wastewaters. *Rsc Advances*, *2*(4), 1248–1263.

Paranthaman, S. R., & Karthikeyan, B. (2015). Bioremediation of heavy metal in paper mill effluent using Pseudomonas spp. *International Journal of Microbiology*, *1*, 1–5.

Peña-Montenegro, T. D., Lozano, L., & Dussán, J. (2015). Genome sequence and description of the mosquitocidal and heavy metal tolerant strain lysinibacillus sphaericus CBAM5. *Standards in Genomic Sciences*, *10*(1), 1–10.

Pereira, P., Enguita, F. J., Ferreira, J., & Leitão, A. L. (2014). DNA damage induced by hydroquinone can be prevented by fungal detoxification. *Toxicology Reports*, *1*, 1096–1105.

Phulpoto, A. H., Qazi, M. A., Mangi, S., Ahmed, S., & Kanhar, N. A. (2016). Biodegradation of oil-based paint by bacillus species monocultures isolated from the paint warehouses. *International Journal of Environmental Science and Technology*, *13*(1), 125–134.

Plan, D., & Van den Eede, G. (2010). The EU legislation on GMOs. *JRC Scientific and Technical Reports, EUR, 24279.*

Ramadevi, A., & Srinivasan, K. (2005). Agricultural solid waste for the removal of inorganics: Adsorption of mercury (II) from aqueous solution by Tamarind nut carbon.

Seigle-Murandi, F., Guiraud, P., Croize, J., Falsen, E., & Eriksson, K. L. (1996). Bacteria are omnipresent on Phanerochaete chrysosporium Burdsall. *Applied and Environmental Microbiology*, *62*(7), 2477–2481.

Sevda, S., Yuan, H., He, Z., & Abu-Reesh, I. M. (2015). Microbial desalination cells as a versatile technology: Functions, optimization and prospective. *Desalination*, *371*, 9–17.

Shah, M. P. (2013). Microbial degradation of textile dye (remazol black B) by Bacillus spp. ETL-2012. *Journal of Applied & Environmental Microbiology*, *1*(1), 6–11.

Shah, M. P. (2020). Microbial Bioremediation & Biodegradation; Springer.

Shah, M. P. (2021). Removal of Emerging Contaminants through Microbial Processes; Springer.

Shah, M. P. (2021). Removal of Refractory Pollutants from Wastewater Treatment Plants; CRC Press.

Singh, A., Kumar, V., & Srivastava, J. N. (2013). Assessment of bioremediation of oil and phenol contents in refinery waste water via bacterial consortium. *Journal of Petroleum & Environmental Biotechnology*, *4*(3), 1–4.

Singh, R., & Majumdar, R. S. (2017). Comparative study on bioremediation for oil spills using microbes. *Research & Reviews: A Journal of Bioinformatics*, *4*(1), 16–24.

Sinha, S. N., Biswas, M., Paul, D., & Rahaman, S. (2011). Biodegradation potential of bacterial isolates from tannery effluent with special reference to hexavalent chromium. *Biotechnology Bioinformatics and Bioengineering*, *1*(3), 381–386.

Stelting, S., Burns, R. G., Sunna, A., Visnovsky, G., & Bunt, C. R. (2012). Immobilization of Pseudomonas sp. Strain ADP: A stable inoculant for the bioremediation of atrazine. *Applied Clay Science*, *64*, 90–93.

Strong, P. J., & Burgess, J. E. (2008). Treatment methods for wine-related and distillery wastewaters: A review. *Bioremediation Journal*, *12*(2), 70–87.

Tang, C. Y., Fu, Q. S., Criddle, C. S., & Leckie, J. O. (2007). Effect of flux (transmembrane pressure) and membrane properties on fouling and rejection of reverse osmosis and nanofiltration membranes treating perfluorooctane sulfonate containing wastewater. *Environmental Science & Technology*, *41*(6), 2008–2014.

U.S. EPA (2007). Treatment Technologies for Mercury in Soil, Waste, and Water; Office of Superfund Remediation and Technology Innovation, Washington, DC, 20460.

Wang, J., & Chen, C. (2009). Biosorbents for heavy metals removal and their future. *Biotechnology Advances*, *27*(2), 195–226.

Wanichthanarak, K., Fahrmann, J. F., & Grapov, D. (2015). Genomic, proteomic, and metabolomic data integration strategies. *Biomarker Insights*, *10*, BMI-S29511.

WHO (2006). Air Quality Guidelines for Particulate Matter, Ozone, Nitrogen Dioxide and Sulfur Dioxide-Global Update 2005; World Health Organization, Geneva.

Wolski, E. A., Barrera, V., Castellari, C., & González, J. F. (2012). Biodegradation of phenol in static cultures by Penicillium chrysogenum ERK1: Catalytic abilities and residual phytotoxicity. *Argentine Journal of Microbiology*, *44*(2), 113–121.

Wu, Y. H., Zhou, P., Cheng, H., Wang, C. S., Wu, M., & Xu, X. W. (2015). Draft genome sequence of Microbacterium profundi Shh49T, an Actinobacterium isolated from deep-sea sediment of a polymetallic nodule environment. *Genome Announcements*, *3*(3), e00642–15.

Wuana, R. A., & Okieimen, F. E. (2011). Heavy metals in contaminated soils: A review of sources, chemistry, risks and best available strategies for remediation. *International Scholarly Research Notices*, *2011*, 20.

Yadav, M., Singh, S. K., Sharma, J. K., & Yadav, K. D. S. (2011). Oxidation of polyaromatic hydrocarbons in systems containing water miscible organic solvents by the lignin peroxidase of Gleophyllum striatum MTCC-1117. *Environmental Technology*, *32*(11), 1287–1294.

Zhang, Y., & Angelidaki, I. (2014). Microbial electrolysis cells turning to be versatile technology: Recent advances and future challenges. *Water Research*, *56*, 11–25.

Zhu, X., & Elimelech, M. (1997). Colloidal fouling of reverse osmosis membranes: Measurements and fouling mechanisms. *Environmental Science & Technology*, *31*(12), 3654–3662.

4 Bioelectrochemical System-Based Removal of Hazardous Pollutants from Wastewater

Pranav Pandya and Sougata Ghosh

4.1 INTRODUCTION

There is a continuous increase in the energy demand due to large-scale urbanization and industrialization. Energy requirements have increased in each and every sector, including textile, agriculture, healthcare, transportation, food, and other associated industries. The economic activities of a country are dependent on energy (Brown et al., 2011). Overexploitation of the nonrenewable sources of energy has resulted in its exhaustion, and hence there is a continuous effort to develop new methods to address the rising energy crisis in different parts of the world. Development of sustainable approaches for energy generation has become the prime objective of the present day research (Winter and Brodd, 2004).

One of the most efficient green approaches is to convert the domestic and industrial wastes into energy. This strategy will not only help in management of the waste disposal and environmental pollution but also reduce the cost of the overall process (Khanal, 2008). Microorganism-associated anaerobic technology used for wastewater treatment can be coupled with energy generation (Gallert et al., 2003). Dark fermentation is employed for the production of biohydrogen production. Use of hydrogen fuel cells for eco-friendly energy production has highlighted the scope of exploration in this area. Feedstocks with high carbohydrate content are generally considered as ideal fermentation substrates for energy generation (Logan, 2004; Li and Fang, 2007).

Certain bacteria have the ability to transfer electrons from reduced electron donors toward an electron acceptor that is a solid anode, which is often made up of mostly graphite (Lies et al., 2005). Connecting it to a cathode through an external circuit in a fuel cell can help in efficient electron flow where bacterial respiration can generate power (Biffinger et al., 2007). Biofilm-producing electroactive bacteria (EAB) can thus remove organic carbon from the effluents and generate energy simultaneously in bioelectrochemical systems (BESs) such as microbial fuel cells (MFCs) or microbial electrolysis cells (MECs). In view of this background, this chapter gives an elaborate account of the strategic removal of various wastes using BES that is summarized in Table 4.1.

DOI: 10.1201/9781003368472-4

TABLE 4.1
Application of BES for Removal of Various Pollutants

Type of BES	Inoculum	Pollutant	Removal Efficiency (%)	References
		Dye		
BES integrated with membrane biofilm reactor	Anaerobic sludge	Acid orange 7	>96	Pan et al. 2018
Membrane free microbial electrolysis cell	Anaerobic and aerobic sludge	Congo red	90	Huang et al. 2018
Combined bioanode-biocathode BES	Activated sludge	Congo red	97.5 ± 2.3	Kong et al. 2014
Microbial fuel cell	*Pseudomonas aeruginosa* and *Pseudomonas fluorescens*	Reactive orange 16 and reactive black 5	78 ± 2.3 and 65 ± 2.8 (RO-16) 31 ± 1.5 and 27 ± 1.3 (RB-5)	Ilamathi et al. 2019
Microbial fuel cell	Activated sludge	Triclosan	94	Xu et al. 2020
		Pharmaceutical wastes		
BES	Anaerobic sludge	Carbamazepine	84	Tahir et al. 2019
Microbial electrolysis cell	-	Erythromycin	99	Hua et al. 2019
MFC-fenton	Mixed microbial culture	Paracetamol	70	Zhang et al. 2015
Microbial BES	Primary wastewater	Sulfonamides	-	Harnisch et al. 2013
		Heavy metals		
Microbial fuel cell	MFC effluent	Cd (II), Cu (II), and Cr (VI)	-	Huang et al. 2015
Osmotic microbial fuel cell	Anaerobic sludge	Cr (VI)	97.6	Cao et al. 2021
BES	Activated sludge	Cr (VI)	72.65	Wang et al. 2018
Graphene-modified graphite paper cathode BES	Electrochemically active bacteria	Cr	81	Yao et al. 2021
BES coupled with thermoelectric generators	*Desulfovibrio*, *Megasphaera*, *Geobacter*, and *Propionibacterium*	Cu^{2+}, Cd^{2+}, and Co^{2+}	99.97 ± 0.004 (Cu^{2+}), and 88.90 ± 2.62 (Cd^{2+})	Ai et al. 2020
BES-assisted microelectrolysis system	activated sludge	Cu	100	Wang et al. 2020
		Aromatic compounds		
BES	Anaerobic sludge	Naphthalene, phenanthrene, and pyrene	97.60 (NAP), 42.60 (PHE), and 22.00 (PYR)	Zhou et al. 2020

(Continued)

TABLE 4.1 ***(Continued)***
Application of BES for Removal of Various Pollutants

Type of BES	Inoculum	Pollutant	Removal Efficiency (%)	References
Microbial electrolysis cell	Anaerobic and aerobic sludge	Nitrobenzene	98	Luo et al. 2019
Membrane-free BES	Sludge	*m*-nitrophenol, *p*-nitrophenol, and *o*-nitrophenol	89.23 ± 1.02 (MNP), 49.40 ± 2.42 (PNP), and 26.23 ± 1.66 (ONP)	Jiang et al. 2016
Microbial fuel cell	*Arcobacter*, *Acinetobacter*, *Azospirillum*, *Azonexus* and *Comamonas*	Oxyfluorfen	77	Zhang et al. 2018
Microbial fuel cell	*Kocuria rosea* (GTPAS76), two strains of *Bacillus circulans* (GTPO28 and GTPAS54), and two strains of *Corynebacterium vitaeruminis* (GTPO38 and GTPO42)	Sodium benzoate	89.8	Mukherjee et al. 2021

4.2 DYES

Various attempts are made to remove hazardous dyes using BES, which are discussed in this section. Pan et al. (2018) studied the enhanced removal of acid orange 7 (AO7) using BES integrated with the membrane biofilm reactor (MBfR), as shown in Figure 4.1. The two equal-sized polycarbonate cylindrical tubes were used for constructing the two-chambered BES. The 38.5 cm^2 cation exchange membrane (CEM) separated the chambers (anode and cathode). The carbon brush (anode electrode) was pretreated with pure acetone for 12 h followed by 30 min of heating at 450°C before utilizing it for BES. The granular graphite (prewashed with 32% HCl) was covered with titanium mash and used as a cathode, which was further connected using titanium wire. The adsorption effect was avoided by soaking the cathode for the period of 36 h in 100 mg/L AO7 solution. The measurement of anodic potential was carried out by the Ag/AgCl reference electrode that was mounted in the anode chamber. The circuit was further connected to the external resistor (Rex) of 10 Ω. An anaerobic sludge containing mixed liquor-suspended solids (5980 mg/L) was further inoculated in the anode of BES. Anode was fed with the synthetic solution (0.5 g glucose, 0.125 g $NaHCO_3$, 0.15 g NH_4Cl, and 1 mL trace elements). On the other

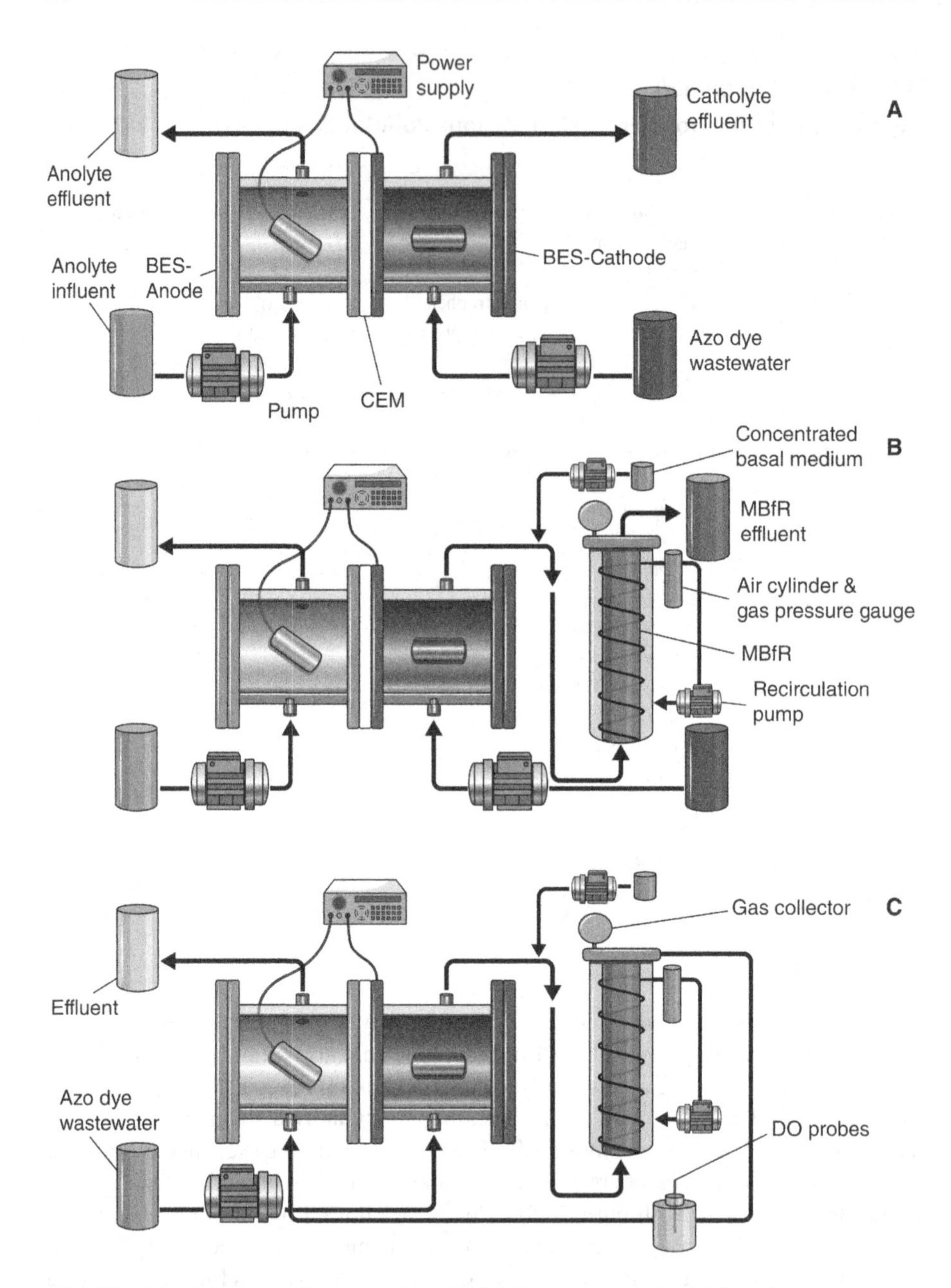

FIGURE 4.1 Schematic diagram of the BES-MBfR system: (A) BES cathode treatment only; (B) BES-MBfR connected by the cathode effluent; and (C) the loop operation of the BES-MBfR system. (Reprinted with permission from Pan et al. (2018). DOI: 10.1021/acs.iecr.8b04725. Copyright © 2018 American Chemical Society.)

hand, the catholyte was mainly composed of 0.1 g/L AO7 and 1 mL of trace elements. The nitrogen gas was purged for around 15 min in both anolyte and catholyte. The BES functioned in a continuous mode with the hydraulic retention time (HRT) of 3 h for each electrode under 0.7 V of external voltage. In the initial 16 days, the hike in the current to 6.1 mA was generated, while the decrease in anode potential was up to −0.5 mV. The pre-enriched aerobic sludge (2200 mg/L) was further used as an inoculum in MBfR. The experiment was performed in three different stages with slight variations in some specific conditions. In the first stage, the study of cathodic removal of AO7 was only carried out in the BES in an open circuit condition (0.7 V supplied voltage) at a constant HRT of 3 h. About 50 mM of phosphate buffer solution (PBS) was poured into the anolyte and catholyte of BES in the first 15 days, which was further replaced with 3 g/L of $NaHCO_3$. Similar conductivity was maintained by supplementing $NaHCO_3$ solution with 2.5 g/L NaCl when PBS was not used. In the second stage, the linkage of MBfR to the BES was fulfilled after the acceptance of effluent from BES cathode. The glucose (co-substrate) was further added to the MBfR at a concentration of 500 mg/L. A loop was created between BES and MBfR for the transfer of MBfR effluent into the BES anode. In BES (open circuit), almost no AO7 decolorization (0.9 ± 0.5%) was observed, while the decolorization capacity of 80.9 ± 1.4% was achieved in a closed circuit condition (0.7 V) which suggested the stimulation of decolorization due to generation of electricity and transfer of electrons. The AO7 was basically reduced to sulfanilic acid (SA) and 1-amino-2-naphthol (1A2N). Due to easy autoxidation of 1A2N even in presence of trace oxygen, the SA concentration was basically monitored in the cathode effluent as it was toxic and not easily biodegradable by the anaerobic bacteria. A significant increase in the concentration of SA was noticed from 1.12 ± 0.27 to 39.19 ± 0.89 mg/L in the cathode effluent. SA formation efficiency reached around 97.2 ± 2.0%, which suggested the efficient reduction of AO7 to SA. The AO7 (18.2 ± 1.2 mg/L) and SA (36.0 ± 0.4 mg/L) concentrations decreased to 3.9 ± 0.6 mg/L and 18.1 ± 2.6 mg/L by MBfR, respectively. BES-MBfR system showed the decolorization efficiency of about 95.9 ± 0.6%, which was higher compared to the BES alone (81.2 ± 0.9%) indicating the advanced removal of AO7 residuals in the MBfR. The SA removal efficiency of BES-MBfR was 63.8 ± 3.1%, which demonstrated its ability to remove the by-products generated during the AO7 decolorization. At different initial concentrations of AO7 (50, 100, 150, and 200 mg/L), the decolorization efficiency by BES were around 89.8 ± 2.2%, 81.2 ± 0.9%, 73.3 ± 2.0%, and 64.6 ± 1.1%, respectively. The overall decolorization efficiency after MBfR treatment for the same different concentrations of AO7 were 98.2 ± 0.6%, 96.5 ± 0.6%, 93.6 ± 0.7%, and 91.3 ± 0.9%, respectively. The decrease in concentration of SA from 34.1 ± 1.0 mg/L to 9.2 ± 0.6 mg/L was accompanied by the increase in the lumen pressure (from 10 to 40 psi), which suggested the role of higher lumen pressure in SA removal. No significant difference in the AO7 removal was noted during the hike in lumen pressure from 10 to 30 psi (3.9 ± 0.2 mg/L), while a sudden increase in the AO7 removal (13.0 ± 1.6 mg/L) was noted at 40 psi lumen pressure. Thus, the BES-MBfR system accomplished greater than 96% removal of AO7 during the loop operation.

Huang et al. (2018) prepared a membrane-free microbial electrolysis cell (MFMEC) to enhance the decolorization of Congo red dye from wastewater. The normal dual-chamber MEC, MFMEC, and MFEC (without biofilm) were constructed with the graphite felts (anode and cathode) presoaked with acetone and nitric acid (1 N) for about 12 h. The circuital current was further monitored by linking the external circuit using titanium wires and 20 Ω resistor. Around 10 mL of sludge mixture (anaerobic and aerobic sludge) was inoculated initially in the anode and cathode followed by 5 mL of bacterial solution from previous MFC used for Congo red treatment. The sludge was filtered before inoculation to remove the impurities followed by three times washing with deionized water in order to eradicate residual chemical oxygen demand (COD). MEC was composed of 500 mg/L of glucose or $NaHCO_3$, 12.5 mL/L of trace metals, 12.5 mL/L of multivitamin, 50 mM/L of phosphate buffer, and 200 mg/L of Congo red. The pH of the electrolyte was adjusted to 7 and 0.6 V of initial voltage was set between the anode and cathode. After completing 5 cycles, the MEC was filled with Congo red for studying the decolorization in new cycle. The cathode potential at the applied voltage of 0.3, 0.6, and 0.9 V was −0.05, −0.16, and −0.22 V for MFEC, while −0.79, −0.95, and −1.20 V in case of MFMEC, respectively, after 500 h of enrichment. The Congo red dye was efficiently reduced at the cathode potential below −0.79. The initial current (1.64 mA) of MFMEC was maintained for 6 h and a sudden drop in current to 1.1 mA was noted at 15 h. In case of MEC, the initial current (1.44 mA), which was maintained only for 1 h, dropped to 0.7 mA at 15 h. This declining current suggested the consumption of carbon sources. Congo red decolorization (CR-DE) abilities in MFMEC (90%) and MFC (below 10%) were evaluated on the completion of 42 h cycle. Decolorization rates of 7.1 mg L^{-1} h^{-1} and 0.5 mg L^{-1} h^{-1} were observed by MFMEC and MFEC at 24 h. The optimum applied voltage was around 0.3 V, which led to desirable decolorization efficiency at low power consumption. The biocathode containing glucose gave decolorization efficiencies of 81.8% (24 h) and 90% (30 h), while biocathode containing $NaHCO_3$ exhibited decolorization efficiencies of 70% (24 h) and 90% (42 h). Lower decolorization efficiency of 81.6% was noted for the anode. The study of degradation product obtained after the decolorization of Congo red using gas chromatography-mass spectroscopy (GC-MS) demonstrated the characteristic peak at 22 min in GC and at m/z 184 in MS, which suggested the formation of benzidines after comparing the data with mass spectral library (NIST). The gas product generated during the decolorization of Congo red was further investigated at different voltages (0.3, 0.6, and 0.9 V) using a GC-thermal conductivity detector (GC-TCD). Two gases dinitrogen (N_2) and methane (CH_4) were produced at 0.6 V and 0.9 V in MFMEC, whereas peak of N_2 was only observed in case of MFMEC at 0.3 V.

Kong et al. (2014) investigated the acceleration in the decolorization of the Congo red using the combined bioanode-biocathode BES. A single-chambered reactor was developed from the cylindrical Plexiglas tube without a membrane. The dual-chamber reactor, having anode and cathode chambers, was parted by the means of CEM. The activated sludge was inoculated in the anode and cathode (10% V/V) followed by feeding the BES reactors with the solution containing 300 mg/L Congo red and 500 mg/L glucose. The anaerobic conditions were maintained continuously by flushing N_2 gas before each batch cycle. The cathode potential of −0.7 V was identified for CR-DE, which was only maintained by applying 0.3 V to the system. The CR-DE of combined bioanode-biocathode single-chamber BES was around

98.3 ± 1.3%, showing 30% higher decolorization than the mixed solution in the dual-chamber BES derived from anode and cathode chambers (67.2 ± 3.5%). The current density of about 25.1 A/m^3 noted in the single-chamber BES occurred to be 3.6-times greater than that of the dual-chamber BES (7.1 A/m^3). Bioanode was the only limiting factor in the CR-DE, which was negatively affected by the Congo red by having the toxic effect on the anode-associated microbes, restricting the glucose oxidation or Congo red reduction in the anode chamber. Additionally, the CR-DE in the cathode chamber was affected indirectly due to the resultant electrons generated from the anode. On the basis of the product benzidine ($CE_{benzidine}$) the Coulombic efficiency for the single- and dual-chamber BES was calculated to be around 29.2% and 41.4%, respectively. In the case of single-chamber BES, an important role in the glucose oxidation was played by the bioanode, which resulted in the generation of electrons, and the biocathode played role in breaking of azo bonds. The generation of electrons due to the oxidation of glucose at bioanode and/or biocathode was taken by the anodophilic and/or cathodophilic microbes in order to reduce Congo red. In the dual-chamber BES, the electrochemically active microbes at the anode oxidized glucose, and the resultant electrons resulted in the Congo red reduction in the anode chamber. On the other hand, the reductive cleavage of azo bonds was carried out by the partial electrons transferred to the cathode through the electric circuit, which were further converted to aromatic amines. Better performance of combined bioanode and biocathode in the single-chamber BES was noted compared to the separated arrangement used in a dual-chamber BES. The CR-DE of biocathode in the presence and absence of glucose was 80.6 ± 2.5% and 88.7 ± 4.4% at 23 h, respectively. Abiotic cathode had 57.6 ± 2.3% CR-DE, which was 30% lower than the biocathode. Different types of electrodes, like horizontal, vertical, and surrounding, in the combined bioanode-biocathode BES were equated depending on their performance during decolorization and electrochemical characteristic. On modifying the horizontal deployment to surrounding deployment within 11 h, the improvement in CR-DE from 87.4 ± 1.3% to 97.5 ± 2.3% was noted, whereas the decrease in internal resistance was from 236.6 to 42.2 Ω.

Ilamathi et al. (2019) studied the comparative evaluation of *Pseudomonas* species (*Pseudomonas aeruginosa* and *Pseudomonas fluorescens*) in the removal of the reactive orange 16 (RO-16) and reactive black 5 (RB-5) dyes by utilizing a single-chamber microbial fuel cell (SCMFC) having a cathode coated with manganese. The manganese-coated membrane cathode assembly was made by coating the carbon cloth with a mixture of 1.5 g manganese sulfate ($MnSO_4.H_2O$), conductive graphite (0.46 g), unsaturated polyester with the accelerator (0.15 g), cobalt oceteneoate (2% w/w), and methyl ethyl ketone peroxide (0.7% w/w), which resulted in the polymerization reaction. A paste was prepared by mixing 60 μm graphite powder and manganese salt. The polymerization was initiated by using the paste of cobalt oceteneoate, unsaturated polyester, and methyl ethyl ketone peroxide catalyst. The reaction mixture was further added to the salt mixture of graphite/manganese (50:50 ratio) followed by the application of the free-flowing slurry thoroughly on the carbon cloth and air-drying it at room temperature for 36 h. The proton exchange membrane (PEM) associated with hot-pressed Mn-coated carbon cloth was denoted as MPEM-MFC. Those altered cathodes were washed with acetone (3 times) and deionized water followed by drying at 120°C for 30 min. The non-altered cathode was taken as a control. The MPEM-MFC I (*P. aeruginosa*) and MPEM-MFC II (*P. fluorescens*) were similar types of SCMFCs utilized in RO-16 and RB-5 dyes degradation. In configuration of cathode in each setup, the membrane faced the side containing water and the PTFE laminated carbon cloth faced the air side. The electron transfer was strengthened by the use of MPEM layered cathode as an electrode. The overnight culture (5 mL) of the organisms was inoculated in the SCMFCs in MFC I and MFC II at pH 7, and the reaction was further carried out in the batch conditions for around 360 h. The anaerobic conditions were ensured by the sparging of nitrogen gas in the anode chamber for 20 min. The decolorization abilities of *P. aeruginosa* and *P. fluorescens* were around 78 ± 2.3% and 65 ± 2.8% in the case of RO-16, respectively, while it was about 31 ± 1.5% and 27 ± 1.3%, respectively, in the case of RB-5 reactive dyes in the MFC containing H-shaped graphite electrodes. The power density in the H-shaped MFC was highest with about 83.04 ± 4.0 μW/m^2 for RO-16 dye along with the highest COD removal efficiency (62 ± 2.1%). The MPEM-MFC I showed near-to-complete decolorization of RO-16 dye compared to the MPEM-MFC II. No decolorization was observed in the case of static culture, while more than 94% reduction in concentrations of both dyes was noted after the evaluation of MPEM-MFC I and MPEM-MFC II, respectively. Around 60–65% decolorization of RO-16 and 45–50% decolorization in RB-5 was observed in 48 h of treatment with dye concentration equivalent to 0.5 mM. The decreasing order of the dyes decolorization rates by MPEM-MFC was: MFC-I, RO-16 (0.68 ± 0.05%/day) > MFC-I, RB-5 (0.51 ± 0.08%/day) > MFC-II, RO-16 (0.45 ± 0.03%/day) > MFC-II, RB-5 (0.29 ± 0.04%/day) at different dye loadings equivalent to 100 mg/L, 200 mg/L, 300 mg/L, and 500 mg/L, respectively. The highest COD removal was observed in *P. aeruginosa* compared to *P. fluorescens*. The mechanism behind the reduction of dye was the formation of oxidative laccase and azoreductase enzyme from the *Pseudomonas* sp., which cleaved the dye molecule asymmetrically and the azo bond (—N=N—) by reductive cleavage.

Xu et al. (2020) investigated the elimination of triclosan (TCS) from the wastewater using MFC that was made up of polycarbonate materials with the anode and cathode chambers separated using CEM. The anode was graphite fiber brush, while cathode was graphite plate. Initially, both anode and cathode were soaked in sulfuric acid (30 mol/L) for 2 h at 80°C to remove the organic matter present on their surface followed by further soaking in 30% hydrogen peroxide (H_2O_2) for a period of 2 h in order to eradicate the impurities. The external resistor (1000 Ω) was connected using the titanium wires in the circuit. The K_3 $[Fe(CN)_6]$ accepted electrons in the chamber containing cathode. Around 20 mL of the activated sludge was inoculated in the anode chamber of MFCs, which contained the artificial wastewater composed of 1 g/L sodium acetate, 20 mM PBS (pH 7), 1.7 mM potassium chloride, 5.8 mM ammonium chloride, 2 mL vitamin solution, and 2 mL mineral solution. The MCFs were operated at constant 26°C under light sheltered condition. The TCS was further added to the anode chamber after one month in order to replace acetate in each experimental cycle. The electrode taken as abiotic control was pre-treated for 30 min at 450°C to control the growth of microbes before initiating the experiments. TCS was significantly removed (94%) within 48 h of treatment in the MFC reactor (aqueous phase). The density of the output current in MFC was dependent directly on the metabolic activity of the microbes present in the system. The current density after the inoculation of sludge into the anode chambers reached 160 mA/m^2 within 7 days that increased up to 170 mA/m^2. The drop and recovery of the signals were determined during several successive cycles (1 cycle = 100 h). The TCS removal in the aqueous phase contributed to both adsorption as well as electrochemical degradation. TCS was successfully adsorbed on the inner walls of MFC as well as on the biofilm and on the anode. The confirmation of TCS adsorption on the anode as well as on the reactor was done by the rapid reduction of TCS in the presence of anode in an aqueous solution. The porosity of graphite fiber brush anodes significantly enhanced the adsorption capacity. The TCS adsorption in the anode chamber was not dependent on the nature of the TCS molecules (neutral or ionic). Approximately 21.73% and 19.92% of TCS were adsorbed on the inner wall (reactor) and on the anode. The total adsorption of 72.52% was noted for TCS in an abiotic reactor. The removal of TCS along with the electricity generation in the MFC was around 79% initially, which was much higher than the abiotic system (27%). This suggested that the biodegradation of the TCS was coupled with the electricity generation. At 10 mg/L initial concentration of TCS and 1000 Ω of external resistance (ER), the highest output voltage of 19.4–26.7 mA/m^2 was noted for the MFC for the period of 40 h.

4.3 PHARMACEUTICAL WASTES

Apart from hazardous dyes, various by-products generated from the pharmaceutical industries can also be removed from BES. Tahir et al. (2019) studied the carbamazepine (CBZ) degradation in BES and also investigated the role of anode potential in it. The dual-chamber BES reactor was constructed with plexiglass. The chambers had a PEM, which separated both the anode and cathode chambers from each other. The membrane was pretreated in order to enhance the performance by taking 0.5 M H_2SO_4, deionized water, and 3% H_2O_2, followed by heating for 1 h. The thick carbon felt (3.18 mm) was considered as anode while Pt carbon cloth (20 wt%) was considered as cathode. The dimensions of both the electrodes were 5 cm × 5 cm and were further autoclaved along with the BES reactor at 121°C for 20 min. The anaerobic conditions required for the growth of microbes were fulfilled by the nitrogen gas bag (0.5 L) embedded in the anode chamber. This resulted in the formation of high static N_2 pressure delaying the access of oxygen present in the atmosphere into the chamber. Further, the anaerobic sludge (20 mL) consisting of total suspended solids (16.4 mg/L) and volatile suspended solids (9.5 mg/L) was injected into the anode part along with a specific medium and acetate (500 mg/L), which fulfilled the requirement of the carbon source needed for the growth of the bacteria. The BES anode potential was around +200 mV with respect to the Ag/AgCl reference electrode. The experiment was carried out at 35 ± 1°C. After setup, the BES was further inoculated with acetate-containing anaerobic sludge. The result indicated that biotic degradation was successfully performed by BES 1 in the absence of power source, whereas BES 2 evaluated the effect of anodic potential on the CBZ degradation. A reduction of around 33% of CBZ was observed after 120 h in BES 1, while the CBZ concentration reduced rapidly with a removal percentage of 84% in BES 2 at + 400 mV.

Hua et al. (2019) developed a new approach for the degradation of erythromycin (ERY) from the wastewater by the use of single-chamber MECs comprising of anodes prepared from the carbon cloth. The anodes were pre-treated before use. Initially, the anodes were soaked in acetone for 12 h followed by ultrasonic washing using anhydrous ethanol for 5 min and eventually rinsed with distilled water three times. The MFCs with 500 mV and 1000 Ω resistance were inoculated with acetate (1.0 g/L) containing PBS (50 mM). The MFCs were further converted into MECs by sealing the air-facing cathode side, and the reactors were further run in batch mode (48 h) with the fresh artificial medium. The MEC system was operated with 10 Ω ERs, and the DC voltage was concurrently supplied in the system. All MECs were monitored at room temperature. After achieving 0.9 V cathode potential, the MECs were filled with different concentrations of ERY-containing wastewater (10, 15, 20, 25, and 30 mg/L). The cyclic voltammetry (CV) and electrochemical impedance spectroscopy (EIS) were utilized for the evaluation of MEC anode electrochemical performance. The electrochemical evaluations were carried out using three electrodes: cathode (counter electrode), anode (working electrode), and Ag/AgCl (reference). The frequency ranging from 100 kHz to 5 MHz was used for the EIS measurement, whereas the evaluation of CV was carried between −1.2 and 0.2 V (50 mV S^{-1} scan rate). High-performance liquid chromatography (HPLC) analysis showed that at 20 mg/L ERY concentration, the removal of ERY reached 99% and the COD removal also increased, which indicated the positive effect of ERY concentration (high) on the microorganisms. The confirmation of biofilm development on anode and cathode electrodes was done by the images derived from scanning electron microscope (SEM). High throughput sequencing of 16s-rDNA gene amplicons identified the *Geobacter* (genus of exoelectrogenic bacteria) as the predominant (77%) organism in the anode biofilm of the reactor, while the *Acetoanaerobium* was observed extensively in biocathode.

Zhang et al. (2015) developed the MFC-Fenton system for the degradation of paracetamol (PAM). The dual-chamber MFC reactor containing cathode (216 mL) and anode (108 mL) was prepared. The anode was basically made up of three porous graphite felts (6.0 cm length × 5.5 cm width × 1.0 cm thickness) to provide better surface area required for the attachment and growth of electricigens. The single plate of graphite having smaller surface area (15 cm^2), less porosity, and small specific surface area compared to anode was used as cathode. A PEM fixed between the electrodes prevented the diffusion of dissolved oxygen (DO), direct aeration from the oxygen present in the air as well as iron (III)/iron (II) (Fe^{3+}/Fe^{2+}) ions (sourced from $FeSO_4 \cdot 7H_2O$) transfer between cathode and anode chambers. The voltage output along with the potentials of anode and cathode were monitored using the reference electrode (Ag/AgCl, CHI 111). Every component present in the circuit was connected with the help of titanium wires (> 99.9%). The complete MFC-Fenton systems were maintained at 25°C. A definite microbial mixture obtained from single-chamber air cathode MFC, which was already running for more than two years, was used to inoculate the anode chamber of dual-chambered MFC. Further, the sealed MFC was placed in an open circuit condition for 48 h to achieve anaerobic condition, which enhanced the growth of electricigens on the anode. After inoculation, a 50 mM PBS-buffered medium (pH = 7.0) containing 5736 mg/L $Na_2HPO_4 \cdot 2H_2O$ and 2452 mg/L

$NaH_2PO_4 \cdot H_2O$ was fed to the electricigens in the batch mode with HRT of 48 h. The anodic PBS buffer medium sprinkled nitrogen, which resulted in the removal of DO. Further, the previously used PBS (from the anode chamber) was poured into the cathode chamber. Air at 1.0 L/h speed was bubbled in the cathode chamber in order to provide DO, which played an important role as an electron acceptor. Various resistors with different values of ERs were loaded in the system for around 48 h in the decreasing order starting from $+\infty$ (open circuit), 1000 Ω, 800 Ω, 600 Ω, 390 Ω to 180 Ω. The highest PAM degradation equivalent to 70% was achieved within 9 h at 5 mg/L iron concentration, 10 mg/L of PAM, pH 2, and 20 Ω ER. The complete mineralization of around 25% PAM was noted, whereas the rest was initially reduced to *p*-aminophenol followed by rapid conversion into PNP that was eventually degraded to smaller dicarboxylic/carboxylic acids.

Harnisch et al. (2013) studied the process of sulfonamides elimination from the wastewater with the help of acetate-grown anodic microbial biofilms. The primary wastewater was taken as the source of the microbial inoculum. The experiment was conducted under potentiostatic control by the use of three different electrodes: the working electrode, a reference electrode (Ag/AgCl), and a counter electrode. The graphite rods were taken as the working and counter electrodes. The experiment was carried out with a potentiostat/galvanostat Model VMP3, prepared with 12 independent potentiostat channels. The CV was conducted under non-turnover and turnover conditions at the rate of 1 mV/s. The formation of primary biofilms was initiated by inoculation with wastewater (1 mL) and substrate solution (30 mL) into sealed electrochemical cells (250 mL). The working electrode was constantly supplied with the 0.2 V followed by its regular refilling with substrate solution. The sulfamethoxazole was completely removed from the solution, whereas the sulfathiazole remained constant. Similarly, sulfadiazine was partially removed and sulfadimidine was unaffected, which indicated a substance-specific biotransformation process by the biofilms. The regeneration study revealed that the N^4-acetyl-metabolites qualitatively showed similarity in the removal behavior of their corresponding sulfonamides. The N^4-acetyl-sulfadiazine and N^4-acetyl-sulfamethoxazole were decreased by 69 and 41%, respectively, whereas sulfathiazole and sulfadimidine were not removed from the solution.

4.4 HEAVY METALS

Toxic heavy metals can enter various aquatic flora and fauna, impairing their cellular metabolism. Once these hazardous heavy metals enter the food chain, they can get biomagnified and bioaccumulated that can result in severe toxicity in higher animals as well. Huang et al. (2015) developed a dual-chamber reactor for the degradation of Cd (II), Cu (II), and Cr (VI). The reactor was setup initially with an ER of 510 Ω and 0.5 V was further applied to the circuit of the MECs. The graphite brushes served as the anode, whereas the graphite felt was used as the cathode. Both the anode and cathode chambers were separated by 47 cm^2 CEM. Further, the anode chamber was inoculated with MFCs' effluent. The anodes were filled with the mixture of 5 mM phosphate buffer medium with acetate (1.0 g/L). After the successful adaptation of the biofilm, each cycle was set with the time period of 4 h. The reduction rates of selective metals Cr (VI), Cu (II), and Cd (II) were 1.24 ± 0.01 mg/L-h, 1.07 ± 0.01 mg/L-h, and 0.98 ± 0.01 mg/L-h, respectively.

Cao et al. (2021) used an osmotic microbial fuel cell (OsMFC) for the real-time removal of the hexavalent chromium [Cr (VI)] and organic contaminants, along with formation of energy and retrieval of water. The two-chambered reactor was mainly comprised of the cathode and anode, which were further linked to the 10 Ω external resistor by the means of titanium wires. The cellulose triacetate (CTA) membrane was utilized as a separator in the OsMFC. The carbon cloth was employed as the anode and cathode. An anaerobic sludge was utilized as the substrate to provide enrichment to the electrochemically active microbes. The medium used as the anolyte was mainly composed of 0.469 g glucose, 0.113 g NH_4HCO_3, 0.025 g K_2HPO_4, 2.150 g $NaHCO_3$, and 0.1 g yeast extract in 500 mL of deionized water. The different concentrations of NaCl containing 12 mg/L Cr (VI) were used as a medium in the cathode chamber. The decrease in open circuit voltage (OCV) from 0.84 to 0.35 V was observed with an increase in catholyte concentration from 0.2 to 1 M. The reaction mixture containing 0.2 M NaCl/Cr (VI) showed the highest reduction ability (44.5%). The oxidation-reduction potential (ORP) of 0.2 M NaCl/Cr (VI) was the highest (256 ± 24 mV), which gradually reduced with an increasing NaCl concentration. The reduction amount of Cr (VI) was highest at pH 2 (97.6%) followed by pH 3 (76%), pH 4 (25%), and pH 5 (24%). The highest power density of 76.7 mW/m^3 (476 mA/m^3 and 0.256 V) was generated at pH 2, while at the same pH, the internal resistance was lowest (140 Ω). Thus, it was proved that the pH of the solution had a strong effect on the internal resistance of the system as well as it also improved reduction of Cr (VI). The highest reduction of Cr (VI) was observed at pH 2 in 0.2 M NaCl/Cr (VI) solution, which also displayed a thick layer of Cr (60 wt%) deposited on the surface of electrodes with the size of particles ranging from 32.8 to 53.7 nm. The change in color of the solution from yellow to colorless confirmed Cr removal, mostly by the electrodes after drying.

Wang et al. (2018) used BES for chromium removal from the wastewater and also studied the effect of the pH on its removal efficiency. The reactor was made up of organic glass containing the graphite rod (anode) and biofilm attached carbon fiber (cathode). Both the electrodes were further connected to the direct power and the synthetic wastewater was propelled from the lower end of the reactor. The activated sludge, which was derived from the wastewater treatment plant, served as the inoculum for the BES. The reactor was operated at room temperature (21 ± 2°C) for 20 h (HRT) and at a wide pH range of 6–8. The average Cr (VI) removal efficiencies at different pH ranges were 58.9 (pH 6), 72.65 (pH 7), and 65.08% (pH 8). Thus, the highest Cr (VI) removal efficiency was observed at pH 7.

Yao et al. (2021) developed a graphene-modified graphite paper cathode BES for the chromium removal. A liquid nitrogen treatment method was used for the development of three-dimensional graphite foam (3DGF). Firstly, the pretreatment of graphite paper (GP) was carried out with 0.1 M HCL, NaOH, and Milli-Q water (1 h for each step) in a sequence with ultrasonication that was followed by drying at 60°C for 2 h. The GP was further soaked in liquid nitrogen for 15 min and quickly shifted to the ethanol (>99.5%) at room temperature (20 ± 2°C) followed by drying for 2 h at 60°C in order to form the final 3DGF. The residual ethanol on the surface of formed 3DGF was further removed by washing with Milli-Q water. The MFC reactor (H-type) was utilized for the experiment. Both the plexiglass chambers were separated using the PEM. GP and 3DGF played a role as cathode, whereas carbon felt ($2 \times 4.5 \times 0.5$ cm) was taken as anode. Both the electrodes were linked externally to the resistor (1000 Ω). The cathode chamber was further filled with Cr (VI) (artificial or real) containing wastewater, and electrochemically active bacteria (MFC) were inoculated in the anode. The treatment of the GP with the liquid nitrogen for the formation of 3DGF improved the Cr (VI) reduction by 17% and resulted in the total Cr removal of around 81% at 30 h of treatment in the MFCs.

Ai et al. (2020) developed BES coupled with thermoelectric generators (TEG) in order to separate heavy metals like copper, cadmium, and cobalt present in the smelting wastewater. The electroactive biofilms obtained from the activated sludge were enriched in the single-chambered MFC reactor made of perspex. The anode (carbon brush) and the cathode (carbon cloth with disk shape) were connected to the 910 Ω external resistor. The simulated smelting wastewater was treated using the dual-chamber MEC reactor. The partitioning of both the chambers was accomplished by the anion exchange membrane. In double-chamber MFCs, the pre-enriched anodes with bioelectroactive biofilms from the single-chamber MFCs were taken as anode, and a rectangular-shaped carbon cloth was used as a cathode. The electrodes of the dual-chamber MFC were linked to the outside resistor of 10 Ω resistance. The inoculation of every single-chamber MFC was done with the activated sludge (20 mL) derived from the wastewater treatment plant. The medium utilized in the enrichment of anodic electroactive biofilms in the single-chamber MFCs was comprised of 20 mM lactate (energy substrate), trace element solution, and Wolfe's vitamins (0.5 mL/L) in phosphate buffer. The simulated smelting wastewater with pH 1.80 contained Cu^{2+} (267.59 mg/L), Cd^{2+} (140.88 mg/L), and Co^{2+} (130.16 mg/L). The dual-chamber MFC was utilized for the recovery of Cu^{2+}. The pretreated carbon cloth (cathode) at 1 V and 2 V output voltage were adjusted for the recovery of Cd^{2+} and Co^{2+}. The catholyte pH was maintained below 2.5. The highest power density (228.31 mW/m^2) was obtained, whereas the open circuit potential and coulombic efficiency were about 632 mV and $21.27 \pm 1.20\%$, respectively. Thus, the electroactive biofilms present on the surface of the anode played a noteworthy role in the generation of electricity. The Cu^{2+} was recovered from the smelting wastewater by its deposition on cathode's surface, which received electrons thermodynamically from the biofilm developed on the anode. The Cu^{2+} concentration reduced rapidly in the catholyte from 267.59 mg/L to 67.39 mg/L in 17 h with greater recovery rate of 11.78 mg/L. The Cu^{2+} in the catholyte was recovered completely ($99.97 \pm 0.004\%$) after 53 h of treatment. During the complete MFC treatment, the highest recovery rate of copper, about 121.17 mg/L.d, was acquired. The increase in catholyte pH from 1.80 to 4.35 ± 0.07 after the recovery of Cu^{2+} suggested anions diffusion from anolyte via anion exchange membrane followed by the reaction of anions with the protons of the catholyte. The maximum amount of Cd^{2+} ($88.90 \pm 2.62\%$) was recovered at the rate of 158.20 mg/L.d. At the input voltage of 1.0 V, much higher Cd^{2+} recovery rate was obtained (6.59 mg/L.h). The Co^{2+} separation rate was around 8.08 mg/L.h. The Cu^{2+}, Cd^{2+}, and Co^{2+} were recovered from the smelting wastewater with the respective rates of 121.17, 158.20, and 193.87 mg/L.d. The Cu^{2+} was recovered bioelectrochemically in the form of Cu^0, whereas the Cd^{2+} and Co^{2+} were recovered as $Cd(OH)_2$, $CdCO_3$, or $Co(OH)_2$ by electrodeposition on the cathodic surface. The prominent genera of the activated sludge present in the BES before treatment were *Desulfovibrio* (17%), *Megasphaera* (11.81%), *Geobacter* (10.36%), and *Propionibacterium* (8.64%). After the successive treatment in the batch conditions, the major genera remained were *Geobacter* (34.76%), *Microbacter* (8.60%), and *Desulfovibrio* (5.33%).

Wang et al. (2020) developed a novel BES-assisted microelectrolysis system to enhance the Cu removal. The BES was comprised of microelectrolysis, electroflocculation, trickle filtration/air contact oxidation bed (TF-ACOB), MFC, and electric-membrane bioreactor (EMBR). The graphite/activated carbon particles were added in the anode chamber, which acted as a micro-cathode and the low potential aluminum acted as the micro-anode. During microelectrolysis process, the cations were released continuously from the sacrificial anode. The cathodic chamber was integrated with the EMBR. The novel catalytic conductive membrane (with FE/Mn/O catalyst, membrane pore size 2–7 nm) was taken as cathode of BES and the filtering membrane of EMBR. The cathode and anode were linked to 1000 Ω resistor. The EMR/cathode chamber was further added with the activated sludge derived from the wastewater treatment plant. In the initial stage of BES, the synthetic wastewater was added in the anode chamber for the anaerobic biological treatment. Further, it was mixed with the copper wastewater in the overflow mixing tank and was flown for the subsequent TF-ACOB treatment process. The influent flow rate of the synthetic wastewater was maintained at 12 L/D with the HRT of 6 h. The copper removal efficiency was near to 100% even at the 176–225 mg/L load of copper in the wastewater. The copper concentration in the effluent was only 0–0.09 mg/L. The high copper removal efficiency by the BES was possible because of the microelectrolysis, electroflocculation, and microbial coupling. The output voltage of the system after two weeks of nonstop operation was around 1.2 V. The highest power density during the treatment of copper ions and organic matter was 2.25 W/m^3 at 10.65 mA/m^2 current density. A reduction in internal resistance (about 40%) was also noted. The analysis of microbial community in the cathode chamber/EMBR revealed the abundance of *Chryseobacterium* sp. and *Comamonadaceae* sp.

4.5 AROMATIC COMPOUNDS

BES can also be used for removal of several toxic aromatic compounds from the industrial effluents. Zhou et al. (2020) utilized BES to study the effective co-degradation of complex polycyclic aromatic hydrocarbons (PAHs) in anaerobic conditions. The reactor used in the PAHs degradation studies was a single-chambered air-cathode reactor comprised of pure carbon cloth anode and Pt/C catalyst loaded carbon cloth cathode. The polytetrafluoroethylene (PTFE) was utilized as a hydrophobic material. The distance between these two electrodes was 2 cm, which were further connected to the external resistor (1000 Ω) using the copper wires in the closed-circuit condition. A separate anaerobic system (AS) was built with an open circuit condition in which there was no connection between both the electrodes. The source for microbial inoculum was the anaerobic sludge extracted from the petrochemical wastewater treatment plant. The biofilm formation on the anode was possible by the addition of anaerobic sludge mixed with glucose supplemented medium into the reactor. The reactors were labeled on the basis of the ratio of the concentration of low molecular weight (LMW) PAHs and high molecular weight (HMW) PAHs such as 1:4 (naphthalene (NAP)-phenanthrene (PHE)), 1:2 (NAP-PHE), 1:1 (NAP-PHE), 2:1 (NAP-PHE), 1:4 (NAP-pyrene (PYR)), 1:2 (NAP-PYR), 1:1 (NAP-PYR), 2:1 (NAP-PYR), 1:4 (PHE-PYR), 1:2 (PHE-PYR), 1:1 (PHE-PYR), and 2:1 (PHE-PYR). The removal efficiencies of NAP, PHE, and PYR observed after 4 h were around 20% in case of NAP and 5% in case of PHE and PYR. The degradation efficiency of NAP was higher compared to PHE and PYR in both AS and BES. The highest removal efficiencies of 97.60% (NAP), 42.60% (PHE), and 22.00% (PYR) were noted at 120 h for the BES. The removal efficiency of NAP (0.2500 mg/L) in the NAP-PHE mixed system (79.40%) was lower compared to the mono-system. The degradation efficiency of PHE at different NAP-PHE ratios always surpassed 42.90% at 120 h. In the case of PHE at NAP-PYR ratios, the increase in degradation half-life from 1.55 to 3.62 d was noted along with the decrease in the rate constant by 0.0106 h^{-1}. The removal efficiency of NAP and PYR in the NAP-PYR mixed system reached a maximum within 120 h of the reaction time. The lower removal rate of PYR was noted compared to the mixture of NAP and PHE, but in the case of mono-system, the removal rate of PYR was always higher. The shortest half-lives were noted for NAP (0.84 d) and PYR (4.81 d). The removal rate of PYR showed inverse relationship with NAP concentration in the NAP-PYR system. The removal efficiencies of PHE and PYR in the PHE-PYR system were 87% and 45.40% (1:4), 70% and 45.81% (1:2), 65.47% and 41.67% (1:1), and 51.67% and 40.90% (2:1). At the 1:4 ratio of PHE and PYR, the shortest half-lives for PHE and PYR were 1.70 and 4.53 d, respectively.

Luo et al. (2019) developed a sulfate-reducer enriched biocathode in the MEC to investigate the efficient reduction of nitrobenzene (NB). The two chambered MECs containing graphite brush anode (graphite fiber) and the three parts of polished graphite plate as a cathode were separated using CEM. The pretreatment of anode was carried out by heating it at 450°C for 3 h and the cathode was pretreated by acetone wash, drying, and immersion in 1 M NaOH and 1 M HCl for 24 h which was followed by rinsing, and storing in deionized water. The 10 Ω resistor was linked externally to both the electrodes and the MEC was further supplied with 0.8 V of voltage. The inoculation of anodes with the mixture of anaerobic and aerobic sludge (1:1, 10 mL) was carried out followed by inoculation of 10 mL acclimated sediment in the cathodes. The media present in both the chambers was replaced with the fresh media after every cycle (36 h). The NB (50 mg/L initial concentration) was added to the cathode medium after the completion of one month of operation. The separate MECs having abiotic cathodes were constructed as control. All the MEC reactors were light-protected and were operated at constant temperature (30 ± 3°C). On applying 0.8 V voltage in the MEC with sulfate-reducing biocathode (lacking NB), the increment in current till 2.62 mA (maximum) was noted, which further showed gradual decrease. The SEM images of the biocathode can be seen in Figure 4.2. The sulfate removal efficiency was around 41% after the operation cycle with the average reductive rate of 0.48 ± 0.04 mM/d. With the addition of 50 mg/L (0.41 mM) NB, the maximum current (2.85 mA) was transferred in the

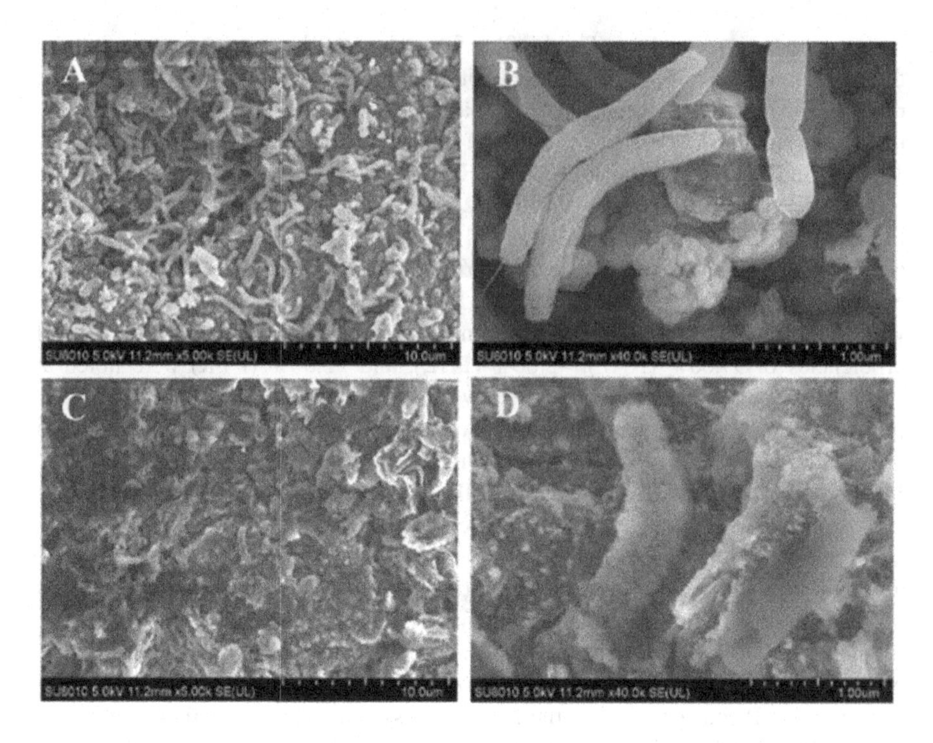

FIGURE 4.2 Scanning electron micrographs of SRB Biocathode surface without nitrobenzene (A and B) and with nitrobenzene (C and D), which were taken under 5.0 and 40.0 k magnification, respectively. (Reprinted with permission from Luo et al. (2019). DOI: 10.1016/j.scitotenv.2019.04.206. Copyright © 2019 Elsevier B.V.)

MEC biocathode. The NB removal efficiency of around 93% was acquired within 16 h and that of 98% at 36 h, while the removal rate of NB for the whole cycle was 0.21 ± 0.01 mM/d. The MEC operated using sulfate lacking catholyte showed the NB removal of 53% (at 4 h) and 98% (at 36 h), while the NB removal in presence of sulfate was 28% (4 h) and 98% (36 h).

Jiang et al. (2016) developed the membrane free-BESs (MFBESs) to investigate the degradation of nitroaromatic contaminants from the wastewater. A tubular bio-contact reactor coupled with MFBESs was filled with granular graphite in the bottom and upper portions were further used as cathode and anode. The electron collectors were the graphite rods inserted in the anodic and cathodic compartments. The MFBES was further inoculated with the sludge (containing chloronitrobenzenes) derived from the upflow anaerobic sludge blanket. The initial mixed liquid suspended solid (MLSS) concentration was around 8.5 g/L in the MFBES. The bottom part of the reactor was continuously filled with the synthetic wastewater, which was further transferred through cathode zones followed by the anode zones. At HRT of 8 h, the complete removal of 0.72 ± 0.01 mM *o*-nitrophenol (ONP) was observed in the cathode zone of MFBES, while the *m*-nitrophenol (MNP) and *p*-nitrophenol (PNP) as high as 0.12 ± 0.01 mM and 0.54 ± 0.01 mM were left. The conversion efficiencies of MNP, PNP, and ONP to their respective aminophenols (AP) were

89.23 ± 1.02%, 49.40 ± 2.42%, and 26.23 ± 1.66%, respectively. Thus, the reduction efficiency of ONP was highest followed by MNP and PNP.

Zhang et al. (2018) analyzed the effective concurrent removal of oxyfluorfen in the MFC along with the electricity generation. The MFC reactor (two-chambered) was designed in which the cationic exchange membrane was used as the separator. After pretreatment using NaOH and HCl, the active carbon felt was used as the anode as well as the cathode. The external resistor of 1000 Ω was connected using a titanium wire between both the electrodes. The bioanode inoculum used in the experiment was the pre-enriched activated sludge derived from the wastewater treatment plant. The reactors were initially autoclaved and were further inoculated with the acclimated activated sludge. The anodes of MFC reactors were injected with oxyfluorfen (50 mg/L) and operated in three different modes: closed circuit mode (MFC), open circuit (OC) mode, and abiotic control mode. The cathode potential was raised using potassium ferricyanide solution as the catholyte. The MFCs exhibited the highest oxyfluorfen removal efficiency followed by OCs and abiotic control. Approximately 77% of oxyfluorfen was eliminated in the MFC group after 24 h of incubation, while it was 46% and 69% after 24 h and 120 h, respectively, in case of the OC control group. The oxyfluorfen removal was significantly affected by the temperature, pH, and initial oxyfluorfen concentration. Thus, the highest degradation rate of 94.95% was achieved at 31.96°C, pH 7.65, and 120.05 mg/L initial concentration of oxyfluorfen. The predominant genera in the anodic biofilm identified after the microbial community analysis were *Arcobacter, Acinetobacter, Azospirillum, Azonexus*, and *Comamonas*.

In another study, Mukherjee et al. (2021) demonstrated the bioremediation of aromatic hydrocarbon along with the bioelectricity generation by utilizing MFC. Five different electrogenic bacteria, *Kocuria rosea* (GTPAS76), two strains of *Bacillus circulans* (GTPO28 and GTPAS54), and two strains of *Corynebacterium vitaeruminis* (GTPO38 and GTPO42) were used individually as well as in consortium in the double-chambered "H" type MFC, which were basically derived from the common effluent treatment plant (CETP). The graphite plates were used as the electrodes, and anoxic situations were maintained in the anodic chamber throughout the operation, whereas oxygen was considered as the terminal electron acceptor in the cathode. The salt bridge prepared using agar slurry (11.6% NaCl with 10% agar) was used to separate both chambers. The modified synthetic wastewater was taken as the anolyte in the anodic chamber, while 0.1 M PBS was taken as the catholyte in the cathode chamber. The anode chamber was further inoculated with 10% active culture followed by operating the MFC at room temperature (30 ± 2°C). The preparation of bacterial consortia was carried out by the combination of equal portions of log-phase cell cultures having the optical density between 0.6 and 0.7 at 600 nm. The consortia that were able to generate output voltage above 0.7 V were further evaluated for their COD and biological oxygen demand (BOD) reduction potential at the end of MFC operation. During the 30 days of MFC operation, a stable generation of electricity in the range of 0.5 ± 0.02 V to 0.7 ± 0.02 V was noted for all the test organisms and *Escherichia coli*. The COD reduction in systems containing all the test organisms was ranging between 40 and 60%, while the reduction in BOD was comparatively low. On the basis of output voltage of 0.8 ± 0.01 V with 81.81% COD and 64% BOD reductions, the consortium comprising

GTPO28, GTPO38, GTPAS54, and GTPAS76 was selected. The MFC anode containing preformed biofilm took around 4–5 days for stabilization giving the average output of 0.75 ± 0.03 V, which was much shorter compared to the MFC inoculated with the fresh anode that took 10 days for stabilization and 0.63 ± 0.02 V average output voltage. The setup with the preformed biofilm generated 11.2 mW/m^2 and 101.85 mA/m^2 of power and current density, respectively which was around 1.74 and 1.2 times higher compared to the MFC with fresh anode (7.5 mW/m^2 and 83.33 mA/m^2) on the 15th day of incubation. The cell viability of the electrode scrapped portion of MFC with biofilm and fresh culture inoculum on the 15th day of MFC setup was $1.13 \pm 0.1 \times 10^5$ and $0.27 \pm 0.1 \times 10^5$ CFU/cm^3, respectively. The highest degradation potential (89.8%), COD reduction potential (88.85 ± 0.28%), and BOD removal (87.15 ± 1.4%) were noted for the 5 mM sodium benzoate (SB) MFC system after 30 days.

4.6 CONCLUSIONS AND FUTURE PERSPECTIVES

Wastewater treatment coupled with electricity generation is attractive due to its low cost, efficiency, and eco-friendly nature. This process allows rapid biomass production at a reduced cost. Thus, MFCs offer a great area for exploring novel materials that can generate easily accessible fuels with high energy density. There is a need to explore and modify the bioanodes and biocathodes to overcome the limitations such as high overpotentials, pH issues, and high ohmic resistance. In order to enhance the overall efficiency of the BES, several working parameters, such as initial concentration of the pollutants, microbial cell density, temperature, time, pH, aeration, light, and external nutrient source, should be carefully optimized. The present wastewater treatment systems can be integrated with advanced MFCs to make them more convenient, economical, and acceptable. In view of the background, BES-based removal of hazardous pollutants can be considered as promising complementary and alternative to wastewater treatment strategy.

REFERENCES

Ai, C.; Yan, Z.; Hou, S.; Huo, Q.; Chai, L.; Qiu, G.; Zeng, W. Sequentially recover heavy metals from smelting wastewater using bioelectrochemical system coupled with thermoelectric generators. *Ecotoxicol. Environ. Saf.* 2020, *205*, 111174. DOI: 10.1016/j.ecoenv.2020.111174

Biffinger, J. C.; Pietron, J.; Ray, R.; Little, B.; Ringeisen, B. R. A biofilm enhanced miniature microbial fuel cell using *Shewanella oneidensis* DSP10 and oxygen reduction cathodes. *Biosens. Bioelectron.* 2007, *22*, 1672–1679. DOI: 10.1016/j.bios.2006.07.027

Brown, J. H.; Burnside, W. R.; Davidson, A. D.; DeLong, J. P.; Dunn, W. C.; Hamilton, M. J.; Mercado-Silva, N.; Nekola, J. C.; Okie, J. G.; Woodruff, W. H.; Zuo, W. Energetic limits to economic growth. *BioScience.* 2011, *61*, 19–26. DOI: 10.1525/bio.2011.61.1.7

Cao, T. N. D.; Chen, S. S.; Chang, H. M.; Ray, S.; Hai, S.; Bui, F. I.; Mukhtar, T. X. Simultaneous hexavalent chromium removal, water reclamation and electricity generation in osmotic bio-electrochemical system. *Sep. Purif. Technol.* 2021, *263*, 118155. DOI: 10.1016/j.seppur.2020.118155

Gallert, C.; Henning, A.; Winter, J. Scale-up of anaerobic digestion of the biowaste fraction from domestic wastes. *Water. Res.* 2003, *37*, 1433–1441. DOI: 10.1016/S0043-1354(02)00537-7

Harnisch, F.; Gimkiewicz, C.; Bogunovic, B.; Kreuzig, R.; Schröder, U. On the removal of sulfonamides using microbial bioelectrochemical systems. *Electrochemistry Commun.* 2013, *26*, 77–80. DOI: 10.1016/j.elecom.2012.10.015

Hua, T.; Li, S.; Li, F.; Ondon, B. S.; Liu, Y.; Wang, H. Degradation performance and microbial community analysis of microbial electrolysis cells for erythromycin wastewater treatment. *Biochem. Eng. J.* 2019, *146*, 1–9. DOI: 10.1016/j.bej.2019.02.008

Huang, L.; Wang, Q.; Jiang, L.; Zhou, P.; Quan, X.; Logan, B. E. Adaptively evolving bacterial communities for complete and selective reduction of cr (VI), cu (II), and cd (II) in biocathode bioelectrochemical systems. *Environ. Sci. Technol.* 2015, *49*, 9914–9924. DOI: 10.1021/acs.est.5b00191

Huang, W.; Chen, J.; Hu, Y.; Zhang, L. Enhancement of Congo red decolorization by membrane-free structure and bio-cathode in a microbial electrolysis cell. *Electrochimica. Acta.* 2018, *260*, 196–203. DOI: 10.1016/j.electacta.2017.12.055

Ilamathi, R.; Sheela, A. M.; Gandhi, N. N. Comparative evaluation of *Pseudomonas* species in single chamber microbial fuel cell with manganese coated cathode for reactive azo dye removal. *Int. Biodeterior. Biodegradation.* 2019, *144*, 104744. DOI: 10.1016/j.ibiod.2019.104744

Jiang, X.; Shen, J.; Lou, S.; Mu, Y.; Wang, N.; Han, W.; Sun, X.; Li, J.; Wang, L. Comprehensive comparison of bacterial communities in a membrane-free bioelectrochemical system for removing different mononitrophenols from wastewater. *Bioresour. Technol.* 2016, *216*, 645–652. DOI: 10.1016/j.biortech.2016.06.005

Khanal, S. K. Anaerobic biotechnology for bioenergy production: Principles and applications. Wiley-Blackwell, Oxford, UK, 2008, 221–246. ISBN: 978-0-813-82346-1

Kong, F.; Wang, A.; Cheng, H.; Liang, B. Accelerated decolorization of azo dye Congo red in a combined bioanode–biocathode bioelectrochemical system with modified electrodes deployment. *Bioresour. Technol.* 2014, *151*, 332–339. DOI: 10.1016/j.biortech.2013.10.027

Li, C.; Fang, H. H. Fermentative hydrogen production from wastewater and solid wastes by mixed cultures. *Crit. Rev. Environ. Sci. Technol.* 2007, *37*, 1–39. DOI: 10.1080/10643380600729071

Lies, D. P.; Hernandez, M. E.; Kappler, A.; Mielke, R. E.; Gralnick, J. A.; Newman, D. K. *Shewanella oneidensis* MR-1 uses overlapping pathways for iron reduction at a distance and by direct contact under conditions relevant for biofilms. *Appl. Environ. Microbiol.* 2005, *71*, 4414–4426. DOI: 10.1128/AEM.71.8.4414-4426.2005

Logan, B. E. Extracting hydrogen and electricity from renewable resources. *Environ. Sci. Technol.* 2004, *38*, 160A–167A. DOI: 10.1021/es040468s

Luo, H.; Hu, J.; Qu, L.; Liu, G.; Zhang, R.; Lu, Y.; Qi, J.; Hu, J.; Zeng, C. Efficient reduction of nitrobenzene by sulfate-reducer enriched biocathode in microbial electrolysis cell. *Sci. Total Environ.* 2019, *674*, 336–343. DOI: 10.1016/j.scitotenv.2019.04.206

Mukherjee, A.; Zaveri, P.; Patel, R.; Shah, M. T.; Munshi, N. S. Optimization of microbial fuel cell process using a novel consortium for aromatic hydrocarbon bioremediation and bioelectricity generation. *J. Environ. Manage.* 2021, *298*, 113546. DOI: 10.1016/j.jenvman.2021.113546

Pan, Y.; Zhu, T.; He, Z. Enhanced removal of azo dye by a bioelectrochemical system integrated with a membrane biofilm reactor. *Ind. Eng. Chem. Res.* 2018, *57*, 16433–16441. DOI: 10.1021/acs.iecr.8b04725

Tahir, K.; Miran, W.; Nawaz, M.; Jang, J.; Shahzad, A.; Moztahida, M.; Kim, B.; Azam, M.; Jeong, S. E.; Jeon, C. O.; Lim, S.-R.; Lee, D. S. Investigating the role of anodic potential in the biodegradation of carbamazepine in bioelectrochemical systems. *Sci. Total Environ.* 2019, *688*, 56–64. DOI: 10.1016/j.scitotenv.2019.06.219

Wang, H.; Wang, H.; Gao, C.; Liu, L. Enhanced removal of copper by electroflocculation and electroreduction in a novel bioelectrochemical system assisted microelectrolysis. *Bioresour. Technol.* 2020, *297*, 122507. DOI: 10.1016/j.biortech.2019.122507

Wang, H.; Zhang, S.; Wang, J.; Song, Q.; Zhang, W.; He, Q.; Song, J.; Ma, F. Comparison of performance and microbial communities in a bioelectrochemical system for simultaneous denitrification and chromium removal: Effects of pH. *Process Biochem.* 2018, *73*, 154–161. DOI: 10.1016/j.procbio.2018.08.007

Winter, M.; Brodd, R. J. What are batteries, fuel cells, and supercapacitors? *Chem. Rev.* 2004, *104*, 4245–4270. DOI: 10.1021/cr020730k

Xu, W.; Jin, B.; Zhou, S.; Su, Y.; Zhang, Y. Triclosan removal in microbial fuel cell: The contribution of adsorption and bioelectricity generation. *Energies.* 2020, *13*, 761. DOI: 10.3390/en13030761

Yao, J.; Huang, Y.; Hou, Y.; Yang, B.; Lei, L.; Tang, X.; Scheckel, K. G.; Li, Z.; Wu, D.; Dionysiou, D. D. Graphene-modified graphite paper cathode for the efficient bioelectrochemical removal of chromium. *Chem. Eng. J.* 2021, *405*, 126545. DOI: 10.1016/j.cej.2020.126545

Zhang, L.; Yin, X.; Li, S. F. Y. Bio-electrochemical degradation of paracetamol in a microbial fuel cell-Fenton system. *Chem. Eng. J.* 2015, *276*, 185–192. DOI: 10.1016/j.cej.2015.04.065

Zhang, Q.; Zhang, L.; Wang, H.; Jiang, Q.; Zhu, X. Simultaneous efficient removal of oxyfluorfen with electricity generation in a microbial fuel cell and its microbial community analysis. *Bioresour. Technol.* 2018, *250*, 658–665. DOI: 10.1016/j.biortech.2017.11.091

Zhou, Y.; Zou, Q.; Fan, M.; Xu, Y.; Chen, Y. Highly efficient anaerobic co-degradation of complex persistent polycyclic aromatic hydrocarbons by a bioelectrochemical system. *J. Hazard. Mater.* 2020, *381*, 120945. DOI: 10.1016/j.jhazmat.2019.120945

5 Microbial Carbon Capture Cell

Carbon Capture, Energy Recovery, and Wastewater Treatment

Thulluru Lakshmi Pathi, Shamik Chowdhury, and Makarand M. Ghangrekar

5.1 INTRODUCTION

Degrading water quality in natural streams, triggered by an influx of untreated or partially treated wastewater, requires urgent attention. This intrusion is perhaps the consequence of point and non-point sources of water pollution. These include urban areas plagued by unprecedented population explosions and the addition of dicey pollutants from industrial discharges. On the other hand, dwindling fossil fuel reserves have resulted in a paradigm shift to develop sustainable energy generation technologies. Microbial fuel cell (MFC), the first derivative of the bioelectrochemical system (BES), is a promising technology for degrading organic pollutants from wastewater while concomitantly generating electricity for onsite applications (Logan et al., 2006).

The anodic and cathodic chambers of MFC are partitioned by a proton exchange membrane (PEM). Electrogens at the anode oxidise the organic matter in the wastewater, transforming it into electrons, protons, and carbon dioxide (CO_2). The BES creates electrical neutrality by directing electrons to the cathode and protons via the PEM. Chemical oxidants, such as oxygen (O_2), are reduced to water (H_2O) at the cathode, accepting the perpetually reaching electrons and protons en masse. The performance of a quintessential MFC depends on the efficiencies of PEM and the reducing ability of encompassed electron acceptors by the cathode (Pant et al., 2010). The deterrents that impede the field-scale functionality of this multifaceted BES are the cost involved in external aeration, besides the exorbitant price of ion exchange membranes.

With high oxidation potential on top of forming a safe product (such as H_2O), O_2 is often the best-suited electron acceptor employed at the cathode. However, the cost involving mechanical aeration for supplying O_2 is outrageous, as it is energy-intensive at the cathode. To this end, microbial carbon cell (MCC) addresses the bottleneck by employing algae. In addition to O_2 liberation at the cathode, it has the additional benefit of serving as feedstock for biodiesel production.

DOI: 10.1201/9781003368472-5

The rate of CO_2 fixation by photosynthetic microorganisms, viz., cyanobacteria and microalgae, is superior to that of traditional photosynthetic plants on land. These photosynthetic microbiotas synthesise CO_2 from the anodic chamber or atmosphere into productive biomass (Costa et al., 2000). Specific systems, including photosynthetic microbial fuel cells (PMFCs), plant-microbial fuel cells, photo-assisted fuel cells (PFCs), and MCC, use this capability to translate CO_2 into valuable products in addition to sequestering carbon (Das et al., 2019). Initially, the gas released during anodic bacteria respiration and metabolism was injected into a cathodic chamber cultivated with the green algae *Chlorella vulgaris* (Wang et al., 2010). Later, cyanobacteria in the photo-biocathode of MCC worked similarly when an exoelectrogen, *Shewanella putrefaciens*, in the anodic chamber was separated by an anion exchange membrane (AEM) (Pandit et al., 2012). It demonstrated the elimination of organic waste by generating electrical energy and sequestering CO_2.

The effectiveness of microalgae introduced at the cathode is strategic, with superior photosynthetic productivity, alongside elevated lipid content (Morita et al., 2000). An investigation by Neethu and Ghangrekar (2017) demonstrated effective wastewater treatment when the photosynthetic biocathode was used in a sediment-type reactor design. Similarly, employing this biocathode effectively provided electron acceptors in microbial desalination cells (MDCs) (Gude et al., 2013). MCC investigations have extensively emphasised components such as reactor configuration (Elmekawy et al., 2014), electrode material (Wang et al., 2010), substrate utilisation (Pant et al., 2010), and algal species (Cui et al., 2014).

MCC is essentially an integration of two distinct and reciprocity processes. Therefore, the individual potency of these processes, operational conditions, and design parameters dictate the performance of MCC. This chapter aims to shed light on MCC, which may be employed for CO_2 sequestration and energy production in the pursuit of sustainability. It examines the current knowledge of functioning, applications, and performance. Subsequently, MCC performance variables and techniques to minimise bioelectrochemical losses have also been discussed. Additionally, the design considerations for practical applications and the existing bottlenecks that restrict its scalability in wastewater treatment and CO_2 capture have been emphasised.

5.2 WORKING PRINCIPLE

The global concern on sequestering captured carbon can be addressed using MCC, a customised form of MFC. It delivers a strategy of manoeuvring CO_2 to grow microalgae in the cathodic chamber. The biomass is then harvested and harnessed as feedstock for biodiesel production. Concurrently, organic matter removal from wastewater and electricity generation occur in the anodic chamber. The microalgae species cultured sequester CO_2 during photosynthesis in the cathodic chamber of MCC. The O_2 liberated during this phenomenon is made available for cathodic reduction, resulting in the formation of H_2O (Figure 5.1). Consequently, the cost associated with external aeration, as witnessed in the case of aqueous cathode MFC, is curbed.

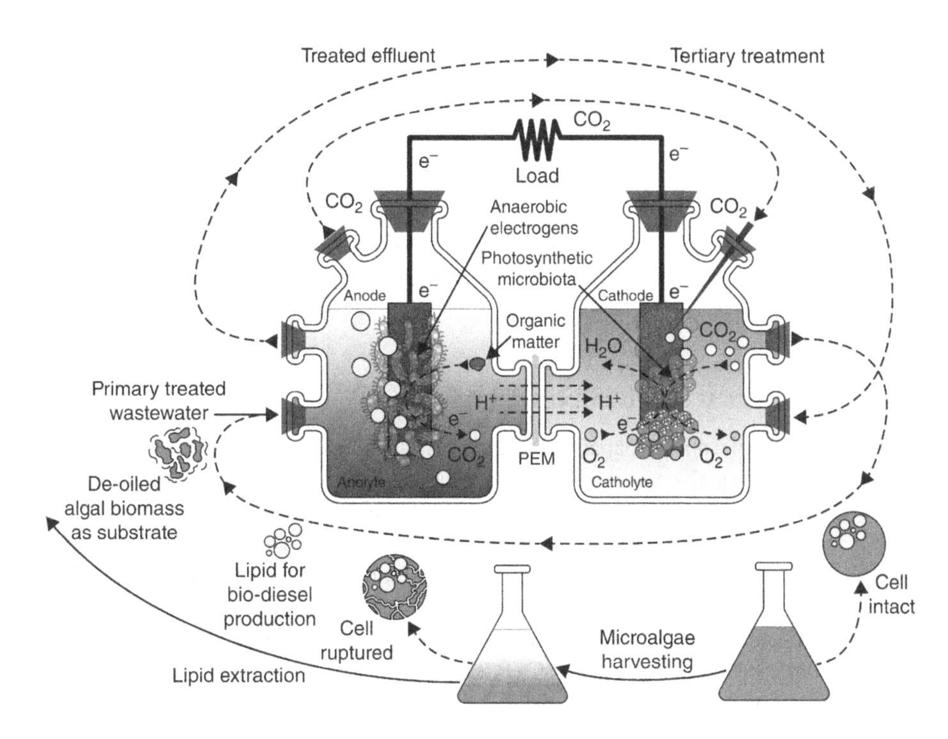

FIGURE 5.1 Schematic demonstrating simultaneous wastewater treatment at the anode and CO_2 capture at the cathode of MCC.

The organic matter degradation at the anode causes anodic off gases such as CO_2 evolution. It can also be introduced in the cathodic chamber and effectively reduced by the algal species during photosynthetic activity. Therefore, it can simultaneously address issues such as wastewater treatment, CO_2 sequestration, biomass production, and electricity generation (Wang et al., 2010). The overall bioelectrochemical reactions witnessed in the anodic (Eqs. 5.1 and 5.2) and cathodic chambers (Eqs. 5.3 to 5.5) are depicted below.

Anode:

$$CH_3COO^- + 2H_2O \rightarrow 2CO_2 + 7H^+ + 8e^- \quad (5.1)$$

Or

$$C_6H_{12}O_6 + 6H_2O \rightarrow 6CO_2 + 24H^+ + 24e^- \quad (5.2)$$

Cathode: Light-dependent photosynthetic reaction resulting in the liberation of O_2.

$$nCO_2 + nH_2O \rightarrow (CH_2O)_n + nO_2 \quad (5.3)$$

$$2O_2 + 8H^+ + 8e^- \rightarrow 4H_2O \quad (5.4)$$

A portion of O_2 liberated during the light phase is consumed during the dark phase for microalgae respiration (González Del Campo et al., 2013).

$$C_2H_4O_2 + 2O_2 \rightarrow 2CO_2 + 4H_2O \quad (5.5)$$

Therefore, optimising the dark and light phase periods is necessary to ameliorate the performance of MCC. The performance of biocathodes containing photosynthetic microorganisms depends on the presence of light and an electron-donating anodic activity.

5.2.1 Microalgae and BES Integrations

A PMFC uses microbiota capable of performing photosynthesis, converting solar energy into electrical energy in either chamber. This advancement in MFC has resolved the external aeration in the cathode by producing O_2 via photosynthesis for the O_2 reduction reaction (ORR). MFC coupled with photo-bioreactor (PBR) operated under continuous mode minimises the fluctuations in the power output. In this context, MCC is a subset of PMFC. Other forms of PMFCs include photosynthetic PMDCs, bio-photovoltaic systems (BPSs), algal microbial fuel cells (AMFCs), and photosynthetic bacteria-assisted MFC (PSB-MFC).

In the case of AMFCs, microalgae serve as electron donors in the anodic chamber, culminating in bioelectricity generation. Microalgae growing in the AMFC cathodic chamber have been demonstrated to augment ORR (Xu et al., 2015). This structure is quite similar to those with a single chamber MCC. PMDC is an advancement of MFC that encompasses brackish water desalination, wastewater treatment, bioelectricity, and biofuel generation. The cathodic chamber is used to culture photosynthetic organisms to speed up the passive ORR. This method consolidates several fixes into a straightforward approach.

5.2.2 Electron Transfer Mechanism of Photosynthetic Microbes in MFC

Photosynthetic microbes introduced in the anodic chamber of PMFCs can generate electrons during photosynthesis. These include anoxygenic photoautotrophic bacteria, oxygenic cyanobacteria, and microalgae. They follow separate electron transfer pathways to the anode, resulting in electricity generation. Unlike microalgae and cyanobacteria, which oxidise water to generate electrons, photosynthetic bacteria oxidise organic matter to generate electrons. Here, the electrons are produced cyclically to promote energy generation in the form of adenosine triphosphate (ATP) in each cycle (Figure 5.2(A)). On the other hand, a linear flow of electrons occurs in photosystem I (PS-I) and photosystem II (PS-II) during photosynthesis in microalgae and cyanobacteria (Figure 5.2(C)).

However, sulphur and organic compounds that function as electron carriers help in providing energy during bacterial photosynthesis by fixing CO_2 (Eq. 5.6).

$$CH_2O + H_2O \rightarrow CO_2 + 4H^+ + 4e^- \quad (5.6)$$

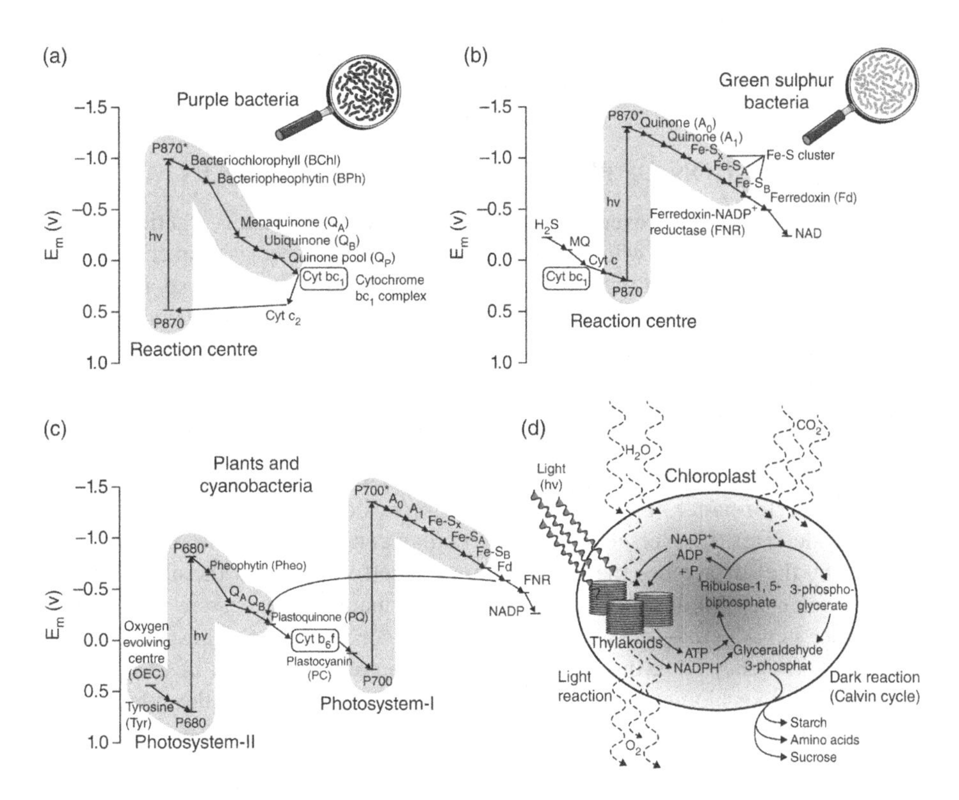

FIGURE 5.2 Electron transport chain diagrams of purple bacteria (A), green sulphur bacteria (B), and plant and cyanobacteria (C). Schematic of light and dark reactions occurring inside the chloroplast (D), in the case of plants and cyanobacteria. (Adopted with permission from Oxford academics (Blankenship et al., 2010.))

In the case of algal photosynthesis, the reaction begins at PS-II, where O_2 and four excited electrons are produced during the water-splitting process using four incident photons (hν). The excited energy is then utilised for ATP synthesis (Eq. 5.7).

$$h\nu + 2H_2O + (\geq 2)ATP \rightarrow 4H^+ + 4e^- + O_2 + (\geq 2)\ ATP \quad (5.7)$$

The electron transfer chain in algal photosynthesis comprises PS-I, cytochrome (Cyt b_6f), PS-II, and the plastoquinone (PQ) free electron carriers (Figure 5.2(C)). The excited electrons transported through PS-II and PS-I help in the reduction of nicotinamide adenine dinucleotide phosphate ($NADP^+$). Therefore, $NADP^+$ can act as an electron carrier and a reducing agent (Rasmussen and Minteer, 2014). Two photons (hν) reduce per molecule of $NADP^+$ to form NADPH (Eq. 5.8).

$$4h\nu + 4e^- + 4H^+ + 2NADP^+ \rightarrow 2NADPH + 2H^+ \quad (5.8)$$

The electron carriers help transport electrons from the chloroplast to the outer membrane of the algae. Water oxidation primarily occurs at PS-II, whereas the reducants for cellular processes are generated at PS-I. They are diffusing carriers, particularly cytochrome b6f complex that acts as a bridge between PS-I and PS-II.

5.3 CULTIVATION OF PHOTOSYNTHETIC MICROBES IN MCC

The efficacy of MCC is governed by determinants such as photosynthetic rate and cell multiplication period, which differ amongst microalgal species (Sun et al., 2016). Investigations have shown that *Chlorella* sp. is superior to *Anabaena* sp. in sequestering CO_2 and releasing photosynthetic oxygen to enhance cathodic efficacy (Jadhav et al., 2017). *Chlorella* sp. is resistant to high CO_2 concentrations, tolerant to urban effluent, and rich in harvestable lipids. The *C. vulgaris* utilised at the biocathode of MCC has exhibited both an elevated CO_2 fixation rate and an increased biomass rate (Hu et al., 2015).

Lower cell concentration translates into lower O_2 evolution, negatively impacting MCC performance. The power output improves as the algae concentration at the cathode rises. Conversely, over a particular concentration, microalgae present nearby the surface use the light resulting in self-shading for the algal cells in deeper water (Ugwu and Aoyagi, 2008). The formation of excessive metabolites is another negative consequence of a high cell concentration. Thus, the photosynthetic efficiency of fixing CO_2 and the lipid output, consisting of unsaturated hydrocarbons for biodiesel synthesis, are critical determinants in the selection of algal species.

5.4 APPLICATION OF MCC

Advancements in MCC have primarily emphasised boosting the performance and biomass yield by optimising the cathodic configuration and operating conditions. Eventually, these present a favourable environment for algal growth and enhance cathodic reaction kinetics.

5.4.1 BIOELECTRICITY GENERATION

Electricity generation fluctuates across MCC operation due to variations in O_2 concentration liberated by the algal species. Algal species demonstrating higher O_2 production rates have resulted in elevated power generation (Jadhav et al., 2017). Despite O_2 concentration in the cathode vicinity, MCC performance depends on anodic pH, electrode materials, and anodic bacteria degrading organic matter. Similarly, cathodic conditions encompass algal and O_2 concentrations.

Based on the quantity of algal biomass yield obtained per unit area of land, it can be considered an excellent choice to convert to electricity. Dry biomass can be utilised as a promising source of substrate introduced at the anode to degrade and generate electricity in MCC. This technique overcomes the issue of potentially being a pollutant in the natural water streams (Cai et al., 2013). The harvested biomass from the cathodic chamber can be used as the carbon substrate to release electrons during anodic oxidation. The biomass employed in the anodic chamber can be pre-treated using heat, enzymatic, chemical, or even untreated (Shukla and Kumar, 2018).

5.4.2 Carbon Capture and Biofuel Generation

Large quantities of CO_2 are added into the atmosphere degrading organic pollutants from numerous wastewater treatment plants (Campos et al., 2016). Feeding captured CO_2 into the cathodic chamber results in the formation of substantial biomass, which has a variety of applications (Gude et al., 2013). Algal species are known to be one of the most prolific biological systems for growing biomass and sequestering carbon while cleaning wastewater (Figure 5.1). *Chlorella* sp. is an algal species known for its high photosynthetic rate and high biomass production, resulting from its robust growth kinetics. Meanwhile, algae biomass production and carbon capture rates depend on process parameters such as nutrient availability, light intensity, and other environmental factors.

Algal biomass is a promising feedstock for lipid extraction and biodiesel production applications, rendering MCC a CO_2-neutral technology. It is rich in carbohydrate, protein, and lipid concentrations.

5.4.3 Wastewater Treatment with Value-Added Products Recovery

Organic matter in wastewater can serve as a carbon substrate for microbial consortia during oxidation at the anode, and MCC can treat it efficaciously. Investigations have demonstrated the adaptability of MFC in degrading carbon substrates ranging from glucose, acetate, starch, and kitchen effluent (Pant et al., 2010). Similarly, it can treat lignocellulosic matter, animal waste, waste-activated sludge, and wastewater from agricultural, domestic, and food processing industries. These are reproducible in the case of MCC accompanying bioelectricity generation. Adopting *Golenkinia* sp. in the MCC cathodic chamber culminated in 6.3 W m^{-3} of power density output and 44% chemical oxidation demand (COD) removal, respectively (Hu et al., 2015). On the other hand, investigations have established the application of anodic oxidation of algal biomass, resulting in high power output (Rajesh et al., 2015). Besides carbon capture, an additional benefit of nutrient removal can be accomplished when wastewater is introduced at the cathodic end (Neethu and Ghangrekar, 2017).

Besides carbon capture, microalgae are effective in nutrient removal by utilising them in cell metabolism (Converti et al., 2009). This was observed when electromigration and diffusion of ions have been occurred from the anodic to cathodic chamber concurred with the nutrient recovery by the biocathode (Colombo et al., 2017). Investigations have revealed the removal of nutrients such as phosphorus (P) in the form of PO_4^{3-} (Y. Huang et al., 2017). Interestingly, substantial (>90%) NH_4^+–N and P removal is achieved when MFC is integrated with a PBR. Similarly, algal cathode-coupled MDCs accomplished complete salt removal (Kokabian and Gude, 2015).

The carbonisation of algae facilitates the development of a prospective electrode material for sodium ion (Na^+) batteries and fuel cell applications. This method combines toxic algae blooms with blue-green algae to produce an inexpensive green electrode (Meng et al., 2015). It is also possible to synthesise nonporous carbon and benefit from its high specific area (Zhou et al., 2012).

5.4.4 Methanogenic Inhibition

While mixed anaerobic consortia are used as an inoculum at the anode, methanogens and other non-electrogenic microbes typically consume the bulk of the substrate. An investigation into the application of algal power observed the suppression of methanogenic activity. This attempt maximised the coulombs recovered from the organic substrate, effectively reducing their consumption by non-electrogenic bacteria (Rajesh et al., 2014). Hexadecatrienoic acid, found in *Chaetoceros*, marine algae, was reported to inhibit methanogens. Methanogen adsorption and cell disruption resulted in increased coulombic efficiency (CE) (45.18%) and power density (21.43 W m^{-3}) (Rajesh et al., 2015).

5.5 FACTORS AFFECTING MCC PERFORMANCE

5.5.1 Algal Biocathode

O_2 is a notable electron acceptor at the cathode, obligated to its high oxidation potential (0.59 V *vs.* standard hydrogen electrode), and reduces itself into a clean end product (H_2O) (Ucar et al., 2017). Most MFC studies have shown considerable energy usage for delivering O_2 in the cathodic chamber. Microalgae may deliver an alternative source of O_2 to the cathodic reduction. During photosynthesis, O_2 liberated comes from water rather than CO_2. The thylakoid membrane of the chloroplast liberates O_2 during the light reaction (Figure 5.2(D)). As a result, the development of O_2 during this phenomenon is influenced by light intensity, cell type, concentration, and operating parameters (Perrine et al., 2012).

5.5.1.1 Light, CO_2, and O_2 Concentration

Environmental factors that directly or indirectly impact algal kinetics are irradiance, temperature, CO_2, pH, aeration, and salinity, either directly or indirectly. Light is a critical component that governs MCC performance based on wavelength, duration, and intensity. Extreme intensity may pose growth inhibition owing to photo-oxidation, whilst low intensity may trigger growth limitation. There have been experiments on altering the light sources, texturing the optical filaments and LEDs for a specific wavelength, and collimating the light beam with a lens (Carvalho et al., 2011). While surplus incoming light transforms into heat energy, red and blue light may improve in maintaining a consistent temperature for algae growth (Michael et al., 2015). Interestingly, an internally illuminating LED can offer advantages, including scattering less heat energy and preventing self-shading. Similarly, maintaining low temperatures enhances microalgae production by increasing CO_2 solubility.

Another critical aspect is the accessibility of carbon in the form of CO_2. The *Chlorella* sp., for example, is resistant to high CO_2 concentrations while still sequestering carbon with high photosynthetic efficiency, converting CO_2, and liberating

O_2 (Singh and Singh, 2014). Though power output, biomass, and lipid content yield improve as CO_2 concentration rises, CO_2 concentration must be optimised based on the consumption rate of particular algae species (Andersen and Andersen, 2006; Ho et al., 2012). The carbon fixation rate in *C. vulgaris* is 6.17 mg L^{-1} h^{-1}, whereas the CO_2 consumption rate in *Scenedesmus* is 59.19 mg L^{-1} h^{-1}. One limitation identified in PBRs is the accumulation of free O_2, which hinders algae growth, but not in MCC. The generated O_2 at the cathode instantly participates in the ORR. When the dissolved oxygen (DO) concentration at the cathodic end was 6.6 mg L^{-1}, the voltage produced by MCC was 706 mV. This inherent function of algae might be a valuable source of O_2 for MCC (Kang et al., 2003).

The efficiency of MCC in generating electricity and treating wastewater is governed by characteristics such as mixing and immobilising biomass. It positively influences the mass transfer of nutrients and separation of metabolites, notably O_2, from the growth medium for the microalgae. It has been reported that a stable, high cell concentration may be generated by immobilising microalgae. Furthermore, due to their logarithmic growth phase, they are resistant to harmful substances and capable of producing steady voltages for extended periods of operation (Jin et al., 2011). The O_2 content acts as a terminal electron acceptor that plays a crucial role in cathodic processes. Nevertheless, the rate at which oxygen is released from algae cultures depends on their rate of development.

5.5.1.2 Nitrate Concentration

Nitrogen (N) availability influences microalgae biomass and its capability to generate lipids. Power output in MCC grew with a rise in nitrate concentration but decreased with a rise in nitrate concentration over a certain threshold. On the other hand, low N content resulted in a more significant percentage of lipids in the dry biomass (Converti et al., 2009). Maximising both lipid content in biomass and power output necessitates optimising N concentration. Therefore, it indicates that wastewater with relatively fewer inorganic nutrients is adequate for biomass synthesis in microalgae.

5.5.1.3 Operation Conditions

Maximum efficiency at the anode and cathode requires operating conditions conducive to the exoelectrogenic bacteria and algae community. On the anodic side, these factors include the organic loading rate (OLR), substrate, inoculum, anolyte pH, hydraulic retention time (HRT), and anodic environment. OLR is managed by the flow rate and concentration of the substrate, which depend on HRT. The COD-loading substrates may vary from complex compounds such as starch to simple acetate. For easy metabolism by bacteria, complex substrates must be broken down into simple molecules before incorporating them into the anodic side (Pant et al., 2010). Acetate, when used as an anodic substrate, produced the highest CE when compared to propionate, butyrate, and glucose, according to the investigation by Chae et al. (2009). Similarly, it was observed that using acetate as a substrate resulted in greater efficiency than using protein-rich effluent (Liu et al., 2009). To generate electricity continuously, running the reactor in a continuous rather than a batch mode is preferable. As reported by several investigations, the HRT estimated primarily relies on substrate degradability in the anodic chamber (Akman et al., 2013; Sharma and Li, 2010). One element that affects how well the anodic chamber functions is the anodic pH. The consortium functions supposedly better in an alkaline environment, following a consensus of findings from many investigations (Behera and Ghangrekar, 2009; Puig et al., 2010; Yuan et al., 2011). The pH imbalance causes voltage loss and thus decreases BES performance. Transporting other cations via the PEM is another possible contributing factor causing a shunting in BES performance (Rismani-Yazdi et al., 2008). Investigations also exhibit the use of an AEM with carbonate and phosphate in the buffer solution, aiding in the efficient transfer of protons (Rozendal et al., 2006). Thus, maximising wastewater treatment and CE involves optimising HRT for a given environment, operating condition, and bacterial community (Table 5.1).

5.5.2 Design Considerations

An ideal reactor design for MCC aims to minimise the internal electron flow resistance and maximise organic degradation, power output, and algal production. These include membrane thickness and area, electrolyte mixing, electrode spacing and material, the volume of chambers, and reactor configuration. The algal culture is integrated into PBRs. It can be integrated into two configurations, connecting PBR to the MFC externally or incorporating PBR in the cathodic chamber of the MFC. With the former design, a peristaltic pump is employed to continuously recycle the algal culture from the PBR to the cathodic chamber of MCC (Jiang et al., 2013). The anodically produced CO_2 is fed into the PBR (Powell and Hill, 2009).

In the latter arrangement, the cathodic chamber of a multi-chamber electrolytic cell is used for growing the algae. The anolyte and catholyte are separated by PEM, while the electrodes are connected via an external circuit. The CO_2 produced in the anodic chamber is exploited as nutrition in the cathodic chamber (Khandelwal et al., 2018). Investigations have been conducted on single-chamber MCC, which eliminates the use of PEM. A symbiosis between algae communities was seen in this case (Fu et al., 2010). However, an airlift MCC system can treat wastewater and sequester significant amounts of carbon (Hu et al., 2015).

Introducing light and fueling microalgae growth is critical in designing a cathodic chamber. In contrast, the MFC does not require a transparent cathodic chamber to allow light penetration. Different investigations have focused on chamber designs that minimise dark regions and light loss for optimum use of incoming light and maximise algal growth. Light trapping, channelling, and dispersion inside the cathodic chamber are pertinent to design considerations. Harnessing Fresnel focal points alongside guiding light beams to collect, transport, and provide it directly into the algal solution has also been investigated (Zijffers et al., 2008). Alternatively, roughening the illumination surface of the distributor to promote horizontal light dispersion has also been demonstrated (Csögör et al., 1999). Therefore, emphasis should be placed on delivering adequate light to ensure algal proliferation. Other variables include the alteration of light intensity and wavelength (red and blue). Variability in light frequency under pervading conditions, homogeneous light distribution throughout the cathodic chamber, and adequate mixing can maximise algal biomass productivity (Zijffers et al., 2008).

TABLE 5.1
Electrical Power Generation, Biomass Productivity, and COD Removal Efficiency Achieved in MCC Investigations

Microbiota Employed						
Cathodic Chamber	Anodic Chamber	Electrical Performance	Biomass Productivity	COD Removal Efficiency/ CO_2 Fixation Rate (•)	Varying/Influencing Parameter	References
Chlorella vulgaris	Anaerobic sludge	2.48 W m^{-3}	18.3 × 10^6 cell mL^{-1}	84.8%	Microalgae (immobilised, suspended)	Zhou et al. (2012)
Desmodesmus sp. A8	-	99.09 W m^{-2}	-	-	Light intensities (0, 1500, 2000, 2500, 3000, and 3500 lx)	Wu et al. (2014)
Chlorella vulgaris	-	0.97 W m^{-3}	0.5 g L^{-1} d^{-1}	86.69%	Airlift MCC	Hu et al. (2015)
Chlorella vulgaris	Consortia from pre-treated cow manure	2.70 W m^{-3}	0.028 kg m^{-3} d^{-1}	70.8% (0.29 kg m^{-3} d^{-1})	Anode substrate (lipid-extracted algae, orange fruit pulp)	Khandelwal et al. (2018)
Chlorella sorokiniana	-	3.2 W m^{-3}	0.27 g L^{-1} d^{-1}	65.97%	PEM (coconut shell, Nafion 117)	Neethu et al. (2018)
Chlorella vulgaris	Anaerobic sludge	5.13 W m^{-3}	0.32 g L^{-1} d^{-1}	683.9 mg L^{-1} d^{-1} •	N source ($NaNO_3$, CH_3COONH_4, CH_2CONH_2, NH_2 $CONH_2$)	Li et al. (2019)
Chlorella sorokiniana	Anaerobic sludge	7.12 W m^{-3}	0.83 g L^{-1} d^{-1}	82%	Nitrate and light period	Neethu et al. (2020)
Synechococcus sp.	Anaerobic sludge	95.63 mW m^{-2}	2.54 g L^{-1}	-	-	Lakshmidevi et al. (2020)
Chlorella sorokiniana	-	2.30 W m^{-3}	812 mg L^{-1}	95%	Inlet pH, light intensity, photoperiod	Varanasi et al. (2020)

Electrode properties, notably biocompatibility, high surface area, and roughness, are highly regarded in MCC while selecting electrode material. These govern the adherence of inoculum to the surface of an electrode resulting in the least electron transfer resistance. Carbon-based graphite felt is used because it has favourable qualities, such as a large surface area that promotes homogeneous biofilm growth and microbial colonisation. Besides lessening the activation losses by selecting a suitable electrode, minimising ohmic losses should be focused on by reducing electrode spacing (Mustakeem, 2015). An increase in power output might occur from a reduction in electrode distance. However, this trade-off could result in the diffusion of substrate and O_2 from either side, leading to cathodic biofouling (Tartakovsky and Guiot, 2006). Overcoming unintended diffusions and attaining maximum power generation both seek optimisation. Investigations have established the influence of electrode spacing on external resistance and maximum power density (Lee and Huang, 2013).

Exoelectrogenic bacteria require anaerobicity in the anodic chamber of MCC to thrive. So, adequate separation from the cathodic chamber is essential owing to its oxygen-rich nature. The critical attributes of an ideal separator are excellent proton conductivity, water absorption, minimum oxygen diffusion, resistance against unnecessary acetate crossover, and biodegradability (Tanaka, 2015). Nafion is the most common membrane, with many cation exchange membranes routinely employed (D. Huang et al., 2017). Meanwhile, unconventional membranes, including bipolar membranes (Kim et al., 2017), chitosan-graphene oxide mixed-matrix membranes (Holder et al., 2017), glass wool membranes (Venkata Mohan et al., 2008), SPEEK membranes (Ghasemi et al., 2016), ceramic membranes (Daud et al., 2018), and clayware membranes (Ghadge et al., 2015), have been explored. Oxygen diffusion, cation aggregation, substrate transfer, and high cost are problems that occur while using a Nafion membrane. These obstacles have prompted research on materials leading to efficient, inexpensive membranes.

5.6 BOTTLENECKS IN UP-SCALING

Despite significant improvements in MCC, a few obstacles must be surmounted before this technology can be scaled up. This chapter encompasses the expanded capabilities of MFC technology once it is combined with microalgae. Efficiency in photosynthetic processes is crucial since O_2 is the terminal electron acceptor. Different algae species have distinct growth kinetics. As a result, a single optimised condition cannot be applied to all algae strains.

Furthermore, the amount of O_2 generated during photosynthesis depends on the CO_2 in the anodic off-gas, which is essential for algae growth. The modifications made to MCC should allow it to process wastewater from various sources with varying compositions, pH levels, and temperatures. Exploration of inexpensive cathodic catalysts that are biocompatible with algae species is also crucial. The reactor design should minimise CO_2 leakage from the anodic to the cathodic chamber. Concurrently, proton transfer from the anode to the cathode is required to accomplish the redox process. The low-cost ion-exchange membrane should prevent undesired O_2 and substrate diffusions from either side. There is a positive correlation between the voltages produced by the reactor and the diurnal variation in DO between day and night. This voltage instability must be addressed immediately. MCC technology has the potential to generate electricity and wastewater remediation, but it faces significant challenges that must be overcome before it can meaningfully compete with established traditional technologies.

5.7 CONCLUSION

MCC is creating a new entity by combining with the biorefinery approach. Microbes in the anodic chamber break down organic matter to release electrons, which flow to the cathode reducing O_2 and completing the circuit. This holistic approach solves a wide range of issues effectively. Algal growth and tertiary treatment are augmented by feeding the treated effluent into the cathodic chamber. It effectively removes nitrates, ammonium, and organic substances from wastewater. The cathodic process is assisted by the O_2 generated during algae growth, which is eventually retained and suspended in the effluent. Understanding and optimising parameters of algal strain selection, photosynthetic efficiency, and substrate properties are essential for achieving maximum performance.

REFERENCES

Akman, D., Cirik, K., Ozdemir, S., Ozkaya, B., Cinar, O., 2013. Bioelectricity generation in continuously-fed microbial fuel cell: Effects of anode electrode material and hydraulic retention time. Bioresour. Technol. 149, 459–464. https://doi.org/10.1016/j.biortech.2013.09.102

Andersen, T., Andersen, F.Ø., 2006. Effects of CO_2 concentration on growth of filamentous algae and *Littorella uniflora* in a Danish softwater lake. Aquat. Bot. 84, 267–271. https://doi.org/10.1016/j.aquabot.2005.09.009

Behera, M., Ghangrekar, M.M., 2009. Performance of microbial fuel cell in response to change in sludge loading rate at different anodic feed pH. Bioresour. Technol. 100, 5114–5121. https://doi.org/10.1016/j.biortech.2009.05.020

Blankenship, R.E., 2010. Early evolution of photosynthesis. Plant Physiology. 154(2), 434–438. https://doi.org/10.1104/pp.110.161687

Cai, P.-J., Xiao, X., He, Y.-R., Li, W.-W., Zang, G.-L., Sheng, G.-P., Hon-Wah Lam, M., Yu, L., Yu, H.-Q., 2013. Reactive oxygen species (ROS) generated by cyanobacteria act as an electron acceptor in the biocathode of a bio-electrochemical system. Biosens. Bioelectron. 39, 306–310. https://doi.org/10.1016/j.bios.2012.06.058

Campos, J.L., Valenzuela-Heredia, D., Pedrouso, A., Val del Río, A., Belmonte, M., Mosquera-Corral, A., 2016. Greenhouse gases emissions from wastewater treatment plants: Minimisation, treatment, and prevention. J. Chem. 2016, 1–12. https://doi.org/10.1155/2016/3796352

Carvalho, A.P., Silva, S.O., Baptista, J.M., Malcata, F.X., 2011. Light requirements in microalgal photobioreactors: An overview of biophotonic aspects. Appl. Microbiol. Biotechnol. 89, 1275–1288. https://doi.org/10.1007/s00253-010-3047-8

Chae, K.-J., Choi, M.-J., Lee, J.-W., Kim, K.-Y., Kim, I.S., 2009. Effect of different substrates on the performance, bacterial diversity, and bacterial viability in microbial fuel cells. Bioresour. Technol. 100, 3518–3525. https://doi.org/10.1016/j.biortech.2009.02.065

Colombo, A., Marzorati, S., Lucchini, G., Cristiani, P., Pant, D., Schievano, A., 2017. Assisting cultivation of photosynthetic microorganisms by microbial fuel cells to enhance nutrients recovery from wastewater. Bioresour. Technol. 237, 240–248. https://doi.org/10.1016/j.biortech.2017.03.038

Converti, A., Casazza, A.A., Ortiz, E.Y., Perego, P., Del Borghi, M., 2009. Effect of temperature and nitrogen concentration on the growth and lipid content of *Nannochloropsis oculata* and *Chlorella vulgaris* for biodiesel production. Chem. Eng. Process. Process Intensif. 48, 1146–1151. https://doi.org/10.1016/j.cep.2009.03.006

Costa, J.A.V., Linde, G.A., Atala, D.I.P., Mibielli, G.M., KrÜger, R.T., 2000. Modelling of growth conditions for cyanobacterium *Spirulina platensis* in microcosms. World J. Microbiol. Biotechnol. 16, 15–18. https://doi.org/10.1023/A:1008992826344

Csögör, Z., Herrenbauer, M., Perner, I., Schmidt, K., Posten, C., 1999. Design of a photobioreactor for modelling purposes. Chem. Eng. Process. Process Intensif. 38, 517–523. https://doi.org/10.1016/S0255-2701(99)00048-3

Cui, Y., Rashid, N., Hu, N., Rehman, M.S.U., Han, J.I., 2014. Electricity generation and microalgae cultivation in microbial fuel cell using microalgae-enriched anode and bio-cathode. Energy Convers. Manag. 79, 674–680. https://doi.org/10.1016/j.enconman.2013.12.032

Das, S, Das, S., Das, I., Ghangrekar, M.M., 2019. Application of bioelectrochemical systems for carbon dioxide sequestration and concomitant valuable recovery: A review. Mater. Sci. Energy Technol. 2, 687–696. https://doi.org/10.1016/j.mset.2019.08.003

Daud, S.M., Daud, W.R.W., Kim, B.H., Somalu, M.R., Bakar, M.H.A., Muchtar, A., Jahim, J.M., Lim, S.S., Chang, I.S., 2018. Comparison of performance and ionic concentration gradient of two-chamber microbial fuel cell using ceramic membrane (CM) and cation exchange membrane (CEM) as separators. Electrochim. Acta. 259, 365–376. https://doi.org/10.1016/j.electacta.2017.10.118

Elmekawy, A., Hegab, H.M., Vanbroekhoven, K., Pant, D., 2014. Techno-productive potential of photosynthetic microbial fuel cells through different configurations. Renew. Sustain. Energy Rev. 39, 617–627. https://doi.org/10.1016/j.rser.2014.07.116

Fu, C.-C., Hung, T.-C., Wu, W.-T., Wen, T.-C., Su, C.-H., 2010. Current and voltage responses in instant photosynthetic microbial cells with *Spirulina platensis*. Biochem. Eng. J. 52, 175–180. https://doi.org/10.1016/j.bej.2010.08.004

Ghadge, A.N., Jadhav, D.A., Pradhan, H., Ghangrekar, M.M., 2015. Enhancing waste activated sludge digestion and power production using hypochlorite as catholyte in clayware microbial fuel cell. Bioresour. Technol. 182, 225–231. https://doi.org/10.1016/j.biortech.2015.02.004

Ghasemi, M., Wan Daud, W.R., Alam, J., Jafari, Y., Sedighi, M., Aljlil, S.A., Ilbeygi, H., 2016. Sulfonated poly ether ether ketone with different degree of sulphonation in microbial fuel cell: Application study and economical analysis. Int. J. Hydrogen Energy. 41, 4862–4871. https://doi.org/10.1016/j.ijhydene.2015.10.029

González Del Campo, A., Cañizares, P., Rodrigo, M.A., Fernández, F.J., Lobato, J., 2013. Microbial fuel cell with an algae-assisted cathode: A preliminary assessment. J. Power Sources. 242, 638–645. https://doi.org/10.1016/j.jpowsour.2013.05.110

Gude, V.G., Kokabian, B., Gadhamshetty, V., 2013. Beneficial bioelectrochemical systems for energy, water, and biomass production. J. Microb. Biochem. Technol. 5, 1–14. https://doi.org/10.4172/1948-5948.S6-005

Ho, S.-H., Chen, C.-Y., Chang, J.-S., 2012. Effect of light intensity and nitrogen starvation on CO_2 fixation and lipid/carbohydrate production of an indigenous microalga *Scenedesmus obliquus* CNW-N. Bioresour. Technol. 113, 244–252. https://doi.org/10.1016/j.biortech.2011.11.133

Holder, S.L., Lee, C.-H., Popuri, S.R., 2017. Simultaneous wastewater treatment and bioelectricity production in microbial fuel cells using cross-linked chitosan-graphene oxide mixed-matrix membranes. Environ. Sci. Pollut. Res. 24, 13782–13796. https://doi.org/10.1007/s11356-017-8839-2

Hu, X., Liu, B., Zhou, J., Jin, R., Qiao, S., Liu, G., 2015. CO_2 fixation, lipid production, and power generation by a novel air-lift-type microbial carbon capture cell system. Environ. Sci. Technol. 49, 10710–10717. https://doi.org/10.1021/acs.est.5b02211

Huang, D., Song, B.-Y., He, Y.-L., Ren, Q., Yao, S., 2017. Cations diffusion in Nafion117 membrane of microbial fuel cells. Electrochim. Acta. 245, 654–663. https://doi.org/10.1016/j.electacta.2017.06.004

Huang, Y., Huang, Yun, Liao, Q., Fu, Q., Xia, A., Zhu, X., 2017. Improving phosphorus removal efficiency and *Chlorella vulgaris* growth in high-phosphate MFC wastewater by frequent addition of small amounts of nitrate. Int. J. Hydrogen Energy. 42, 27749–27758. https://doi.org/10.1016/j.ijhydene.2017.05.069

Jadhav, D.A., Jain, S.C., Ghangrekar, M.M.M., 2017. Simultaneous wastewater treatment, algal biomass production and electricity generation in clayware microbial carbon capture cells. Appl. Biochem. Biotechnol. 183, 1076–1092. https://doi.org/10.1007/s12010-017-2485-5

Jiang, H., Luo, S., Shi, X., Dai, M., Guo, R., 2013. A system combining microbial fuel cell with photobioreactor for continuous domestic wastewater treatment and bioelectricity generation. J. Cent. South Univ. 20, 488–494. https://doi.org/10.1007/s11771-013-1510-2

Jin, J., Yang, L., Chan, S.M.N., Luan, T., Li, Y., Tam, N.F.Y., 2011. Effect of nutrients on the biodegradation of tributyltin (TBT) by alginate immobilised microalga, *Chlorella vulgaris*, in natural river water. J. Hazard. Mater. 185, 1582–1586. https://doi.org/10.1016/j.jhazmat.2010.09.075

Kang, K.H., Jang, J.K., Pham, T.H., Moon, H., Chang, I.S., Kim, B.H., 2003. A microbial fuel cell with improved cathode reaction as a low biochemical oxygen demand sensor. Biotechnol. Lett. 25, 1357–1361. https://doi.org/10.1023/A:1024984521699

Khandelwal, A., Vijay, A., Dixit, A., Chhabra, M., 2018. Microbial fuel cell powered by lipid extracted algae: A promising system for algal lipids and power generation. Bioresour. Technol. 247, 520–527. https://doi.org/10.1016/j.biortech.2017.09.119

Kim, C., Lee, C.R., Song, Y.E., Heo, J., Choi, S.M., Lim, D.-H., Cho, J., Park, C., Jang, M., Kim, J.R., 2017. Hexavalent chromium as a cathodic electron acceptor in a bipolar membrane microbial fuel cell with the simultaneous treatment of electroplating wastewater. Chem. Eng. J. 328, 703–707. https://doi.org/10.1016/j.cej.2017.07.077

Kokabian, B., Gude, V.G., 2015. Sustainable photosynthetic biocathode in microbial desalination cells. Chem. Eng. J. 262, 958–965. https://doi.org/10.1016/j.cej.2014.10.048

Lakshmidevi, R., Gandhi, N.N., Muthukumar, K., 2020. Carbon neutral electricity production from municipal solid waste landfill leachate using algal-assisted microbial fuel cell. Appl. Biochem. Biotechnol. 191, 852–866. https://doi.org/10.1007/s12010-019-03160-5

Lee, C.-Y., Huang, Y.-N., 2013. The effects of electrode spacing on the performance of microbial fuel cells under different substrate concentrations. Water Sci. Technol. 68, 2028–2034. https://doi.org/10.2166/wst.2013.446

Li, M., Zhou, M., Luo, J., Tan, C., Tian, X., Su, P., Gu, T., 2019. Carbon dioxide sequestration accompanied by bioenergy generation using a bubbling-type photosynthetic algae microbial fuel cell. Bioresour. Technol. 280, 95–103. https://doi.org/10.1016/j.biortech.2019.02.038

Liu, Z., Liu, J., Zhang, S., Su, Z., 2009. Study of operational performance and electrical response on mediator-less microbial fuel cells fed with carbon- and protein-rich substrates. Biochem. Eng. J. 45, 185–191. https://doi.org/10.1016/j.bej.2009.03.011

Logan, B.E., Hamelers, B., Rozendal, R., Schröder, U., Keller, J., Freguia, S., Aelterman, P., Verstraete, W., Rabaey, K., 2006. Microbial fuel cells: Methodology and technology. Environ. Sci. Technol. 40, 5181–5192. https://doi.org/10.1021/es0605016

Meng, X., Savage, P.E., Deng, D., 2015. Trash to treasure: From harmful algal blooms to high-performance electrodes for sodium-ion batteries. Environ. Sci. Technol. 49, 12543–12550. https://doi.org/10.1021/acs.est.5b03882

Michael, C., del Ninno, M., Gross, M., Wen, Z., 2015. Use of wavelength-selective optical light filters for enhanced microalgal growth in different algal cultivation systems. Bioresour. Technol. 179, 473–482. https://doi.org/10.1016/j.biortech.2014.12.075

Morita, M., Watanabe, Y., Saiki, H., 2000. High photosynthetic productivity of green microalga *Chlorella sorokiniana*. Appl. Biochem. Biotechnol. 87, 203–218. https://doi.org/10.1385/ABAB:87:3:203

Mustakeem, 2015. Electrode materials for microbial fuel cells: Nanomaterial approach. Mater. Renew. Sustain. Energy. 4, 22. https://doi.org/10.1007/s40243-015-0063-8

Neethu, B., Bhowmick, G.D., Ghangrekar, M.M., 2018. Enhancement of bioelectricity generation and algal productivity in microbial carbon-capture cell using low cost coconut shell as membrane separator. Biochem. Eng. J. 133, 205–213. https://doi.org/10.1016/j.bej.2018.02.014

Neethu, B., Ghangrekar, M.M., 2017. Electricity generation through a photo sediment microbial fuel cell using algae at the cathode. Water Sci. Technol. 76, 3269–3277. https://doi.org/10.2166/wst.2017.485

Neethu, B., Tholia, V., Ghangrekar, M.M., 2020. Optimising performance of a microbial carbon-capture cell using Box-Behnken design. Process Biochem. 95, 99–107. https://doi.org/10.1016/j.procbio.2020.05.018

Pandit, S., Nayak, B.K., Das, D., 2012. Microbial carbon capture cell using cyanobacteria for simultaneous power generation, carbon dioxide sequestration and wastewater treatment. Bioresour. Technol. 107, 97–102. https://doi.org/10.1016/j.biortech.2011.12.067

Pant, D., Van Bogaert, G., Diels, L., Vanbroekhoven, K., 2010. A review of the substrates used in microbial fuel cells (MFCs) for sustainable energy production. Bioresour. Technol. 101, 1533–1543. https://doi.org/10.1016/j.biortech.2009.10.017

Perrine, Z., Negi, S., Sayre, R.T., 2012. Optimisation of photosynthetic light energy utilisation by microalgae. Algal Res. 1, 134–142. https://doi.org/10.1016/j.algal.2012.07.002

Powell, E.E., Hill, G.A., 2009. Economic assessment of an integrated bioethanol–biodiesel–microbial fuel cell facility utilising yeast and photosynthetic algae. Chem. Eng. Res. Des. 87, 1340–1348. https://doi.org/10.1016/j.cherd.2009.06.018

Puig, S., Serra, M., Coma, M., Cabré, M., Balaguer, M.D., Colprim, J., 2010. Effect of pH on nutrient dynamics and electricity production using microbial fuel cells. Bioresour. Technol. 101, 9594–9599. https://doi.org/10.1016/j.biortech.2010.07.082

Rajesh, P.P., Jadhav, D.A., Ghangrekar, M.M., 2015. Improving performance of microbial fuel cell while controlling methanogenesis by *Chaetoceros* pretreatment of anodic inoculum. Bioresour. Technol. 180, 66–71. https://doi.org/10.1016/j.biortech.2014.12.095

Rajesh, P.P., Noori, T., Ghangrekar, M.M., 2014. Controlling methanogenesis and improving power production of microbial fuel cell by lauric acid dosing. Water Sci. Technol. 70, 1363–1369. https://doi.org/10.2166/wst.2014.386

Rasmussen, M., Minteer, S.D., 2014. Photobioelectrochemistry: Solar energy conversion and biofuel production with photosynthetic catalysts. J. Electrochem. Soc. 161, H647–H655. https://doi.org/10.1149/2.0651410jes

Rismani-Yazdi, H., Carver, S.M., Christy, A.D., Tuovinen, O.H., 2008. Cathodic limitations in microbial fuel cells: An overview. J. Power Sources. 180, 683–694. https://doi.org/10.1016/j.jpowsour.2008.02.074

Rozendal, R.A., Hamelers, H.V.M., Buisman, C.J.N., 2006. Effects of membrane cation transport on pH and microbial fuel cell performance. Environ. Sci. Technol. 40, 5206–5211. https://doi.org/10.1021/es060387r

Sharma, Y., Li, B., 2010. Optimising energy harvest in wastewater treatment by combining anaerobic hydrogen producing biofermentor (HPB) and microbial fuel cell (MFC). Int. J. Hydrogen Energy. 35, 3789–3797. https://doi.org/10.1016/j.ijhydene.2010.01.042

Shukla, M., Kumar, S., 2018. Algal growth in photosynthetic algal microbial fuel cell and its subsequent utilisation for biofuels. Renew. Sustain. Energy Rev. 82, 402–414. https://doi.org/10.1016/j.rser.2017.09.067

Singh, S.P., Singh, P., 2014. Effect of CO_2 concentration on algal growth: A review. Renew. Sustain. Energy Rev. 38, 172–179. https://doi.org/10.1016/j.rser.2014.05.043

Sun, Z., Liu, J., Zhou, Z.G., 2016. Algae for biofuels: An emerging feedstock. An emerging feedstock, in: Handbook of Biofuels Production: Processes and Technologies: Second Edition. Elsevier, pp. 673–698. https://doi.org/10.1016/B978-0-08-100455-5.00022-9

Tanaka, Y., 2015. Electrodialysis, in: Ion Exchange Membranes. Elsevier, pp. 255–293. https://doi.org/10.1016/B978-0-444-63319-4.00012-2

Tartakovsky, B., Guiot, S.R., 2006. A comparison of air and hydrogen peroxide oxygenated microbial fuel cell reactors. Biotechnol. Prog. 22, 241–246. https://doi.org/10.1021/bp050225j

Ucar, D., Zhang, Y., Angelidaki, I., 2017. An overview of electron acceptors in microbial fuel cells. Front. Microbiol. 8, 1–14. https://doi.org/10.3389/fmicb.2017.00643

Ugwu, C.U., Aoyagi, H., 2008. Influence of shading inclined tubular photobioreactor surfaces on biomass productivity of *C. sorokiniana*. Photosynthetica. 46, 283–285. https://doi.org/10.1007/s11099-008-0049-1

Varanasi, J.L., Prasad, S., Singh, H., Das, D., 2020. Improvement of bioelectricity generation and microalgal productivity with concomitant wastewater treatment in flat-plate microbial carbon capture cell. Fuel. 263, 116696. https://doi.org/10.1016/j.fuel.2019.116696

Venkata Mohan, S., Veer Raghavulu, S., Sarma, P.N., 2008. Biochemical evaluation of bioelectricity production process from anaerobic wastewater treatment in a single chambered microbial fuel cell (MFC) employing glass wool membrane. Biosens. Bioelectron. 23, 1326–1332. https://doi.org/10.1016/j.bios.2007.11.016

Wang, X., Feng, Y., Liu, J., Lee, H., Li, C., Li, N., Ren, N., 2010. Sequestration of CO_2 discharged from anode by algal cathode in microbial carbon capture cells (MCCs). Biosens. Bioelectron. 25, 2639–2643. https://doi.org/10.1016/j.bios.2010.04.036

Wu, Y., Wang, Z., Zheng, Y., Xiao, Y., Yang, Z., Zhao, F., 2014. Light intensity affects the performance of photo microbial fuel cells with *Desmodesmus* sp. A8 as cathodic microorganism. Appl. Energy. 116, 86–90. https://doi.org/10.1016/j.apenergy.2013.11.066

Xu, C., Poon, K., Choi, M.M.F., Wang, R., 2015. Using live algae at the anode of a microbial fuel cell to generate electricity. Environ. Sci. Pollut. Res. 22, 15621–15635. https://doi.org/10.1007/s11356-015-4744-8

Yuan, Y., Zhao, B., Zhou, S., Zhong, S., Zhuang, L., 2011. Electrocatalytic activity of anodic biofilm responses to pH changes in microbial fuel cells. Bioresour. Technol. 102, 6887–6891. https://doi.org/10.1016/j.biortech.2011.04.008

Zhou, M., He, H., Jin, T., Wang, H., 2012. Power generation enhancement in novel microbial carbon capture cells with immobilised *Chlorella vulgaris*. J. Power Sources. 214, 216–219. https://doi.org/10.1016/j.jpowsour.2012.04.043

Zijffers, J.W.F., Salim, S., Janssen, M., Tramper, J., Wijffels, R.H., 2008. Capturing sunlight into a photobioreactor: Ray tracing simulations of the propagation of light from capture to distribution into the reactor. Chem. Eng. J. 145, 316–327. https://doi.org/10.1016/j.cej.2008.08.011

6 Sewage Treatment and Energy Recovery by Bioelectrochemical Oxidation System

Priyanka Devi, Prasann Kumar, and Joginder Singh

6.1 INTRODUCTION

The world's population is expected to surpass 9 billion by 2050, which would only increase the already significant problems of energy scarcity and environmental degradation (Xie et al., 2016). Population increase isn't the only factor putting a strain on water supplies; rising standards of life and urbanization (Bdour et al., 2009). Reduced carbon, nutrient, and pathogen discharge from wastewater treatment plants (WWTPs) protect water resources and human health (Cherchi et al., 2015). Because the technologies typically used at WWTPs were established in the early 20th century. And have low levels of efficiency and high levels of energy consumption, an improvement in the quality of treated effluent discharged from WWTPs, for reuse or recycling, has become more important due to the worldwide problems of fossil-fuel depletion, environmental pollution, and water shortages (Panepinto et al., 2016). The thousands of emerging pollutants (EPs) that are introduced and emitted annually may overwhelm conventional WWTPs. Most conventional WWTP procedures have never come across these EPs before, and because of their relatively stable chemical structure, the vast majority of them can exit WWTPs untreated. The discharge from WWTPs has been pinpointed as a significant contributor to environmental aquatic contamination (Reemtsma et al., 2006). As a result, cutting-edge techniques such as membrane processes (reverse osmosis and nanofiltration), membrane bioreactors, and advanced oxidation processes (AOPs) such as Fenton, ozonation, photocatalytic, and radiation are gaining popularity. These cutting-edge treatments, however, are either prohibitively expensive or fall short. More and more people are paying attention to electrochemical technologies such as electro-oxidation (EO), electrochemical coagulation (EC), and electrochemical flotation (EF) for advanced wastewater treatment, water disinfection, and improved remediation of contaminated soils (Särkkä et al., 2015; Shah, 2020). Electrochemical systems provide several benefits over other processes, including the capacity to function at room temperature and pressure, high performance, and responsiveness to changes in influent composition and flow rate. Reducing the electrical power consumption of electrochemical technologies and creating innovative electrical power supply for large-scale applications are both critical

DOI: 10.1201/9781003368472-6

because of the substantial amounts of electricity needed to run these systems. Due to its flexibility and applicability, adsorption has become a standard technique in the field of advanced wastewater treatment (Mor et al., 2007). With its high adsorption capacity that stems from its microporous structure, commercially available activated carbon (AC) is a common adsorbent in water treatment operations, where it is very effective for heavy metals, organic matter, and other wastes. A commercially available AC system, however, is prohibitively costly and can easily be saturated, rendering it ineffective (Luiz et al., 2002). As a result, it is essential to find solutions to the challenges of AC regeneration before the widespread application of AC to the field of wastewater treatment may occur. In contrast to the conventional practice, where AC is discarded together with waste-activated sludge, regeneration is an innovative approach to lowering economic expenses.

Electrochemical regeneration is the most cost-effective and efficient method of AC regeneration; nevertheless, its applicability is constrained for wastewater treatment due to its high power consumption (Hou et al., 2014; Shah, 2021). WWTPs today are extremely power-hungry. The issue of energy shortages may be greatly mitigated if WWTPs improve their energy management practices. Over 21 million metric tonnes of greenhouse gas (GHG) emissions are produced each year in the United States due to municipal wastewater treatment, which uses 2.8–4.1% of the country's total power consumption (30.2 billion kWh) (Shen et al., 2015). In the United States, electrical power consumption accounts for more than 30% of overall operating and maintenance expenditures and up to 80% of GHG emissions in WWTPs (USEPA, 2008). According to data from China's Ministry of Housing and Urban-Rural Development, the country's sewage treatment capacity is second only to that of the United States (Jin et al., 2014). Energy consumption from municipal wastewater treatment accounts for almost 30% of China's overall power usage; in 2012, there were 3340 municipal WWTPs situated in Chinese cities (Jin et al., 2014). Because of the foregoing, modern WWTP design and operation must prioritize the efficient removal of pollutants while also lowering the facility's energy and financial needs (Gude, 2016). Sustainable wastewater treatment and reuse is the emerging paradigm in this space (SWTR). Carbon reduction may be greatly aided by SWTR. Furthermore, SWTR is considered a technique to recover and reuse materials and energy from wastewater (Guest et al., 2009), which may contribute to addressing the major issues of energy scarcity and environmental pollution. As well as tackling the massive problems of energy scarcity and pollution, the new paradigm of recovering energy and minerals from wastewater is within reach thanks to several existing technologies. Water quality, discharge requirements, location, and other variables determine the optimal treatment method for municipal WWTPs (Zhang et al., 2015). One of the most common types of municipal WWTP is the activated sludge system; however, anaerobic treatment units are also common (Dos Santos et al., 2016). Up until recently, most efforts at recovering energy from municipal WWTPs had concentrated on the anaerobic biological treatment of wastewater and anaerobic digestion (AD) of waste sludge.

The anaerobic biological process in municipal WWTPs results in the release of biogas. Methane (CH_4), carbon dioxide (CO_2), and small amounts of other gases make up the majority of the biogas that are produced from digested sludge (Shen et al., 2015).

Biogas is a powerful GHG that contributes to climate change and should be avoided as an energy source for electricity generation. Therefore, to reduce carbon emissions, biogas emissions must be reduced (Daelman et al., 2012). Improved energy and material recovery from wastewater are possible with the use of bioelectrochemical technologies (BETs), which can supplement conventional methods. BETs, such as microbial fuel cells (MFCs), have been widely explored over the last decades because of the wastewater's high level of organic contaminants, which provide an attractive alternative source for energy generation. BETs can immediately transform the energy in organic matter found in wastewater into electricity and other valuable energy and molecules such as hydrogen, ozone, methane, and a variety of other compounds (Xu et al., 2015). The literature indicates that WWTPs act as energy storage facilities. Both electrochemical technology and alternating current adsorption are cutting-edge, efficient approaches for WWTPs; nevertheless, they require substantial electrical power inputs. Biological therapy and BETs, however, can generate electricity. These systems may be integrated to boost electricity recovery, allowing for greater energy independence and reduced carbon emissions. There has not been enough research on the connection between electrochemical and biological technologies and BETs, and this lack of study is reflected in the dearth of scholarly articles on the topic. This article's goal is to assess the viability of WWTPs for achieving energy independence and carbon mitigation by the integration of electrochemical, biological, and BETs (Machado et al., 2007, Memon et al., 2007, Uggetti et al., 2010, Yildirim & Topkaya 2012).

The fast depletion of fossil fuel reserves brought about by the world's rising energy consumption is a key reason for alarm. Excessive usage of these fossil fuels has led to a rise in atmospheric GHGs, which in turn has triggered widespread climate change and harsh weather. Because of their finite supply, ever-increasing prices, and harmful impacts on the environment, nonrenewable energy sources have prompted the exploration of alternative, renewable, and sustainable energy options (Mohanakrishna et al., 2018). Direct discharge of untreated or partially treated wastewater to the neighbouring wetlands and/or water bodies is harmful to both aquatic and terrestrial ecosystems, making wastewater treatment a top priority for most developing nations alongside the energy crises. Energy storage of 4.92–7.97 kW h kg COD^{-1} has been measured in wastewater, making it a possible source for energy generation and resource recovery due to its high content of organic matter (Dong et al., 2015). Activated sludge, oxidation ditches, trickling filters, lagoons, and aerobic digestion are just a few of the conventional biological techniques used to remove organic contaminants from wastewater. However, the energy requirements and financial burdens of these procedures are substantial (Daverey et al., 2019). In this context, membrane fuel cells (MFCs) have emerged as cutting-edge sustainable technology for producing electricity while treating wastewater. This method uses microorganisms to efficiently generate electricity from wastewater. While conventional treatments like activated sludge processes and trickling filter plants consume around 1322 kW h per million gallons of water, MFC takes only a fraction of that amount to run (Khan et al., 2017). For an influent flow of 318 m^3 h^{-1}, the cost of treating the wastewater by MFC is expected to be just 9% ($6.4 million) of the total cost of treating the wastewater by a conventional WWTP ($68.2 million). Compared to traditional AD, MFC offers various benefits, including the capacity to handle low substrate concentrations even at temperatures below 20°C, the production of water molecules, and the direct conversion of organic substrate into electrical energy (Pham et al., 2006). In a nutshell, MFC's capacity to generate power from wastewater makes it a sustainable, eco-friendly option. MFC's infrastructure for bioenergy generation and wastewater treatment has undergone significant and visible upgrades in recent years. Researchers worldwide are working hard on fixing the scale-up problems and making MFC more effective. More than half of all publications were published between 2015 and 2019, as shown in Figure 6.1. Synthesizing cost-effective parts (electrodes and membranes) and tackling MFC scale-up issues have been the focus of recent studies aimed at enhancing MFC performance.

Using exoelectrogenic microorganisms, a microbial electrolysis cell (MEC) oxidizes organic materials at the anode and produces a chemical of interest at the cathode (Logan et al., 2008). If wastewater is treated with MECs, the organic molecules contained within can have their chemical energy converted into hydrogen gas or another usable fuel. Adding voltage to produce hydrogen is far less demanding than electrolyzing water. Unlike water electrolysis, which requires just 1.2 V of power, the acetate-based MEC requires a much higher voltage of >0.2 V, with >0.5 V generally used. It has also been shown that additional compounds, like hydrogen peroxide and methane, may be produced at the cathode (Cheng et al., 2009). In the lab,

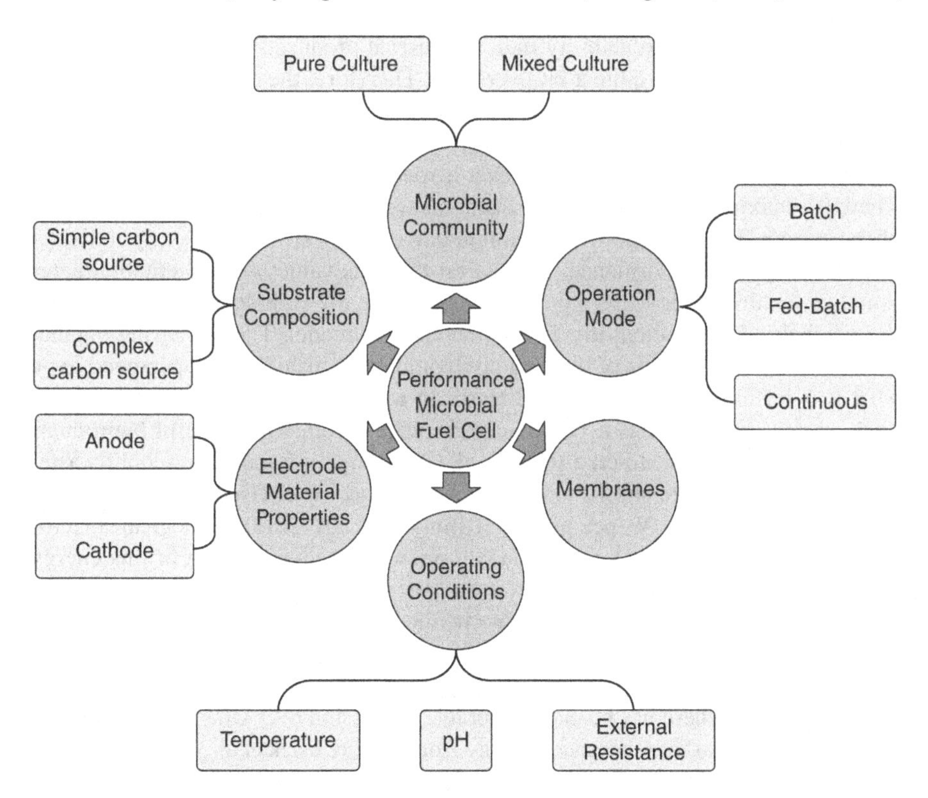

FIGURE 6.1 Microbial fuel cell efficiency in wastewater treatment and power generation is affected by several factors.

MECs are often run in fed-batch mode, where the liquid is replenished if the current or voltage drops below a predetermined threshold. In real-world applications, however, continuous flow operation is the norm. Initial testing and characterization of MEC reactor designs, electrode materials (Hu et al., 2009), and treatability of wastewaters may be done quickly and effectively with fed-batch experiments (Lu et al., 2010). However, as the current profiles generated during a fed-batch cycle tend to be fairly varied, it can be tricky to deduce how they apply to reactors operated under continuous flow conditions from batch test data. Potentiostatic (constant voltage) and galvanostatic (constant current) modes are both possible during fed-batch cycle MEC testing. The constant current and hydrogen production rates at the cathode are only possible under galvanostatic circumstances; however, the anode might have issues. At the end of a fed-batch cycle, when the substrate is no longer available in the anode, the potentiostat or power supply will continue to draw current from the anode, causing a higher voltage and destroying biofilm activity by forcing a less thermodynamically favourable reaction there, such as oxygen evolution or oxidation of the electrode (dissolution). Potentiostatic mode, in which the end of a batch cycle is indicated by the decrease of current, is recommended for batch-fed MECs to prevent such undesirable responses. Different current profiles can emerge during the fed-batch cycle since the end of a cycle is defined by a fixed current threshold. The overall parameters of the batch cycle (initial and final) are more significant for evaluating wastewater treatability than the current produced using constant polarization, which can vary quite a bit over time. Therefore, the cumulative Coulombs or recovered volume of hydrogen gas is more valuable than the peak current produced instantly. We propose that the major performance criteria for MECs should be a cumulative product and charge over a specified period, in contrast to MFCs, where the maximum power is a significant measure for measuring the performance of the system. Even though the maximum current may serve as an early indicator of treatability and performance, we suggest that this value is not a suitable representative feature of the MEC's performance during the complete cycle (equivalent to a set hydraulic retention time in continuous flow mode). The increasing population and the energy needs of modern civilizations have shifted our perspective on what was formerly considered to be trash. In recent years, people have begun to view wastewater (WW) as a "misplaced resource" from which useful byproducts and energy might be extracted (Gao et al., 2014). For instance, household wastewater (dWW) has been shown to contain as much as 7.6 kJ (Heidrich et al., 2010), which translates to 23 W per person (Rittmann et al., 2013). Biological anaerobic treatments, such as AD, are the ideal option for recovering part of this energy since they are strong and flexible. Over the past 10 years, a new generation of bio-based technologies known as bioelectrochemical systems (BESs) has evolved with considerable potential for wastewater treatment and resource recovery. Based on their manner of operation, bacterial electrolysis systems (BESs) are either MFCs or MECs. Although there has been considerable discussion over whether an approach is the most practical for recovering energy from WW (Cusick et al., 2010), it appears that MECs provide major benefits over MFCs from economic, environmental, and technological vantage points (Dominguez-Benetton et al., 2012). The reliability of this claim, however, is highly dependent on the MEC product (hydrogen, methane, ethanol, hydrogen peroxide, etc.). Hydrogen in particular is a promising prospect because it serves as a key input in a wide range of important industries (including the metals and minerals, fertilizer, chemical, and petrochemical sectors, among others) and because its high energy yield has prompted predictions that it will become the dominant form of energy transport shortly. In comparison to the sheer volume of published academic research, the number of reviews devoted to MECs is quite limited. These reviews have mostly concentrated on topics like substrates for MECs (Kadier et al., 2014), BESs as a versatile platform for a wide range of engineering functions, and the generation of high-value chemicals and bio-inspired nanomaterials, cathodic catalysts (Liu et al., 2014), ion exchange membranes (IEMs), electrode materials, and electroactive biofilms (Borole et al., 2011). Instead, this research digs into how MECs are being put to use in the real world, namely in the areas of wastewater treatment and hydrogen generation, and it offers a thoughtful examination of the obstacles and possibilities that lie ahead. We begin by quickly discussing the merits and disadvantages of the electrode materials and MEC designs most likely to be of practical use. We then discuss the parameters that would be most important at a commercial scale and thus have the greatest impact on the techno-economic viability of this technology as we review experiments in which real WWs were used as substrates for hydrogen production at the laboratory, semi-pilot, and pilot level. We then briefly explore some environmental and economic concerns before concluding with a critical study of MEC technology's prospects by looking at things like application niches, hydrogen alternatives, and MEC integration into energy transportation networks (Zhi et al., 2014).

6.2 THE FUNDAMENTALS OF MFC

The MFC is a BES that converts the chemical energy existing in the form of organic matter in wastewater into electrical energy through the cellular respiration power of microorganisms (Chandrasekhar et al., 2017). In general, it has three parts: a cathode chamber, an anode chamber, and an IEM. The IEM (such as anion, cation, and proton exchange membranes) separates the aerobic cathode chamber with a cathode electrode from the anaerobic anode chamber with an anode electrode. Glass wool or beads (porous proton exchange device) replace IEM in membraneless MFC. Bacteria cannot decompose organic waste and generate electricity in the anodic chamber unless oxygen is severely limited. MFC efficiency is affected by the substrate's characteristics, which in this case means the nature of the organic waste being used. Acetate, lignocellulose waste, brewery waste, starch processing waste, cellulose, chitin, sludge, and wastewater are all examples of waste materials that can be utilized as substrates (Parkash, 2016).

6.3 FUNCTIONING MFC MECHANISMS

Microorganisms in MFC oxidize organic substrate to provide energy and support their development and metabolic processes. Bacterial metabolism splits its energy between cellular processes and bioelectricity production. The oxidative pathway (respiration) and the fermentation pathway are the two primary routes used by microbes to convert energy (Chandrasekhar et al., 2017). Microbes use an electron transport chain (ETC) through their cell walls to generate energy through cellular respiration (ETC). The ETC begins with the release of a high-energy electron and proton from nicotinamide adenine dinucleotide (NADH). In this process, electrons are transferred from donor molecules like flavin adenine dinucleotide (FADH2) and electron-carrier molecules like NADH and H+, which are located in the intercellular space, to the terminal electron acceptor, which is often oxygen (Guang et al., 2020). The electron travels to the mitochondrial membrane through a predetermined pathway, including protein components. Both electrons and protons are pushed over the membrane and through the protein molecules. The goal of ETC is to generate byproducts like protons, electrons, and carbon dioxide. Electrons in an MFC go along a predetermined course until they are picked up by a mediator molecule and transported to the anode, where they are absorbed by adenosine triphosphate (ATP) and the byproducts (e, H+, and CO_2) are utilized to generate electricity. As the number of protons increases in the anode chamber relative to the cathode chamber, an electrochemical gradient is formed. Protons are forced through the semi-permeable membrane from the anode to the cathode by the electrochemical gradient. The anode generates electrons, which are sent to the cathode through an external circuit, where they join with oxygen and protons to make water, which then passes through a membrane. An electric current is produced by the movement of electrons across the external circuit, and this current has several uses (Dimuro et al., 2014, Dixon et al., 2003, Larsen et al., 2018).

6.4 MICROORGANISMS INTERACT WITH THE ELECTRODE SURFACE

Microorganism fuel cells (MFCs) use a wide variety of bacteria to generate electricity. Electrode/anode surface microbial interactions affect electrochemical reaction efficiency. This means that the electrode surface is crucial to the MFC process. *Geobacter sulfurreducens*, *Pichia fermentans*, and *Pseudomonas aeruginosa* are examples of exoelectrogenic microorganisms that exhibit extracellular electron transfer (EET) behaviours (Slate et al., 2019). Through direct and indirect processes, the exoelectrogenic microbes aided in the transport of electrons (Almatouq et al., 2020). Using nanowires and redox-active proteins, a process known as direct electron transfer (DET) brings bacteria into direct touch with the electrode surface. Nanowires (flagella and pili)/biofilm are developed by electrogenic microorganisms close to the electrode surface, allowing for DET from the outer cell membrane. DET also takes advantage of redox-active proteins like cytochromes located on the plasma membrane. Nanowire-forming bacteria like Shewanella and Geobacter are favoured in MFC because of their excellent electric power production capacity; nevertheless, the rate of electron transfer is slowed due to indirect electron transfer (IET) and the buried active sites of electron transport proteins (Slate et al., 2019). Without direct cytochrome-to-electrode contact, electrons can be transferred to the electrode surface by IET, which uses low molecular weight organic compounds known as endogenous redox mediators/electron shuttles. Oxidation-reduction processes can be catalyzed by endogenous redox mediators. It is the low molecular weight molecules or redox-active proteins that aid in the first transport of "e" to the cell surface during metabolic processes. Then they made their way to the outer membrane, where the cytochromes were waiting for them. Once "e" has been shuttled to the cell's outside media, it can go to specific external acceptors. Through these intermediates, electrons created by metabolic processes are collected and transported to the anode. MFC mediators should have high chemical stability and be non-biodegradable. Mediators utilized in MFC range from naturally occurring to synthetically produced and include phenazine, pyocyanin, anthracene-dione, thionine, neutral red, humic acid, riboflavin, and methylene blue (Bilal et al., 2018). Since mediators have an aromatics group, this may alter the make-up of microorganism communities and lower electric current densities by damaging cell proteins and membranes and impairing their fluidity and functioning (Kastury et al., 2015). For this reason, exogenous mediators are not recommended for use in MFC in the long term since they are both costly and technically infeasible (Cao et al., 2019). Microbes' negatively charged exterior surfaces cause them to be drawn to the anode (Palanisamy et al., 2019). And thus produces the microbial biofilm over the outer surface of the electrode, which also defines the surface charge idea owing to the adherence of bacteria towards the anode. Different surface modification techniques also aid the surface charge attraction in the MFC system. Increasing the EET rate is possible through a variety of means, one of which is by modifying the electrodes. Electrode surfaces are often treated with ammonia gas to allow the readily attaching of negatively charged microorganisms to the anode surface. Ammonia gas treatment of electrodes raises the positive charge throughout the electrode surface, which, at higher current densities, enhances electrode performance relative to untreated electrodes. However, ammonia treatment of electrodes is an expensive technique for large-scale applications (Choudhury et al., 2017). Using the electrons received by microorganisms at the cathode requires different methods than using electrons donated at the anode. Although, at the cathode end, DET and IET methods also occurred. Bacteria, such as *Acinetobacter calcoaceticus*, *Pseudomonas aeruginosa*, etc., use soluble redox shuttles to move electrons from the cathode to their cells, a process called DET or IET. A wide variety of abiotic (carbon, graphite, Pt, stainless steel, etc.) and biotic (biocathode) cathodes have been used in MFC (Kalathil et al., 2018). With the use of an abiotic catalyst (Pt, Au, AC, and transition metal compounds), oxygen-reducing materials that can create water in the cathodic chamber have been produced as abiotic cathodes (Breheny et al., 2019). However, ferricyanide as a catholyte might reduce the need for high-priced catalysts like Pt, Au, etc. at the cathode (Kalathil et al., 2018). However, ferricyanide usage cannot be maintained indefinitely. Some research has demonstrated that the use of a bacterial biocatalyst to promote microbial growth in the cathodic chamber and the subsequent development of biocathodes improve the current generation efficiency of MFC. Biocathodes are electrodes that promote microbial growth and accept the last electron by a process analogous to that which occurs in an anodic chamber, using oxygen, carbon dioxide, inorganic salts, and transition metal compounds (Erable et al., 2012). They could need oxygen or not. Transition metals, such as iron(II) and manganese(II), were oxidized by aerobic microorganisms at aerobic biocathodes in the presence of oxygen. However, non-oxygen oxidants (carbon dioxide (CO_2), manganese (Mn), iron (Fe), arsenic

(AsO_3), nitric oxide (NO_2), sulphate $(SO_4)_2$, selenium (SeO_2), etc.) were reduced by anaerobic or facultative microbes (Guo et al., 2012). Carbon- and nitrogen-rich wastewaters (anoxic and denitrifying biocathode) and sewage (Al-Mamun et al., 2020) can be treated, electricity can be generated, carbon dioxide can be captured, and microalgal biomass can be produced for use in a biorefinery, all at the same time, thanks to the use of biocathodes (Yadav et al., 2020).

6.5 ELECTROCHEMICAL ENERGY CONVERSION IS THE SECOND MOST COMMON METHOD OF ENERGY UTILIZATION

6.5.1 Electrochemical Technology: Advantages and Drawbacks

It takes more energy to run the EO process than it does to run EF or EC, even though it is one of the most widely used electrochemical technologies and an effective approach for pollutant degradation. Direct oxidation of pollutants in wastewater by hydroxyl radicals produced at anodes, as well as indirect oxidation by other means, are both possible while the EO is operating (Eryuruk et al., 2018). Direct oxidation is a process in which macromolecular organic contaminants are absorbed on the anode surface and subsequently eliminated by the anodic electron transfer reaction. Indirect oxidation makes use of the electrochemical generation of strong oxidants like chlorine, ozone, hypochlorite, and hydrogen peroxide. The oxidant produced from the oxidation process then degrades the contaminants in a large volume of solution (Nguyen and Ahn, 2018). In addition, "EC" refers to the method of using an electric current to cause the contaminants in wastewater to become more labile. A typical EC setup is an electrolytic cell with one or more sets of plates functioning as the anode and cathode. The electrodes, which might be the same or different materials, are sometimes referred to as "sacrificial electrodes." To neutralize the electric charge and remove contaminants, the anode can be resolved to create flocs. More gas bubbles (H_2 and O_2) of uniform size and size distribution can be generated at the electrode surface during an EF than during conventional flotation (Mook et al., 2014). Conditions such as current density, plate material qualities, and electrode spacing can be adjusted to regulate gas bubble production. The concentration of bubbles is influenced by the frequency and intensity of collisions between bubbles, particles, and oil droplets, all of which increase as the current density rises (Mook et al., 2014). Electrochemical methods have made significant strides in wastewater treatment over the last few decades because of their high efficiency, environmental friendliness, and adaptability, particularly in the abatement of bio-refractory chemicals in industrial wastewater. For perfluorinated compound removal in the lab, electrochemical technology uses significantly less energy than other AOPs such as ozonation, UV photolysis, sonochemical breakdown, and homogeneous photochemical oxidation (Niu et al., 2016). Electrochemical technologies can be considered "green technology" (Särkkä et al., 2015) due to the little chemical input required for water purification and the high removal efficiency and clean energy produced during the conversion process. However, limited power production and high costs have thus far hampered the widespread use of electrochemical technology in WWTPs. Some obstacles prevent the widespread use of electrochemical technologies, such as the limited durability of electrode materials and the inefficiency of their current output (resulting in excessive energy consumption) (Mahvi et al., 2011).

6.5.2 Consumption of Energy in Electrochemical Technology

The viability of electrochemical technologies depends on the conversion of electrical current that results in reductions in solute concentrations to employ current efficiency, such as columbic efficiency, through the elimination of COD or TOC target organic contaminants. According to Radjenovic and Sedlak, technologies depend on the conversion of electrical current that results in reductions in solute concentrations to employ current efficiency, such as columbic efficiency, through the elimination of COD or TOC target organic contaminants. According to Radjenovic and Sedlak (2015), the efficiency of a current is defined as the fraction of total electricity used by the targeted electrode reaction. This simplified equation may be used to determine the electrical energy usage in kilowatt-hours: E ()/1000 kWh U I t = ×× (1) (1) In this equation, U represents the voltage (V) across the cell, I represents the current (A), and t represents the period (t) over which the electrochemical process occurs (h). When considering the use of electrochemical technology for wastewater treatment, energy consumption is crucial. Water quality, discharge requirements, electrode type, and other factors all affect how much energy is needed to treat wastewater. Palahouane et al. (2015) used EC to remove fluoride from processed photovoltaic wastewater at an energy cost of 3.43 kWh/m^3 and an area flux of 18.51 A/m^2. According to Lacasa et al. (2013), a low value of 1.0 kWh/m^3 was achieved while treating 1 m^3 of seawater using an electrochemical oxidation technique for 45 minutes at 6.7 V and 255 A m^2.

6.5.3 Renewable Energy Generation Using Biological Engineering

The capacity of biological technology to generate energy from wastewater is widely recognized. Wastewater contains both thermal energy and the chemically bonded energy of organic compounds. The biodegradation of organics can release chemical energy stored in them in either anaerobic or aerobic environments (Nowak et al., 2011). Half of the biodegradable COD may be converted anaerobically to CH_4 at optimal operating circumstances, while the other half must be decomposed aerobically, necessitating more energy for aeration (Nowak et al., 2011). Activated sludge, anaerobic/anoxic/aerobic (A2/O), and AD are the three most common biological technology applications. Activated sludge techniques, which have been around for a while, currently form the backbone of most WWTPs (Gude, 2016). Activated sludge systems are inexpensive to build and run, yet effective in treating wastewater with high organic contents, resulting in biogas (Dos Santos et al., 2016). The A2/O process is an evolution of the activated sludge biological treatment technique. Some 25% of China's urban WWTPs utilize A2/O biological treatments to purify 33.2% of the country's municipal wastewater. With proper sludge AD, WWTPs may optimize energy output while reducing total treatment costs. The energy requirements of anaerobic technologies are also often lower.

Technologies that don't rely on oxygen include anaerobic ponds, upflow anaerobic sludge beds (UASBs), and other digesters. More than 60% of Brazil's current WWTPs employ anaerobic treatment (Dos Santos et al., 2016). Anaerobic application in WWTPs is more common in Latin America than in Brazil because of the region's warmer temperatures. It has been shown that WWTPs using an AD unit may save energy by producing energy-rich biogas. Biogas, which is made up of methane and carbon dioxide, offers significant environmental benefits since it may be used as a sustainable replacement for fossil fuels to generate power and fuel for vehicles, thereby lowering the carbon footprint of WWTP operations (Shen et al., 2015). With proper sludge AD, WWTPs may optimize energy output while reducing total treatment costs. While the operation of the active sludge system has a greater impact on energy consumption than the concentration of organic matter in the original wastewater, energy generation is highly dependent on organic matter levels in the water. Total WWTP CH_4 production may be standardized in several ways, including wastewater flow, population served, and incoming chemical oxygen demand (Daelman et al., 2012). High initial investment and ongoing operating and maintenance expenses are associated with chemical and energy-intensive processes. Based on studies conducted by the Electric Power Research Institute (EPRI), AD with biogas usage may provide 350 kWh of power per million gallons of wastewater treated at the WWTP. Produced biogas may supply between 39% and 75% of the WWTPs' entire electrical energy needs (Silvestre et al., 2015). If all the biodegradable materials in municipal wastewater were converted into biogas, then 30–60 L/d of CH_4 per capita might be produced. Emissions of CH_4 were found to be between 0.53% and 1.20% of the incoming COD at Dutch WWTPs. By 2040, an anaerobic WWTP's maximum energy output might reach 1 TWh (Dos Santos et al., 2016). Yet Cai et al. (2015) found that in China, around 4.4% of the entering COD was released as CH_4. A total of 39,921 t/a of CH_4 was produced at China's municipal WWTPs in 2012. More than 800 German WWTPs are now using biogas to create electricity and heat, with the potential to produce 900 GWh/electric year and 1800 GWh/thermal year (Dos Santos et al., 2016). Existing biological technologies have many benefits, including the ability to reduce organic and nitrogen levels in wastewater, to be applied on very small and very large scales, to have low investment costs, to require little space, to use little energy, to produce little excess sludge, to have a high loading capacity, and to have high treatment efficiencies. Unlike some other renewable fuels, biogas may be produced anywhere in the world, and the technology needed to convert it into usable energy is simple and not monopolized (Taleghani and Kia, 2005). This means its creators can earn renewable identification number credits (Shen et al., 2015). Many financial, power, and ecological advantages accrue from using biomethane as a transportation fuel. However, few WWTPs convert biogas into usable forms of energy like electricity or heat. About 90% of WWTPs in the United States only burn their biogas in boilers or let it go up in smoke, while the remaining 10% use combined heat and power technology (USEPA, 2011). CH_4 has 20–25 times the radiative forcing per gram of CO_2 as a GHG, depending on the evaluated emission time horizon, implying that improper management or use of biogas from WWTPs can increase carbon emissions and waste a valuable energy source (Shen et al., 2015).

Anaerobic biological application in WWTPs is impeded by several factors. Direct anaerobic biological treatment in WWTPs is sometimes hampered by factors such as low temperature, high organic content, and high conductivity (McCarty et al., 2011). To get the most out of AD, it's best to minimize the time and money spent on cleaning and improving biogas, which is contaminated with carbon dioxide and other substances (McCarty et al., 2011). Some strategies for improving the quality of biogas have been used in recent years. These include water scrubbing and pressure swing adsorption (PSA). Biogas generation at WWTPs enhances energy supply variety, but energy cannot be effectively stored in electrical or thermal forms for quick dispatch at WWTPs, which needs innovative technologies to overcome (Law et al., 2012).

6.6 BIOELECTROCHEMICAL TECHNOLOGY FOR ENERGY GENERATION

6.6.1 The Benefits of Bioelectrochemical Technology for Energy Generation

Current WWTPs rely heavily on the chemically and energy-intensive activated sludge process, which is costly to set up and maintain (Gude, 2016). More than 40%, and often 75%, of WWTP energy expenses are associated with aeration, which is necessary for the activated sludge process. However, sludge treatment and disposal can account for as much as 60% of the total energy consumed in WWTP operations (Gude, 2016). WWTPs use biological treatment systems to transform chemical energy (in the form of organic components in wastewater) into electrical energy or other valuable goods. BESs are what scientists refer to as "BETs" (Kumar et al., 2012). Bacteria can play a catalytic role in BETs, allowing for the oxidation of inorganic or organic materials and the subsequent production of electricity. The bacteria use these substrates to create electrons, which are then transported to the anode and flow to the cathode through a conductive substance with a resistor or when operating under a load (Logan and Regan, 2006). MFCs, electrolysis cells (MECs), and desalination cells (MDCs) are the most common types of BESs (Kelly and He, 2014). MFCs generate energy from both organic and inorganic sources through the bioelectrochemical catalytic activity of microorganisms (Wei et al., 2011). MDC is an expansion of MFC technology that allows MECs to create additional value products such as ethanol, methane, hydrogen, and hydrogen peroxide (Gude, 2016). In contrast, MDCs that do not employ a mediator can make use of the naturally occurring, electrochemically active bacteria already present in the sludge without the need for any bacterial additions. Municipal, hospital, brewery, and animal wastewater, as well as synthetic wastewater, have all been processed using lab-scale BETs. By generating clean power from organic materials in wastewater without any further procedures for separation and purification, BETs are generally regarded as an ecologically benign technology. BETs, in particular, can operate effectively in moderate operating conditions, notably at ambient temperatures, while still producing clean power directly (Gude, 2016). However, more effective BET designs with low costs and sustainable materials must be created to facilitate a successful scale-up effort (Fornero et al., 2010).

6.6.2 Obstacles to Generating Power with Bioelectrochemical Processes

Using BETs to cleanse wastewater while also generating electricity is one of their most promising uses. There are, however, several technological hurdles that must be overcome before this technology can be broadly adopted. For example, power densities should be considerably enhanced, the cost of building materials should be minimized, and the architecture developed to generate electricity should be scalable so that enormous wastewater flows may be treated (Logan and Regan, 2006). The low power densities provided by BETs are the primary obstacle to their widespread use among these problems. It was claimed that the power densities varied widely, from 0.0017 to 2.0 W/m^2 (0.2–200 W/m^3) (Janicek et al., 2014). A power density of 2.87 kW/m^3 was published recently; this is twice as high as what might be generated by AD. Swine wastewater and landfill leachate, but not municipal wastewater, were able to attain significant power densities. The high cost of building BETs is another obstacle to their wider use. Economic assessments show that power production from BETs alone may not be sufficient to make wastewater systems energy-neutral, energy-positive, or even cost-effective, according to past research (Gude, 2016). There is also a need to enhance the design of BETs so that they can generate electricity while processing high volumes of wastewater. Only 2% of the research included in BET-related literature mentioned using a reactor bigger than 1 L, and only 30% of those used a continuous operation mode (Zhang et al., 2013a). Fornero et al. (2010) identified three barriers to BETs' expansion that need to be addressed on the path to widespread adoption: (1) to produce more low-resistance, electrochemically active biomass; (2) to fine-tune BET's reactor layouts; and (3) to find better ways of dividing the anode and cathode. High electrolyte resistance of the anode and cathode solutions, or the IEM, causes power loss in BETs because of the sluggish passage of ions between the electrodes. Thus, researchers need to shrink BETs, which decrease electrolyte resistances by increasing the membrane surface area to volume ratio and shortening the distance between electrodes (Liu et al., 2013). However, this method cannot be used in wastewater treatment systems that require reacting quantities.

6.7 INTEGRATION OF CARBON-REDUCING AND ENERGY-GENERATING TECHNOLOGIES

6.7.1 Technology Integration for Municipal Wastewater Treatment Plants with Combined Energy Independence

There's no denying that today's WWTPs are among the highest energy users in the industry. Energy is required for the collection, physical treatment, chemical treatment, sludge treatment, and discharging processes in WWTPs (Rothausen and Conway, 2011). Since current density is a crucial factor in electrochemical wastewater treatment procedures, its implementation results in higher power consumption and operational expenses for more advanced treatment (Mook et al., 2014). There have been several attempts to supply the WWTP with renewable energy sources such as geothermal, solar, wind, and hydro (Galil and Levinsky, 2007). However, industrial processes often need very little energy, and that energy may be recovered in a far shorter time frame than was originally anticipated. Furthermore, when in use, these renewable sources have their own energy requirements. Furthermore, the manufacturing process for some renewable energy components, including solar panel materials, requires a great deal of energy (García-García et al., 2015). So far, biogas has only been produced through the AD of sludge and the treatment of wastewater in anaerobic conditions. But is it possible to generate electricity from the entire process of treating wastewater? That's what a few investigators are trying to find. McCarty et al. (2011), for instance, noted that BET is an innovative technique for harvesting energy from wastewater. The anaerobic biological units and BETs in WWTPs can generate electricity to be transformed into an energy-self-sufficient unit; this hybrid system has the potential to be more stable and sustainable in the voltage generated; the overall treatment efficiency is improved; and energy savings are realized (Tee et al., 2016). The normal treatment procedure for municipal WWTPs in dense metropolitan areas involves a large amount of water and a lot of time. Current municipal WWTPs can be retrofitted to become energy-neutral or even energy-positive. The treatment procedure of a hypothetical WWTP that generates all of its energy to use in therapy is depicted. Some of the steps in the process include the generation of electricity from biological anaerobic sludge digestion, BET wastewater treatment, and biological anaerobic wastewater treatment. There is no energy loss in transmission because the electricity generated may be used directly in the WWTPs. The created electrical energy can supply electricity for electrochemical advanced treatment in WWTPs in addition to powering the water pump, blower, and other power equipment in WWTPs. Electricity for the AC electrochemical regeneration process can be generated if AC adsorption is employed in the advanced treatment unit of WWTPs. More and more municipal WWTPs will need to undergo upgrades to bring them up to date with current effluent regulations as wastewater effluent discharge standards become more stringent. Stricter criteria for water quality and environmental protection have led to a significant rise in the energy required for wastewater collection and treatment (Cano et al., 2012). AC adsorption and electrochemical technologies are frequently employed for advanced treatment in WWTPs because of their low initial investment and ongoing operational expenses (Zhang et al., 2014). With an estimated electrical energy usage of 650 kW h/tonne of AC, Weng and Hsu's (2008) electrochemical regeneration system costs around 4% as much as buying new AC. With the addition of electrochemical regeneration technology, AC adsorption can gain a competitive edge in the rapidly expanding sector of advanced wastewater treatment. Figure 6.1 depicts a cutting-edge WWTP treatment method that employs both AC adsorption and electrochemical technologies. After the standard secondary treatment, this mostly comprises an AC adsorption unit for the removal of organic contaminants and suspended materials. Electronic commutation (EC) technology is used in AC regeneration. With the aforementioned in mind, it is apparent that the technology used in municipal WWTPs may save tonnes of money on energy costs since the BETs can provide power and AC adsorption is an advanced treatment process. Until now, the potential for WWTPs to become energy-neutral by using physical, chemical, and biological technology has been highlighted by Tee et al. (2016). The idea of WWTPs becoming energy-neutral through the incorporation of biological technology and BETs was considered by McCarty et al. (2011). This is the first study of its kind to analyze the energy requirements and carbon emissions of BETs alongside biological and electrochemical technology. Possibilities for municipal WWTPs are to achieve energy independence and balance.

Since WWTPs are energy-intensive machines, their daily operating expenses are heavily influenced by their electricity usage. Some recorded figures show how much energy is used by municipal WWTPs. WTTPs are used between 0.2 and 0.305 kWh/m^3 for general secondary treatment (CAS). When comparing the energy requirements of general secondary treatment (CAS) and secondary treatment coupled with advanced treatment and secondary treatment (MBR) in WWTPs, the latter comes out on top, at 1.184 kWh/m^3 compared to 0.64–1.39 kWh/m^3 for CAS. The two largest consumers of electricity in WWTPs are the aeration and pump lifting systems. For their case study, Xie & Wang (2012) looked at a municipal WWTP in Beijing with a treatment capacity of 600,000 m^3/d. A2/O biological technology was used to remediate this municipal wastewater. Various WWTP units' relative energy usage approximately 69.34% of the WWTP's total energy usage is attributable to the aeration and pump lifting units. Energy usage for electrochemical methods ranged from 1.0 to 3.43 kWh/m^3 for both salt water and industrial wastewater, as detailed in Section 2.2 of this book. The energy required by electrochemical technology used to treat municipal wastewater is substantially lower than that required to treat seawater or industrial wastewater. For the sake of the following contents, the electrochemical technology's energy consumption is determined to be 1.0 kWh/m^3. However, the electrochemical approach is not cost-effective at this time because of the high energy consumption of 1.0 kWh/m^3 required by large-scale municipal WWTPs with a significant volume of treatment capacity. Therefore, the future of electrochemical technology lies in figuring out how to significantly reduce its dependence on external power sources. Analyzing the energy balance of municipal WWTPs requires secondary treatment (CAS), biological technologies (BETs), and electrochemical technology for AC regeneration. Energy consumption of 0.305 kWh/m^3 and energy recovery of 0.758 kWh/m^3 result in a net surplus of 0.453 kWh/m^3 when general secondary treatment (CAS) is used in conjunction with biological technology (sludge digestion and AD) and BETs. When comparing this to the total energy consumption of 1.385 kWh/m^3 and the total energy recovery of 0.758 kWh/m^3 achieved by the secondary treatment (CAS) combined with AC adsorption advanced treatment (PAC-sand filter), biological technologies (sludge digestion and AD), BETs, and electrochemical technology for AC regeneration, a net insufficient energy of 0.627 kWh/m^3 is achieved. Therefore, an extra 0.627 kWh/m^3 is required to produce an energy-neutral system. To that end, advances in electrochemical technology and better BETs for energy production are needed in the not-too-distant future. The daily economic consumption of a sewage treatment plant processing 100,000 m^3 and 500,000 m^3 of waste was roughly calculated based on the tables above and then compared with that of a system of biological technology and BET as well as a system of electrochemical and biological technology for a comprehensive economic cost analysis of the various processing systems and levels, using secondary and advanced treatment technology plants as examples. After doing the math, it turns out that it costs around 150,000 yuan per day to power a facility that treats 100,000 m^3 per day of sewage with conventional secondary treatment equipment. The daily expenditures of modern treatment methods, particularly the electrochemical process, vary depending on the processing method used. In contrast, the use of biological technology in conjunction with BET may significantly reduce daily energy use through the generation of biogas for self-production, which in turn delivers steady economic advantages with the long-term operation. On the other hand, carbon emissions in WWTPs need to be reduced due to factors such as pollution and increased regulations on the release of wastewater effluent (Rothausen and Conway, 2011). Reduced GHG emissions are a common side effect of energy efficiency strategies. The quantity of GHG emissions avoided is not calculable as a whole since it is tied to the specifics of the energy production market in the nation where the WWTP is located (Nowak et al., 2011). After AD, environmental impacts are drastically reduced (Dong et al., 2014). Methane emissions from WWTPs account for 75% of the total; however, biogas use has reduced that number significantly in recent years. Sludge processing plants may drastically cut their methane emissions by improving their ventilation systems. The methane in the sewage was mostly oxidized aerobically in the activated sludge tanks. This has the potential to be used to further reduce methane emissions from sewage treatment facilities (Daelman et al., 2012).

6.7.2 Modern Wastewater Treatment Using MFC with Energy Recovery

Energy generation and the recovery of useful products (methane, biofuel, biohydrogen, etc.) and chemicals (ammonium, acetate, glycerol, 2-isobutyrate, etc.) during treatment are only two of the reasons why MFC has emerged as a viable technique (Jabeen and Farooq, 2017). Modifying MFC to use electrohydrogenesis and electromethanogenesis to generate hydrogen and methane, respectively, is one possibility. Organic material, nutrients, medicines, fats/oil, heavy metals, pesticides, toxic colours, and other contaminants may all be effectively removed from wastewater by using MFC. Domestic wastewater (Koffi and Okabe, 2020), dairy wastewater (Tanikkul et al., 2019), distillery wastewater (Jayashree et al., 2016), slaughterhouse wastewater (Jayashree et al., 2016), food-processing wastewater (Jayashree et al., 2016), tannery wastewater (Jayashree et al., 2016), pulp/paper wastewater (Jayashree et al., 2016). Table 6.1 displays the efficiency with which MFC recovers energy and treats wastewater from a variety of sources. Table 6.1 shows that the energy recovered by the MFC system is highly sensitive to the chemical make-up of the wastewater. Recent studies have concentrated on exploring various strategies for maximizing energy salvaged from wastewater (Babanova et al., 2017; Liu et al., 2005).

Water from three separate sources (sewage, sugar, and dairy) was treated effectively by a dual-chambered MFC, with 92% of BOD removed and 75.5% of COD reduced. Sewage (0.64 mA) produced more current than sugar (0.54 mA) and dairy wastewater (0.34 mA) during the day (Xiao et al., 2020). Five MFC modules were stacked by Dong et al. (2015) to create a 90 L pilot plant that successfully cleaned brewery wastewater and generated power with no additional fuel. The system was able to remove up to 87.6% of COD, and 86.3% of SS, and collect 0.034 kW h m^{-3} of energy in two phases that ran for more than 6 months; however, the power performance of each module degraded over time. Reversal current was shown to impair the electrochemical performance of a stacked MFC (72 L)

TABLE 6.1
Different Biochar-Based Materials Used in Direct Carbon Fuel Cells and Microbial Fuel Cells

S.No.	Biomass	Methods of Biochar Synthesis	Application	Performance
1	Wheat Straw	The biomass of wheat straw was pyrolyzed at 700°C for 2 hours, then Ca $(NO_3)_2$ was added, and the mixture was heated at 400°C for another 2 hours	Direct carbon fuel cells (DCFCs)	Its maximum power density is 258 mW/cm^2. The open circuit voltage is 0.98 V.
2	Corn Straw	The biomass of maize straw is pyrolyzed at 700°C for 1 hour	Direct carbon fuel cells (DCFCs)	Measured a maximum power density of 218.5 mW/cm^2. 1.06 volts is the open circuit voltage.
3	Biomass of Bamboo	The bamboo biomass was pyrolyzed at 700°C for 2 hours	Direct carbon fuel cells (DCFCs)	Maximum 121 mW/cm^2 power density. At 1.10 volts, an open circuit is considered to be "good."
4	Pinewood	Biomass of pine wood pyrolyzed at 1000°C for 1 hour	Microbial fuel cells (MFCs)	Get a maximum power density of 6 W/m^2. Value of open circuit: 0.72 V.
5	Alfalfa Leaf	Alfalfa leaf biomass was pyrolyzed at 250°C for 1 hour, then activated at 900°C for 2 hours with KOH, $FeCl_3$, or $ZnCl_2$	Microbial fuel cells (MFCs)	The open circuit voltage is 0.59 V, and the maximum power density is 1328.9 mW/m^2.
6	Coconut Shell	H_2SO_4 pyrolysis of coconut shell biomass at temperatures between 100 and 500°C for 2 hours	Microbial fuel cells (MFCs)	The maximum current is 1.057 mA, and the open circuit voltage is 0.722 V.

operating in parallel configuration while in continuous mode. Under batch mode, the same system demonstrated good COD removal (97%), in addition to a power output of 50.9 1.7 W m^{-3}. There has been some speculation that a stacked MFC with electrodes made of granular activated carbon (GAC) as a packed bed might be a viable arrangement for future expansion. Poor pollutant removal efficiency (COD 24%, nitrogen 28%, and TSS 40%) was observed after 9 months of operation of a membrane-less single-chamber MFC utilizing effluent supplied from a primary clarifier of the wastewater treatment facility. It indicated that the configuration, when combined with other WWTP technologies, might be appropriate for large-scale operations (Hiegemann et al., 2016). Successful operation of a set of five flat panel air-cathode MFCs containing a nitrifying and denitrifying microbial population for eight months resulted in 85% and 94% COD and TN removal efficiencies, respectively, for the treatment of household wastewater (Park et al., 2017). Using a hybrid lab-scale MFC system (MFC combined with algal biofilm, AB-MFC), the nutrient removal efficiency (TN 96%, TP 91.5%) from home wastewater was enhanced. Energy output from the AB-MFC was 18.1% greater (62.93 mW m^{-2}) than that of the MFC (52.33 mW m^{-2}) (Wu et al., 2016). Using nitrifying-denitrifying bacteria like Nitrosomonas, Clostridium, Pseudomonas, Arcobacter, etc., a recently published research found that air-cathode MFC could remove 99% ammonia, 91% COD, and up to 95% TN with continuous generation of 100 mW m^{-3} of electricity. Due to the increased oxidation capacity of contaminants accessible in wastewater, the pilot plant (65 L) MFC combined with a horizontal subsurface-created wetland demonstrated improved performance (approximately 33%) to synthetic sewage of greater strength for a period of three months. Although there have been some studies in this area, researchers have found that scaling up MFCs is one of the greatest obstacles they face. By stacking six modules, we were able to create a field-scale MFC (720 L) that removed 87% of COD from sewage sludge in under 36 hours (HRT) and produced 61 milliwatts (mW) of electricity, which was then utilized to power low-voltage light-emitting diode (LED) lights (Das et al., 2020). To treat low strength wastewater (60–100 mg L^{-1} COD) with 70–90% COD removal efficiency and a power density of 0.033–0.005 kW h m^{-3}, a modularized MFC system with a 1000 L capacity and 50 modules was run for almost a year at a short HRT (2 h) (Liang et al., 2018). The electron transfer efficiency, open circuit potential (OCP), and current density of this modularized MFC were all improved by adding N, P co-doped catalyst for the cathode, bringing the total to 1603.6 80 mW m^{-2} (Liang et al., 2019). Small-size MFCs, built with the aid of 3D bioanode and air-cathode, have also been used to generate stable power output (2000 mW m^2) for 45 d (Jiang et al., 2020). Overall, these investigations support the idea that air-cathode MFCs, hybrid MFCs (AB-MFCs, built wetland MFCs), cathode-doped modularized MFCs, etc., have the potential to be scaled-up for effective wastewater treatment together with bioelectricity production for long-term operation.

The potential of MFC in wastewater treatment has been acknowledged by researchers all around the world. Therefore, the development of MFC technology in recent years has been accomplished via the use of a wide range of novel, efficient, and cost-effective designs and building materials. Much research has shown improving MFC performance using different membranes, electrode materials, optimal operating conditions, inoculums, and biocatalysts; however, most assessments are limited to small-scale laboratory systems. Some MFCs pilot-scale units have also been implemented in WWTPs in several different nations. To treat the dye industry effluent, "JSP Enviro" and the Indian Institute of Technology Chennai have recently installed a 200 L prototype MFC at a WWTP in Tirupur (India) (IIT Madras et al., 2020). In Okinawa (Japan), the MFC system at the Awamori distillery has been in use for around five years. Okinawa Environment Science Center and Okinawa Prefectural Livestock and Grassland Research Center are collaborating to make MFC operational for the treatment of swine wastewater and pig farm effluent (Aelterman et al., 2006). Increased electrochemical voltage losses with increased volume, high material costs (electrodes, membranes, catalysts, etc.), increased internal resistance, long-term operational stability problem, etc. are only a few of the major problems that limit the practical deployment of MFCs. Because of this, research into a remedy for such practical issues is essential for the future. Maximum energy extraction from MFCs calls for careful monitoring of their internal resistance. Resistance in charge transfer (active resistance), ohmic resistance, and diffusion resistance all contribute to the slowed current flow in an MFC while it operates (concentration resistance). Problems including active surface fouling, degradation, activity loss, and corrosion, as well as the electrode's physical and biological features like ETP, interspacing, biocompatibility, etc., all have an impact on electrode performance. Unwanted biomass development, microbial processes like methanogenesis, and the complicated composition of genuine wastewater are only a few of the primary difficulties that restrict its usage. Therefore, additional research into the degradation kinetics involved in the biochemical process of MFC and biofilm development from the initiation stage to a stable stage is necessary to improve the performance of MFC. To take MFC from a pilot plant to a commercial facility, it is necessary to understand the interplay between the force, electricity, concentration, and scalability of biological processes. The MFC system has been scaled up in several ways, including by stacking MFCs in parallel/series connections and by making use of a high-volume reactor. Some key issues that will need to be addressed in the future include the assessment and monitoring of a large number of stacked MFCs, as well as decreased power production owing to the development of reverse voltage during the stacked arrangement. It has been noticed that a buffer solution is commonly utilized in laboratory-scale designs to keep the system's pH stable, but this approach is not practical in a full-scale reactor due to the high amount of external input and cost involved. When discussing the feasibility of expanding MFC infrastructure, the cost naturally arises as a key and pressing issue. Therefore, the ideal design and configuration should be chosen while expanding the MFC system to lessen the financial and time commitment required. Another challenge that must be overcome when expanding the use of MFC technology is its weak and variable power output. Unfortunately, the current density produced by a single MFC unit is insufficient to allow wastewater treatment to function autonomously. To address this challenge and conform to the rigorous water discharge criteria mandated by the Environmental Protection Agency, MFC units are increasingly being combined with other technologies, such as hybrid MFC, wetland-integrated MFC, stacked MFCs, etc. In any case, the fully flashing commercialized MFC system on a larger scale is still awaited, and much more study is required to bring it about.

6.8 CONCLUSIONS

Protecting human health and the environment through reduced pollution while minimizing energy use is a problem for municipal wastewater treatment. Recovering energy and minerals from WWTPs are becoming a priority as part of a paradigm shift towards sustainability. Sustainable wastewater treatment methods that generate energy, such as biological technologies and BETs, were discussed in this literature review. An additional energy-intensive but potentially effective advanced wastewater treatment option is electrochemical technology. While electrochemical technology has a low environmental impact for advanced treatment in WWTPs, it is not yet cost-effective for treating huge volumes of wastewater or for municipal wastewater on a large scale. In addition to achieving energy neutrality, this review shows that generic CAS paired with biological technologies (sludge digestion and AD) and BETs may provide a surplus of energy. By utilizing sludge digestion, AD, BETs, and electrochemical technology for AC regeneration, a CAS may be operated on a completely renewable energy basis. By recovering biogas and putting BETs to use in WWTPs, we can drastically cut back on GHG emissions. More study is needed; however, integrating WWTPs with BETs and electrochemical technologies might boost power recovery to the point where energy independence is achieved and GHG emissions are reduced. MFCs are a novel approach to generating electricity from wastewater. This environmentally friendly technology has been successfully used to remove pollutants and generate electricity from both model and real wastewater in laboratory settings. As a result, it is important to optimize the MFC's operating parameters for the intended effluent. Without sacrificing MFC performance, cheaper electrodes and membranes are needed. Hybrid systems, which include MFC, have demonstrated improved performance but present challenges for process optimization and control. Overall, for its effective application in self-energy adequate full-scale wastewater treatment, future studies should try to simplify its operation, reduce its cost through inexpensive building materials, and improve its performance. MECs have a long way to go before they can replace traditional wastewater treatment methods, but the promise is there. From the standpoint of actual implementations, we have reviewed the current state of MEC technology and its potential future developments. The following inferences are possible: Many different electrode materials have been produced, but their practical usefulness is limited by biocompatibility and cost concerns. The use of MECs has proven that they are a reliable technology capable of processing various organic substrates. Energy usage for MECs treating household wastewater is lower than that for comparable conventional technologies, and they can remove over 75% of COD. Hydrogen recovery is typically poor, though. Hydrogen generation rates are often greater in industrial wastewater due to the increased concentration of organics present in these fluids. However, they often need to be altered in some way before being given to the MEC. By acting as a hub connecting the electricity and gas grids, MECs can ease pressure on the power transmission system. MEC technology has significant future possibilities, with the development of many pilot-scale reactors over the past few years and the initiation of initial commercial experiences. There are still significant obstacles to overcome; the high initial investment required for MECs is the primary impediment to their widespread adoption. In addition, hydrogen itself can be a significant technical obstacle since, if not properly controlled on the cathode side, it can give rise to undesirable microorganisms that may reduce the reactor's overall performance. It has been suggested that methane-producing MECs might be used as a viable alternative to hydrogen-producing MECs to speed up the commercialization of this technology. Standardized methods of characterization, laboratory practices, and data reporting on MECs might all speed up their commercial development.

ACKNOWLEDGEMENT

We would like to acknowledge the Department of Agronomy at the Lovely Professional University, Phagwara, Punjab, India, for their consistent moral support and encouragement throughout the writing process.

Conflicts: None

REFERENCES

Aelterman, P., Rabaey, K., Clauwaert, P., Verstraete, W., 2006. Microbial fuel cells for wastewater treatment, water. Sci. Technol. 54, 9–15, https://doi.org/10.2166/wst.2006.702.

Al-Mamun, A., Jafary, T., Baawain, M.S., Rahman, S., Choudhury, M.R., Tabatabaei, M., Lam, S.S., 2020. Energy recovery and carbon/nitrogen removal from sewage and contaminated groundwater in a coupled hydrolytic-acidogenic sequencing batch reactor and denitrifying biocathode microbial fuel cell. Environ. Res. 183, 109273.

Almatouq, A., Babatunde, A.O., Khajah, M., Webster, G., Alfodari, M., 2020. Microbial community structure of anode electrodes in microbial fuel cells and microbial electrolysis cells. J. Water Process Eng. 34, 101140.

Babanova, S., Carpenter, K., Phadke, S., Suzuki, S., Ishii, S., Phan, T., GrossiSoyster, E., Flynn, M., Hogan, J., Bretschger, O., 2017. The effect of membrane type on the performance of microbial electrosynthesis cells for methane production. J. Electrochem. Soc. 164, H3015–H3023.

Bdour, A.N., Hamdi, M.R., Taraeneh, Z., 2009. Perspectives on sustainable wastewater treatment technologies and reuse options in the urban areas of the mediterranean region. Desalination. 237, 162–174.

Bilal, M., Wang, S., Iqbal, M.N., Zhao, Y., Hu, H., Wang, W., Zhang, X., 2018. Metabolic engineering strategies for enhanced shikimate biosynthesis: Current scenario and future developments. Appl. Microbiol. Biotechnol. 102, 7759–7773.

Borole, A.P., Reguera, G., Ringeisen, B., Wang, Z., Feng, Y., Kim, B.H., 2011. Electroactive biofilms: Current status and future research needs. Energy Environ. Sci. 4, 4813–4834.

Breheny, M., Bowman, K., Farahmand, N., Gomaa, O., Keshavarz, T., Kyazze, G., 2019. Biocatalytic electrode improvement strategies in microbial fuel cell systems. J. Chem. Technol. Biotechnol. 94, 2081–2091.

Cai, B.F., Gao, Q.X., Li, Z.H., Wu, J., Wang, J.X., 2015. Estimation of methane emissions of wastewater treatment plants in China. Chin. Environ. Sci. 35, 3810–3816.

Cano, A., Cañizares, P., Barrera-Díaz, C., Sáez, C., Rodrigo, M.A., 2012. Use of conductive diamond electrochemical-oxidation for the disinfection of several actual treated wastewaters. Chem. Eng. J. 211–212, 463–469.

Cao, Y., Mu, H., Liu, W., Zhang, R., Guo, J., Xian, M., Liu, H., 2019. Electricigens in the anode of microbial fuel cells: Pure cultures versus mixed communities, microb. Cell Fact. 18, 1–14.

Chandrasekhar, K., Kadier, A., Kumar, G., Nastro, R.A., Jeevitha, V., 2017. Challenges in microbial fuel cell and future scope. Microb. Fuel Cell. 483–499.

Cheng, S.A., Xing, D.F., Call, D.F., Logan, B.E., 2009. Direct biological conversion of electrical current into methane by electromethanogenesis. Environ Sci. Technol. 43, 3953–3958.

Cherchi, C., Badruzzaman, M., Oppenheimer, J., Bros, C.M., Jacangelo, J.G., 2015. Energy and water quality management systems for water utility's operations: A review. J. Environ. Manag. 153, 108–120.

Choudhury, P., Prasad Uday, U.S., Bandyopadhyay, T.K., Ray, R.N., Bhunia, B., 2017. Performance improvement of microbial fuel cell (MFC) using suitable electrode and bioengineered organisms: A review. Bioengineered. 8, 471–487.

Cusick, R.D., Kiely, P.D., Logan, B.E., 2010. A monetary comparison of energy recovered from microbial fuel cells and microbial electrolysis cells fed winery or domestic wastewaters. Int. J. Hydrog. Energy. 35, 8855–8861.

Daelman, M.R.J., van Voorthuizen, E.M., van Dongen, U.G.J.M., Volcke, E.I.P., van Loosdrecht, M.C., 2012. Methane emission during municipal wastewater treatment. Water. Res. 46, 3657–3670.

Das, I., Ghangrekar, M.M., Satyakam, R., Srivastava, P., Khan, S, Pandey, H.N., 2020. Onsite sanitary wastewater treatment system using 720-l stacked microbial fuel cell: Case study. J. Hazardous, Toxic, Radioact. Waste. 24, 1–7.

Daverey, A., Pandey, D., Verma, P., Verma, S., Shah, V., Dutta, K., Arunachalam, K., 2019. Recent advances in energy efficient biological treatment of municipal wastewater, bioresour. Technol. Reports. 7, 100252.

Dimuro, J.L., Guertin, F.M., Helling, R.K., Perkins, J.L., Romer, S., 2014. A financial and environmental analysis of constructed wetlands for industrial wastewater treatment. J. Ind. Ecol. 18, 631–640.

Dixon, A., Simon, M., Burkitt, T., 2003. Assessing the environmental impact of two options for small-scale wastewater treatment: Comparing a reedbed and an aerated biological filter using a life cycle approach. Ecol. Eng. 20, 297–308.

Dominguez-Benetton, X., Sevda, S., Vanbroekhoven, K., Pant, D., 2012. The accurate use of impedance analysis for the study of microbial electrochemical systems. Chem. Soc. Rev. 41, 7228–7246.

Dong, J., Chi, Y., Tang, Y.J., Wang, F., Huang, Q.X., 2014. Combined life cycle environmental and exergetic assessment of four typical sewage treatment techniques in China. Energy Fuels. 28, 2114–2122.

Dong, Y., Qu, Y., He, W., Du, Y., Liu, J., Han, X., Feng, Y., 2015. A 90-liter stackable baffled microbial fuel cell for brewery wastewater treatment based on energy selfsufficient mode. Bioresour. Technol. 195, 66–72.

Dos Santos, I.F.S., Barros, R.M., Tiago Filho, G.L., 2016. Electricity generation from biogas of anaerobic wastewater treatment plants in Brazil: An assessment of feasibility and potential. J. Clean. Prod. 126, 504–514.

Erable, D., F´eron, A., Bergel, 2012. Microbial catalysis of the oxygen reduction reaction for microbial fuel cells: A review. ChemSusChem. 5, 975–987.

Eryuruk, K., Un, U.T., Ogutveren, U.B., 2018. Electrochemical treatment of wastewaters from poultry slaughtering and processing by using iron electrode. J. Clean. Prod. 172, 1089–1095.

Fornero, J.J., Rosenbaum, M., Angenent, L.T., 2010. Electric power generation from municipal, food, and animal wastewater using microbial fuel cells. Electroanalysis. 22, 832–843.

Galil, N.I., Levinsky, Y., 2007. Sustainable reclamation and reuse of industrial wastewater including membrane bioreactor technologies: Case studies. Desalination. 202, 411–417.

Gao, H., Scherson, Y.D., Wells, G.F., 2014. Towards energy neutral wastewater treatment: Methodology and state of the art. Environ Sci.: Process Impacts. 16, 1223–1246.

García-García, A., Martínez-Miranda, V., Martínez-Cienfuegos, I.G., Almazán-Sánchez, P.T., Castañeda-Juárez, M., Linares-Hernández, I., 2015. Industrial wastewater treatment by electrocoagulation–electrooxidation processes powered by solar cells. Fuel. 149, 46–54.

Guang, L., Koomson, D.A., Jingyu, H., Ewusi-Mensah, D., Miwornunyuie, N., 2020. Performance of exoelectrogenic bacteria used in microbial desalination cell technology. Int. J. Environ. Res. Public Health. 17, 10–12.

Gude, V.G., 2016. Wastewater treatment in microbial fuel cells – An overview. J. Clean. Prod. 122, 287–307.

Guest, J.S., Skerlos, S.J., Barnard, J.L., Beck, M.B., Daigger, G.T., Hilger, H., Jackson, S.J., Karvazy, K., Kelly, L., Macpherson, L., Mihelcic, J.R., Pramanik, A., Raskin, L., Van Loosdrecht, M.C., Yeh, D., Love, N.G., 2009. A new planning and design paradigm to achieve sustainable resource recovery from wastewater. Environ. Sci. Technol. 43, 6126–6130.

Guo, K., Hassett, D.J., Gu, T., 2012. Microbial fuel cells: Electricity generation from organic wastes by microbes.

Heidrich, E., Curtis, T., Dolfing, J., 2010. Determination of the internal chemical energy of wastewater. Environ. Sci. Technol. 45, 827–832.

Hiegemann, H., Herzer, D., Nettmann, E., Lübken, M., Schulte, P., Schmelz, K.G., Gredigk-Hoffmann, S., Wichern, M., 2016. An integrated 45 L pilot microbial fuel cell system at a full-scale wastewater treatment plant. Bioresour. Technol. 218, 115–122.

Hou, P., Byrne, T., Cannon, F.S., Chaplin, B.P., Hong, S.Q., Nieto-Delgado, C., 2014. Electrochemical regeneration of polypyrrole-tailored activated carbons that have removed sulfate. Carbon. 79, 46–57.

Hu, H.Q., Fan, Y.Z., Liu, H., 2009. Hydrogen production in singlechamber tubular microbial electrolysis cells using nonprecious-metal catalysts. Int. J. Hydrogen Energy. 34, 8535–8542.

IIT Madras alumni-founded start-up the microbial fuel cell tech to generate electricity by treating textile wastewater into use, 2020 (n.d.).

Jabeen, G., Farooq, R., 2017. Microbial fuel cells and their applications for cost effective water pollution remediation. Proc. Natl. Acad. Sci. India Sect. B – Biol. Sci. 87, 625–635.

Janicek, A., Fan, Y., Liu, H., 2014. Design of microbial fuel cells for practical application: A review and analysis of scale-up studies. Biofuels. 5, 79–92.

Jayashree, C., Tamilarasan, K., Rajkumar, M., Arulazhagan, P., Yogalakshmi, K.N., Srikanth, M., Banu, J.R., 2016. Treatment of seafood processing wastewater using upflow microbial fuel cell for power generation and identification of bacterial community in anodic biofilm. J. Environ. Manage. 180, 351–358.

Jiang, M., Xu, T., Chen, S., 2020. A mechanical rechargeable small-size microbial fuel cell with long-term and stable power output. Appl. Energy. 260, 114336, https://doi.org/10.1016/j.apenergy.2019.114336.

Jin, L.Y., Zhang, G.M., Tian, H.F., 2014. Current state of sewage treatment in China. Water Res. 66, 85–98.

Kadier, A., Simayi, Y., Kalil, M.S., Abdeshahian, P., Hamid, A.A., 2014. A review of the substrates used in microbial electrolysis cells (MECs) for producing sustainable and clean hydrogen gas. Renew Energy. 71, 466–472.

Kalathil, S., Patil, S.A., Pant, D., 2018. Microbial fuel cells: Electrode materials. Encycl. Interfacial Chem. 309–318.

Kastury, F., Juhasz, A., Beckmann, S., Manefield, M., 2015. Ecotoxicity of neutral red (dye) and its environmental applications. Ecotoxicol. Environ. Saf. 122, 186–192.

Kelly, P.T., He, Z., 2014. Nutrients removal and recovery in bio-electrochemical systems: A review. Bioresour. Technol. 153, 351–360.

Khan, M.D., Khan, N., Sultana, S., Joshi, R., Ahmed, S., Yu, E., Scott, K., Ahmad, A., Khan, M.Z., 2017. Bioelectrochemical conversion of waste to energy using microbial fuel cell technology. Process Biochem. 57, 141–158.

Koffi, N.J., Okabe, S., 2020. Domestic wastewater treatment and energy harvesting by serpentine up-flow MFCs equipped with PVDF-based activated carbon aircathodes and a low voltage booster. Chem. Eng. J. 380, 122443.

Kumar, A.K., Reddy, M.V., Chandrasekhar, K., Srikanth, S., Mohan, S.V., 2012. Endocrine disruptive estrogens role in electron transfer: Bio-electrochemical remediation with microbial mediated electrogenesis. Bioresour. Technol. 104, 547–556.

Lacasa, E., Tsolaki, E., Sbokou, Z., Rodrigo, M.A., Mantzavinos, D., Diamadopoulos, E., 2013. Electrochemical disinfection of simulated ballast water on conductive diamond electrodes. Chem. Eng. J. 223, 516–523.

Larsen, J.D., Hoeve, M., Nielsen, S., Scheutz, C., 2018. Life cycle assessment comparing the treatment of surplus activated sludge in a sludge treatment reed bed system with mechanical treatment on centrifuge. J. Clean Prod. 185, 148–156.

Law, Y., Ye, L., Pan, Y., Yuan, Z., 2012. Nitrous oxide emissions from wastewater treatment processes. Philos. Trans. R. Soc. Lond. Ser. B Biol. Sci. 30, 441–444.

Liang, K., Li, Y., Liu, X., Kang, 2019. Nitrogen and phosphorus dual-doped carbon derived from chitosan: An excellent cathode catalyst in microbial fuel cell. Chem. Eng. J. 358, 1002–1011.

Liang, P., Duan, R., Jiang, Y., Zhang, X., Qiu, Y., Huang, X., 2018. One-year operation of 1000-L modularized microbial fuel cell for municipal wastewater treatment. Water Res. 141, 1–8.

Liu, J., Liu, L., Gao, B., Yang, F., 2013. Integration of bio-electrochemical cell in membrane bioreactor for membrane cathode fouling reduction through electricity generation. J Membr Sci. 430, 196–202. https://doi.org/10.1016/j.memsci.2012.11.046

Liu, H., Grot, S., Logan, B.E., 2005. Electrochemically assisted microbial production of hydrogen from acetate. Environ. Sci. Technol. 39, 4317–4320.

Liu, X., Li, W., Yu, H., 2014. Cathodic catalysts in bioelectrochemical systems for energy recovery from wastewater. Chem. Soc. Rev. 43, 7718–7745.

Logan, B.E., Call, D., Cheng, S., Hamelers, H.V.M., Sleutels, T., Jeremiasse, A.W., et al., 2008. Microbial electrolysis cells for high yield hydrogen gas production from organic matter. Environ. Sci. Technol. 42, 8630–8640.

Logan, B.E., Regan, J.M., 2006. Microbial fuel cells – Challenges and applications. Environ. Sci. Technol. 40, 5172–5180.

Lu, L., Xing, D.F., Xie, T.H., Ren, N.Q., Logan, B.E., 2010. Hydrogen production from proteins via electrohydrogenesis in microbial electrolysis cells. Biosens Bioelectron. 25, 2690–2695.

Luiz, C.A.O., Rachel, V.R.A.R., Jose, D.F., Karim, S., Garg, V., Rochel, M.L., 2002. Activated carbon/iron oxide magnetic composites for the adsorption of contaminants in water. Carbon. 40, 2177–2183.

Machado, A.P., Urbano, L., Brito, A.G., Janknecht, P., Salas, J.J., Nogueira, R., 2007. Life cycle assessment of wastewater treatment options for small and decentralized communities. Water Sci. Technol. 56, 15–22.

Mahvi, A.H., Ebrahimi, S.J., Mesdaghinia, A., Gharibi, H., Sowlat, M.H., 2011. Performance evaluation of a continuous bipolar electrocoagulation/electrooxidation-electrofluotation (ECEO-EF) reactor designed for simultaneous removal of ammonia and phosphate from wastewater effluent. J. Hazard. Mater. 192, 1267–1274.

McCarty, P.L., Bae, J., Kim, J., 2011. Domestic wastewater treatment as a net energy products – Can this be achieved. Environ. Sci. Technol. 45, 7100–7106.

Memon, F.A., Zheng, Z., Butler, D., Shirley-Smith, C., Lui, S., Makropoulos, C., Avery, L., 2007. Life cycle impact assessment of greywater recycling technologies for new developments. Environ. Monit. Assess. 129, 27–35.

Mohanakrishna, G., Abu-Reesh, I.M., Kondaveeti, S., Al-Raoush, R.I., He, Z., 2018. Enhanced treatment of petroleum refinery wastewater by short-term applied voltage in single chamber microbial fuel cell. Bioresour. Technol. 253, 16–21.

Mook, W.T., Aroua, M.K., Issabayeva, G., 2014. Prospective applications of renewable energy based electrochemical systems in wastewater treatment: A review. Renew. Sustain. Energy. Rev. 38, 36–46.

Mor, S., Ravindra, K., Bishnoi, N.R., 2007. Adsorption of chromium from aqueous solution by activated alumina and activated charcoal. Bioresour. Technol. 98, 954–957.

Nguyen, V.K., Ahn, Y., 2018. Electrochemical removal and recovery of iron from groundwater using non-corrosive electrodes. J. Environ. Manag. 211, 36–41.

Niu, J.F., Li, Y., Shang, E.X., Xu, Z.S., Liu, J.Z., 2016. Electrochemical oxidation of perfluorinated compounds in water. Chemosphere. 146, 526–538.

Nowak, O., Keil, S., Fimml, C., 2011. Examples of energy self-sufficient municipal nutrient removal plants. Water Sci. Technol. 64, 1–6.

Palahouane, B., Drouiche, N., Aoudj, S., Bensadok, K., 2015. Cost-effective electrocoagulation process for the remediation of fluoride from pretreated photovoltaic wastewater. J. Ind. Eng. Chem. 22, 127–131.

Palanisamy, G., Jung, H.Y., Sadhasivam, T., Kurkuri, M.D., Kim, S.C., Roh, S.H., 2019. A comprehensive review on microbial fuel cell technologies: Processes, utilization, and advanced developments in electrodes and membranes. J. Clean. Prod. 221, 598–621.

Panepinto, D., Fiore, S., Zappone, M., Genon, G., Meucci, L., 2016. Evaluation of the energy efficiency of a large-scale wastewater treatment plant in Italy. Appl. Energy. 161, 404–411.

Park, Y., Park, S., Nguyen, V.K., Yu, J., Torres, C.I., Rittmann, B.E., Lee, T., 2017. Complete nitrogen removal by simultaneous nitrification and denitrification in flat-panel air-cathode microbial fuel cells treating domestic wastewater. Chem. Eng. J. 316, 673–679.

Parkash, A., 2016. Microbial fuel cells: A source of bioenergy. J. Microb. Biochem. Technol. 8, 247–255.

Pham, B.T.H., Rabaey, K., Aelterman, P., Clauwaert, P., De Schamphelaire, L., Boon, N., Verstraete, W., 2006. Microbial fuel cells in relation to conventional anaerobic digestion technology. Eng. Life Sci. 6, 285–292.

Radjenovic, J., Sedlak, D.L., 2015. Challenges and opportunities for electrochemical processes as next-generation technologies for the treatment of contaminated water. Environ. Sci. Technol. 49, 11292–11302.

Reemtsma, T., Weiss, T., Mueller, J., Petrovic, M., Gonzalez, S., Barcelo, D., Ventura, F., Knepper, T.P., 2006. Polar pollutants entry into the water cycle by municipal wastewater: A European perspective. Environ. Sci. Technol. 40, 5451–5458.

Rittmann, B.E., 2013. The Energy Issue in Urban Water Management. Source Separation and Decentralization for Wastewater Management; IWA Publishing, London, pp. 13–27.

Rothausen, S.G.S.A., Conway, D., 2011. Greenhouse-gas emissions from energy use in the water sector. Nat. Clim. Change. 1, 210–219.

Särkkä, H., Bhatnagar, A., Sillanpää, M., 2015. Recent developments of electro-oxidation in water treatment – A review. J. Electroanal. Chem. 754, 46–56.

Shah, M.P., 2020. Microbial Bioremediation & Biodegradation; Springer.

Shah, M.P., 2021. Removal of Refractory Pollutants from Wastewater Treatment Plants; CRC Press.

Shen, Y.W., Linville, J.L., Urgun-Demirtas, M., Mintz, M.M., Snyder, S.W., 2015. An overview of biogas production and utilization at full-scale wastewater treatment plants (WWTPs) in the United States: Challenges and opportunities towards energy-neutral WWTPs. Renew. Sustain. Energy Rev. 50, 346–336.

Silvestre, G., Fernández, B., Bonmatí, A., 2015. Significance of anaerobic digestion as a source of clean energy in wastewater treatment plants. Energy Convers. Manag. 101, 255–262.

Slate, A.J., Whitehead, K.A., Brownson, D.A.C., Banks, C.E., 2019. Microbial fuel cells: An overview of current technology. Renew. Sustain. Energy Rev. 101, 60–81.

Taleghani, G., Kia, A.S., 2005. Technical economical analysis of the Saveh biogas power plant. Renew. Energy. 30, 441–446.

Tanikkul, P., Boonyawanich, S., Pisutpaisal, N., 2019. Bioelectricity recovery and pollution reduction of distillery wastewater in air-cathode SCMFC. Int. J. Hydrogen Energy. 44(11), 5481–5487.

Tee, P.F., Abdullah, M.O., Tan, I.A.W., Rashid, N.K.A., Amin, M.A.M., Nolasco-Hipolito, C., Bujang, K., 2016. Review on hybrid energy systems for wastewater treatment and bio-energy production. Renew. Sustain. Energy. Rev. 54, 235–246.

Uggetti, E., Ferrer, I., Molist, J., Garcı, J., 2010. Technical, economic and environmental assessment of sludge treatment wetlands. Water Res. 45, 573–582.

USEPA, 2008. Ensuring a Sustainable Future: An Energy Management Guidebook for Wastewater and Water Utilities. US Environmental Protection Agency, Washington, DC.

USEPA, 2011. Opportunities for Combined Heat and Power at Wastewater Treatment Facilities: Market Analysis and Lessons from the Field. US Environmental Protection Agency, Washington, DC.

Wei, J., Liang, P., Cao, X., Huang, X., 2011. Use of inexpensive semicoke and activated carbon as biocathode in microbial fuel cells. Bioresour. Technol. 102, 10431–10435.

Weng, C.H., Hsu, M.C., 2008. Regeneration of granular activated carbon by an electrochemical process. Separ. Purif. Technol. 64, 227–236.

Wu, S., Li, H., Zhou, X., Liang, P., Zhang, X., Jiang, Y., Huang, X., 2016. A novel pilot-scale stacked microbial fuel cell for efficient electricity generation and wastewater treatment. Water Res. 98, 396–403.

Xiao, N., Selvaganapathy, P.R., Wu, R., Huang, J.J., 2020. Influence of wastewater microbial community on the performance of miniaturized microbial fuel cell biosensor. Bioresour. Technol. 302, 122777.

Xie, D., Wang, S. 2012. Consensus of second-order discrete-time multi-agent systems with fixed topology. J Math Anal Appl. 387(1), 8–16.

Xie, M., Shon, H.K., Gray, S.R., Elimelech, M., 2016. Membrane-based processes for wastewater nutrient recovery: Technology, challenges, and future direction. Water Res. 89, 210–221.

Xu, X., Zhou, B., Ji, F.Y., Zhou, Q.L., Yuan, Y.S., Jin, Z., Zhao, D.Q., Long, J., 2015. Nitrification, denitrification, and power generation enhanced by photocatalysis in microbial fuel cells in the absence of organic compounds. Energy Fuels. 29, 1227–1232.

Yadav, G., Sharma, I., Ghangrekar, M., Sen, R., 2020. A live bio-cathode to enhance power output steered by bacteria-microalgae synergistic metabolism in microbial fuel cell. J. Power Sources. 449, 227560.

Yildirim, M., Topkaya, B., 2012. Assessing environmental impacts of wastewater treatment alternatives for small-scale communities. Clean. 40, 171–178.

Zhang, C., Jiang, Y.H., Li, Y.L., Hu, Z.X., Zhou, L., Zhou, M.H., 2013a. Three-dimensional electrochemical process for wastewater treatment: A general review. Chem. Eng. J. 228, 455–467.

Zhang, C.H., Peng, C., Li, J., 2015. Past and future trends of wastewater treatment in Beijing. Pol. J. Environ. Stud. 24, 917–921.

Zhang, C.H., Peng, Y., Ning, K., Niu, X.M., Tan, S.H., Su, P.D., 2014. Remediation of perfluoroalkyl substances in landfill leachates by electrocoagulation. Clean. – Soil, Air, Water. 42, 1740–1743.

Zhi, W., Ge, Z., He, Z., Zhang, H., 2014. Methods for understanding microbial community structures and functions in microbial fuel cells: A review. Bioresour Technol. 171, 461–468.

7 Constructed Wetland-Bioelectrochemical Oxidation Systems

A Hybrid System for Wastewater Treatment

Prasann Kumar, Priyanka Devi, and Joginder Singh Panwar

7.1 INTRODUCTION

Since their initial large-scale implementation in the late 1960s, constructed wetlands (CWs) have seen extensive use. Due to their low cost and ease of operation and maintenance, CWs have seen widespread use in the last several decades to clean up anything from household sewage to industrial effluent, agricultural wastewater to mine drainage, landfill leachate to urban runoff, and dirty river water (Wu et al., 2015). It also provides a safe and effective method of therapy for those living in rural and underdeveloped locations. However, CWs have a few flaws that might restrict their usefulness and longevity. Substrate clogging is a major challenge when using CWs for wastewater treatment with high organic and SS input rates (Ruiz et al., 2010). Additionally, with high nitrogen loading rates, CWs may be nitrification restricted due to poor oxygen transfer or denitrification limited due to low levels of accessible organics. Additionally, several stubborn contaminants and heavy metals in industrial wastewater might hinder the efficiency of CWs (Wu et al., 2014). Despite advancements in design and operational strategies and the use of intensified systems like flow direction reciprocation, artificial aeration, tidal flow CWs, etc., CW systems operating as standalone technologies are sometimes unable to meet the requirements of these new guidelines due to the deteriorating environment and stringent discharge standards, including the emphasis on effluent reuse. Therefore, in recent years, there has been a rise in the practice of combining or integrating CWs with other existing or emerging technologies to maximize the individual advantages in terms of wastewater treatment. These technologies include membrane bio-reactors (MBRs), electrochemical oxidation (EO), microbial fuel cells (MFCs), etc. It is well established in the literature that these methods have proved effective in the treatment of particular contaminants and/or as green processes for energy recovery, albeit still having significant limits (Dalmau et al., 2015).

DOI: 10.1201/9781003368472-7

Over the past decade, the prevalence of POPs in water has become a major environmental issue. The presence of these pollutants, even at low concentrations, generates harmful consequences due to their toxicity, carcinogenicity, and mutagenicity. They are often discovered at trace concentration levels (i.e., ng up to mg) in water bodies (UNESCO 2012). These contaminants are notoriously difficult to eliminate using standard wastewater treatment methods (Subedi and Kannan, 2015). Therefore, effective and innovative solutions for POP removal are needed so that vital water supplies can be restored and reused. Electrochemical advanced oxidation processes (EAOPs) have been singled out as a promising subset of alternative water treatment technologies that might help mitigate this environmental threat. Since EAOPs have the potential to mineralize extremely recalcitrant organic contaminants such as medicines (Cavalcanti et al., 2013), insecticides, azo dyes, and even carboxylic acids, they have garnered growing attention (Garcia-Segura et al., 2016). Not only do EAOPs effectively remove POPs from the environment but they also exhibit several environmentally significant features, including (i) mild operation conditions under ambient temperature and pressure, (ii) compact reactors of smaller physical footprint that require lesser land space, (iii) no additional requirement of auxiliary chemicals, which eliminates the need for transportation and storage of these chemicals, and (iv) no production of secondary waste streams. With these positive qualities, EAOPs are a carbon-light technology that benefits the environment. Due to its adaptability and scalability, anodic oxidation (AO), or EO, is the most researched electrochemical process among the EAOPs (Scialdone et al., 2011). Only a small number of research have focused on synthetic water matrices and actual wastewater effluents, whereas the great majority of experimental investigations on EO and other EAOPs deal with the oxidation of POPs in synthetic wastewater. The basics of EO technology are briefly covered so that readers may better understand its functioning and its ramifications, the foundation for environmental remediation of POPs, and the potential conjunction with other water technologies employed as pre- or post-treatment (Xu et al., 2017).

Because ensuring safe supplies of water, energy, and the natural environment has a direct impact on people's quality of life, the interdependence of these three sectors is sometimes referred to as the "water-energy-environment nexus" (WEEN). In recent years, several ideas have been proposed all across the world for building nexus-based management techniques and methodologies (Yazdandoost and Yazdani, 2019). For sustainable development to be achieved, progress must be made in the development of technologies that effectively weave together water, energy, and the environment. In this sense, wetlands may be viewed as a nexus approach, as they generate and govern features while attenuating consequences in numerous ways. A sustainable and cost-effective wastewater treatment method, an engineered treatment wetland (ETW) CW or reed bed treatment system, is a ground equivalent to a natural wetland (Vymazal, 2010). A better investigation of CW technology is warranted in light of the pressing necessity to concentrate on cutting-edge technologies and innovations for WEEN's security. Sustainable Development Goal (SDG) 6 of the United Nations (UN) calls for a greater focus on reclaimed water, greater recycling, and a worldwide decrease in wastewater production (Vymazal, 2010). This fascination is related to the UN "Nature-based solutions" for bettering water quality while also generating extra benefits and the "Water Action Decade 2018-2028," which aims to organize action on water reuse worldwide. For the treatment of conventional contaminants in a variety of wastewaters (municipal, industrial, farm, and agro-industrial), CW has matured as an environmentally preferable alternative to conventional energy and/or chemical-intensive treatment systems in a variety of climates and environmental settings (Pardee et al., 2021). Physically, CW is an artificial wetland, with substrate, biofilm, and macrophyte as the main components that play a crucial role in degrading pollutants in wastewater. Researchers have noticed that until around halfway through this decade, most studies on CW focused on technical questions including flow pattern, design, pollutant removal efficiency, and associated mechanisms (Kataki et al., 2021). However, researchers have recently been interested in improved CW designs with more competent qualities to promote technological efficiency in terms of treatment efficacy, cost-effectiveness, and energy results. Our investigation into the topic's history revealed that researchers have focused their attention on improving the CW system's performance by expanding its design requirements and increasing its efficiency.

Reflecting the main trends in the broader water sector, the CW community is beginning to address innovative paradigms such as the potential for energy harvesting from the technology in addition to nutrient removal, the solution to the problem of temperature inhibition on its performance, and the environmental competence of the technology compared to other parallel treatment technologies. Improved design components are being included in inefficient and complex treatment technologies like CW in response to the pressing need to address the interconnected issues of water, energy, and the environment. This is why recent studies have placed such tremendous emphasis on integrating CW with MFCs, a highly developed and efficient design aspect. Intending to increase the wastewater treatment capacity of wetlands while also providing electrical power, CW-linked MFC (CWMFC) has arisen in recent years (Guadarrama-P'erez et al., 2019). Even though there is evidence that incorporating MFC technology into a CW may increase treatment efficiency by 27–49%, there are still several obstacles that must be overcome before the technology can be effectively used at a large scale. In the past, reviews of CW-MFC have mainly provided an overview of the technology's operation, different configurations and mechanisms of action and its advantages, elimination of conventional pollutants, and its potential for bioelectricity production (Guadarrama-P'erez et al., 2019). The effectiveness of CW-MFC in removing emerging contaminants is evaluated, as is the impact of macrophytes on the system's performance and microbial profile. There is also significant room for comprehensive compilation and critical analysis of the information regarding its treatment efficiency in comparison to standard CW, improving cathodic and anodic efficiency for higher nutrient removal and bioelectricity generation. Competence of the technology in climatically difficult conditions and the essential strategy for efficiency development are two major issues surrounding CW that are receiving more and more attention (Varma et al., 2020). Since the effectiveness of wastewater treatment in CW is directly related to local weather patterns, it stands to reason that the technology for treating wastewater in chilly climates has not spread as quickly as it could in warmer regions. Many thousands of studies have been conducted over the past three decades on the topic of wastewater treatment utilizing CWs, but only a small fraction of these have addressed the challenges and concerns of a cold environment and the essential strategy for dealing with them. It has been noted that the scattered nature of the few published accounts of seasonal changes in CW performance, difficulties encountered by CW in cold temperatures, and the several methods for regaining lost treatment efficiency. This calls for a thorough analysis to guide future actions and research towards achieving sufficient CW performance at low temperatures for further technological penetration in harsh climates. Our research shows that there is growing interest in using Life Cycle Assessment (LCA) and other systematic methods to compare and contrast competing technologies that have comparable life cycle consequences (Fuchs et al., 2011). Considering CW technology's widespread application, life cycle assessment (LCA) methods would help determine which treatment strategies would have the least negative impact on consumers and the environment over time. In light of the lack of a comprehensive compilation of research on the comparative life cycle impact of the technology and conventional treatment units using standard LCA, the present work also provided a comprehensive review of concise information on the state of the art of knowledge. With so much work being done on CW-MFC, but so little information readily available on its environmental footprint in comparison to other related technologies as assessed by the LCA approach, there is an urgent need to compile the state of the art. Improvements in these areas would make CW a net energy generator, improve the system's ability to withstand extreme weather, and provide information on whether or not the technology is preferred over more conventional treatment methods. Finally, we've tried to point out several research gaps in linked CWMFC, cold temperature-operated CW and LCA-based CW studies that are worth exploring to go to the next level of CW performance. Scientific direction for making well-informed judgements concerning CW's practical implementation and R&D effort would be provided by the future views and difficulties mentioned in this part (Jenssen et al., 2005) (Table 7.1).

Environmental engineering, electrochemistry, biochemistry, and physics all have a stake in the emerging discipline of bioelectrochemistry, which combines microbiology with electrochemistry. Bioelectrochemical systems (BESs) encompass the research and application of a wide variety of subfields, including but not limited to microbial, enzyme, protein, DNA, and neuro-electrochemistry. Power production, chemical synthesis, and environmental services including soil bioremediation, desalination, and wastewater treatment are all examples of the kinds of goods that BES has been developed to supply (Arends and Verstraete, 2015). In a bacterial electron transport system (BES), bacteria interact with insoluble electron donors and acceptors, transferring metabolic electrons from the donor to the acceptor (electrode) via an electroconductive substance (Rabaey et al., 2007). Early experiments in microbial electrochemistry can be traced back to the first half of the twentieth century, when Potter discovered that certain species of bacteria, *Escherichia coli*, could generate electricity through substrate oxidation processes. Later, Hooker attempted to use the oxidation-reduction reactions to create an electrical cell (Desloover et al., 2012). Different names have been developed to explain their concepts and applications due to their quick evolution and multidisciplinary approach. Schröder distinguished between microbial electrochemistry, microbial electrocatalysis, and microbial electrosynthesis. The initial attempts to use oxidation-reduction reactions to produce an electrical cell can be traced back to the first half of the twentieth century with the discovery that some species of bacteria, such as *Escherichia coli*, could generate

TABLE 7.1
Comparative Performance of CW Concerning Other Water Treatment Technology through LCA Analysis

S.No	Different Methods of Constructed Wetland Comparison	Category Impact	Boundary System	Environmental Remarks	Reference
1	Plant that uses both CW and biofiltration	Exertion, Carbon Dioxide Emissions, Area Utilization, and Particulate Emissions	Phases of Production, Assembly, and Use	Comparable energy use over the course of a product's entire life cycle. The majority of energy consumption and carbon dioxide emissions come from the transportation sector, while CW has the lowest CO_2 emission and may lessen its environmental impact by recycling the soil it excavates.	Dixon et al., 2003
2	Membrane bio (MB) and chemical (MC) reactors Green roof water recycling system (GROW)	Eutrophication and Acidification	Different operation and construction	Both GROW and CW have minimal ecological impact; however, GROW is noticeably less so. MCR, on the other hand, has the most effect. 80% of the entire effect of CW occurs during the construction phase.	Memon et al., 2007
3	The three most common methods of transporting sludge from a STW to a WTP (AST with aeration) are: (i) direct land application; (ii) compost post-treatment; and (iii) centrifugation with compost post-treatment	Eutrophication	Treatment and transportation of sludge (not including the water treatment line)	STWs are most beneficial when used in conjunction with direct land application; STWs are less favourable when compost post-treatment is necessary, such as using mechanical dewatering processes, due to the impact associated with sludge transport; and Gases emissions from STWs are negligible.	Uggetti et al., 2010

(Continued)

TABLE 7.1 *(Continued)*
Comparative Performance of CW Concerning Other Water Treatment Technology through LCA Analysis

S.No	Different Methods of Constructed Wetland Comparison	Category Impact	Boundary System	Environmental Remarks	Reference
4	Slow rate infiltration using coagulant-waste (CW) and activated sludge (AST) (SRI)	Eutrophication and Acidification	Production, manufacture, assembly, use, maintenance, and eventual disposal of components	Due to its lower global warming potential (GWP) from carbon sequestration, negative carbon balance, and lower aquatic toxicity and eutrophication (compared to AST), CW is better suited to rural settings. A 10% increase in CW's operational lifetime can reduce CO_2 emissions by 1% and abiotic depletion by 5%. Lessening of the energy needs of CW and SRI. The CW phase is the most consequential stage of demolition.	Machado et al., 2007
5	CW and Vegetated land treatment (VLT) Activated sludge treatment	Acid Rain, Eutrophication, and Climate Change	Construction and operation phase	AST with the P-removal option has the most consequences, with the exception of eutrophication. Small towns should prioritize VLT and CW since they are the least harmful to the environment. Because CW uses an impermeable liner, it is more practical than VLT in areas with greater groundwater susceptibility. VLT is lower because plants absorb more CO_2.	Yildirim et al., 2012
6	Constructed wetland and Sequencing batch reactor (SBR)	Acidification potential	A research project that spans from beginning to end	Since CW uses less energy and materials, it has a less overall impact on the environment. Despite popular belief, CW actually has a smaller footprint on the land it occupies.	Dimuro et al., 2014

(Continued)

TABLE 7.1 *(Continued)*
Comparative Performance of CW Concerning Other Water Treatment Technology through LCA Analysis

S.No	Different Methods of Constructed Wetland Comparison	Category Impact	Boundary System	Environmental Remarks	Reference
7	CW and Municipal wastewater treatment plant (WWTP)	Greenhouse gases	Use of combustion by-products in plant operations	The total environmental effect of CW is smaller than that of WTP because: The total environmental effect of CW is smaller than that of WTP because: CW emits 38.34 gCO_2e per tonne of influent wastewater compared to 60.63 gCO_2e in WTP (the primary difference is the low energy consumption of CW).	Huang et al., 2016
8	Combination centrifuge and CW sludge treatment, sludge treatment followed by stockpiling	Ecotoxicity	Weed control, fertilizer replacement, water reuse, and sewage treatment	Equal to or cheaper in cost than mechanical treatment (greater N emission on soil was observed after mechanically treating sludge, likely due to increased N retention). Adding post-treatment on a stockpile area to CW does not result in significant environmental benefits. CW was more effective in removing carbon and nitrogen than other methods. In terms of human toxicity and ecotoxicity, the effects of the three different treatment scenarios were roughly the same.	Larsen et al., 2018

electricity through substrate oxidation processes. As a result of their dynamic nature and the interdisciplinarity of their growth, several names have been coined to characterize their underlying concepts and practical implementations. Microbial electrochemistry, microbial electrocatalysis, and microbial electrosynthesis are all unique concepts, as outlined by Schröder et al. (2015).

Microbial electrochemical technologies refer to the use of electrochemical processes powered by microorganisms to create chemicals, energy, and environmental remedies (MET). Primary and secondary METs are defined by the operational circumstances in which they function. Direct extracellular electron transfer (EET) routes from cell to acceptor or electron shuttle mediation are required for primary MET in microbial electrochemical activities (exclusively Faraday processes). However, in secondary MET, such as in microbial electrolysis of hydrogen or soil remediation, the bioelectrochemical processes are regulated by the modification of the microbial ambient conditions (pH, oxygen pressure, metabolite concentrations, etc.) (Schröder et al., 2015). MET can be roughly divided into power producers (where electrons are generated from oxidized organic matter and transferred to the cathode via an external circuit), power consumers (where external power is required to achieve bioelectrochemical cathodic reactions because small or negative potential differences prevent electron flow from anode to cathode), and intermediate systems (which neither produce nor consume power but require stable electrochemical conditions). Microbial electrolysis cells (MECs) synthesize H2 or other compounds by using external power to reduce cathode potential; microbial desalination cells (MDCs) purify water; microbial remediation cells (MRCs) cathodically reduce oxidized pollutants like uranium, perchlorate, and chlorinated solvents in polluted environments; and MFCs generate electricity. Microbial electroremediating cells (MERCs) are another type of MET system that aims to optimize the biodegradation of contaminants such as the herbicides isoproturon and atrazine by overcoming electron acceptor constraints. Since 2012, a variety of research has investigated the technological capabilities and benefits of integrating MET with built wetlands to clean wastewater and generate electricity at the same time (CW). The presence of plants and microorganisms, as well as the interplay of physical, chemical, and biological processes, and various removal techniques, are the backbone of a CW, which is a biologically engineered wastewater treatment system (Kadlec and Wallace, 2009). The prominent macrophytes in a CW may be used to determine its hydrology (surface vs. subsurface flow) and flow direction (horizontal vs. vertical) (Vymazal, 2010). The effectiveness of CWs in treating the home, industrial, drainage mining, runoff, and agricultural effluents has been extensively studied (Paredes et al., 2007). CWs are widely utilized as a mature option for decentralized wastewater treatment because they are reliable, cost-effective, and need little in the way of operation and maintenance. However, compared to other compact wastewater treatment technologies, the land footprint of CW implementation is substantially bigger. Therefore, CW has developed from passive to intensified systems, with some relatively modern designs combining CW with MET, to lessen the surface requirements (Yadav et al., 2012). Therefore, this study examines the foundations and current practices of MET as it relates to wastewater treatment, with an eye towards the possible gains, difficulties, and unique configurations that may come from including CW.

7.2 MICROBIAL EXTRACELLULAR ELECTRON TRANSFER (EET)

Synergistic consortiums of fermentative and bioelectronic bacteria can form in anaerobic settings in aqueous solutions thanks to a process called microbial EET (Reimers et al., 2001). Bioelectronic microbes can readily oxidize the simpler structures produced by fermentative microorganisms, such as acetate, ethanol, glucose, hydrogen gas, amino acids, and polymers (polysaccharides, proteins, and cellulose) (Logan and Rabaey, 2012). Electroactive microorganisms can get their energy from the transfer of electrons to a terminal electron acceptor or from an electron donor that is extracellular and conductive but insoluble due to their metabolic makeup. In these ideal circumstances, microorganisms may develop colonies, which eventually form electrochemically competent biofilms on solid-state electrodes (Rabaey et al., 2007). It is well established that electron transfer is made possible by the formation of electroactive biofilms. However, research is still needed to fully understand the kinetics involved in the transmission of electron acceptors to electroactive microorganisms. Direct extracellular electron transfer (DEET) and mediated extracellular electron transfer (MET) are the two primary mechanisms by which bioelectrochemical microorganisms execute EET, with the former being affected by the potential difference between the final electron carrier and the anode (Mao and Verwoerd, 2013).

7.2.1 Electron Transport from the Extracellular Space to the Cell

The bacterium and electrode must come into direct contact for DEET to work; typically, the bacteria will cling to the electrode, creating a biofilm. Therefore, the catalysis is constrained by the maximal cell density in this bacterial monolayer (Mao and Verwoerd, 2013), suggesting that only bacteria in the first monolayer at the anode surface are electrochemically active. Nanowires and pili are used by certain species to access and use far-off, insoluble electron acceptors and to interconnect internal biofilm layers. Nanowires, also called electroconductive pili, are vesicular extensions of a cell's periplasm and outer membrane (2–3 m in length) that are optimized for direct electron transmission between cells and electron acceptors (Kracke et al., 2015). This progress allows for the formation of thick electroactive layers and the establishment of contact with far-off electron acceptors as a direct response to the scarcity of these resources. In the case of Geobacter and Shewanella, this tactic has been seen.

7.3 DIRECT EXTRACELLULAR ELECTRON TRANSFER (DIET)

By exchanging electrons across species boundaries, many different kinds of microbial communities may harvest energy from processes that no one organism could catalyze on its own. This process occurs during syntrophic metabolism and involves the transfer of electrons between different organisms. Electrically conducting aggregates are formed by anode biofilms (Malvankar et al., 2012). In mineral-mediated DIET, nano-mineral particles or conductive surfaces like activated carbon granules, coke, or biochar are used by several organisms as electron conduits (Liu et al., 2012).

7.4 EXTRACELLULAR ELECTRON TRANSFER THROUGH A MEDIATING PROTEIN (MEET)

Bacteria that lack the necessary metabolic processes for DEET or DIET have created a third option, MEET, which involves EET mediated by electron shuttles. Flavins and phenazine compounds are two examples of endogenous redox-active chemicals that some bacteria, including *Escherichia coli*, Pseudomonas, Proteus, and Bacillus, may produce and excrete (Erable et al., 2010). Mediating compounds can also originate from the outside world, such as synthetic chemicals or naturally occurring humic substances (Voordeckers et al., 2010). Both the inside and exterior of the bacterial cell can provide electrons for the mediators in their oxidized state, allowing them to cycle between reduced and oxidized states. They should be able to be transferred easily and have a positive redox potential without disrupting cellular metabolism (Patil et al., 2012). The lack of transparency in the stoichiometry of converting electron donors into electricity is a drawback of this method. Energy production may be negatively impacted if fermentation processes continue, leading to methanogenesis activities, which would decrease electron transport to the electrodes (Bonanni et al., 2012). Constant addition is required, the chemicals are unstable, and they may be hazardous to the environment, all of which restrict their usefulness in large-scale MET applications. Secondary metabolites, such as phenazines, pyocyanine, ACNQ (2-amino-3-carboxy-1, 4-naphthol-quinone), and flavins, are produced by MEET-dependent bacteria and are responsible for the transport of MEET's low-molecular weight cargo (Voordeckers et al., 2010). Their synthesis is energy expensive, thus they'll only work in tight conditions with minimum substrate renewal, like batch setups, despite the possibility of reuse as an electron carrier (Lovley, 2006). Secondary chemicals, however, will be used in dynamic settings like wastewater treatment. Pseudomonas, Shewanella putrefaciens, and Geothrix fermentans are all examples of bacteria that are capable of producing their mediators to raise the EET rate (Kato, 2015).

7.5 MEDIATED EXTRACELLULAR ELECTRON TRANSFER (MEET)

DIET has created metabolic processes mediated by electron shuttles to enable extracellular electron transfer; this process is known as MEET. Flavins and phenazine compounds are two examples of endogenous redox-active chemicals that some bacteria, including *Escherichia coli*, Pseudomonas, Proteus, and Bacillus, may produce and excrete (Erable et al., 2010). Mediating compounds can also originate from the outside world, such as synthetic chemicals or naturally occurring humic substances (Kotloski and Gralnick, 2013). Both the inside and exterior of the bacterial cell can provide electrons for the mediators in their oxidized state, allowing them to cycle between reduced and oxidized states. They should be able to be transferred easily and have a positive redox potential without disrupting cellular metabolism (Patil et al., 2012). The lack of transparency in the stoichiometry of converting electron donors into electricity is a drawback of this method. Energy production may be negatively impacted if fermentation processes continue, leading to methanogenesis activities, which would decrease electron transport to the electrodes (Lovley, 2006). Constant addition is required, the chemicals are unstable, and they may be hazardous to the environment, all of which restrict their usefulness in large-scale MET applications. Low-molecular-weight molecules used for transport, including phenazines, pyocyanine, ACNQ (2-amino-3-carboxy-1, 4-naphthol-quinone), and flavins, are produced by bacteria and are essential for MEET. Their synthesis is energy-intensive, thus they'll only work in tight conditions with minimum substrate change, like batch setups, despite the possibility of reuse as an electron carrier. Secondary chemicals, however, will be used in dynamic settings like wastewater treatment. Pseudomonas, Shewanella putrefaciens, and Geothrix fermentans are all examples of bacteria capable of producing their mediators to raise the EET rate (Hernandez et al., 2004).

7.6 WASTEWATER TREATMENT USING MICROBIAL ELECTROCHEMICAL TECHNOLOGIES

Wastewaters store a significant amount of potential energy in the form of biodegradable organic materials. Although a clear relationship between wastewater energy and COD levels does not exist, estimates suggest that wastewaters contain between 17.7 and 28.7 kJ gl COD (Heidrich et al., 2011). It is estimated that wastewater treatment plants in the United States use around 15 GW of electricity (3% of national power generation), whereas 17 GW can be contained in wastewater of varying origins. Due to their high COD removal rates (up to 90%) and high coulombic efficiency (per cent of electrons recovered as current vs. maximal recovery), MET has shown promise in wastewater treatment (Rahimnejad et al., 2015). Since it is estimated that 1 kg of COD can be converted to 4.16 kWh of power in anaerobic digestion (1 kWh of usable energy in the form of electricity), a MET must achieve a substrate conversion rate comparable to anaerobic digestion if it is to compete as an alternative wastewater treatment technology with simultaneous energy generation. There is no way for a MET to create affordable energy at this time. However, it represents a new method of directly extracting energy from wastewater (Corbella et al., 2015). Installation costs, pricey electrode materials, and poor energy density generated are only a few of the drawbacks of METs that prevent them from being widely used in the field (Zhang et al., 2016). MET systems are a promising new option for treating wastewater since they can efficiently oxidize biodegradable organic materials. MET are adaptable platforms that treat wastewater at a rate of 7.1 kg COD m^3 (reactor volume day 1) using oxidation and reduction processes, with the added benefit of reducing expenses associated with aeration and sludge disposal (Wang and Ren, 2013). However, it is suggested that MET be supplemented with other methods to yield fermented products that can be oxidized by electroactive bacteria in high organic loaded/complex effluent (Rosenbaum et al., 2010).

7.7 METHODS AND CUTTING-EDGE FACILITIES FOR WASTEWATER TREATMENT

By connecting a microbial anode to a cathode that catalyzes a reduction process, bioelectrochemical wastewater treatment is possible. When the anode and cathode are connected, electrons can more easily flow from one to the other (Rozendal et al., 2008). Since many reactions may be catalyzed by EAB, METs have been implemented for wastewater treatment, and the MET operation can be divided into two categories: systems based on non-spontaneous reactions and spontaneous reactions.

7.8 NON-SPONTANEOUS REACTION SYSTEMS

When the Gibbs free energy of the reaction is positive and the theoretical cell voltage or electromotive force is negative, we have a non-spontaneous reaction that requires an external source of energy to proceed. Some METs, like the MEC and fluidized systems, don't rely on spontaneous responses. The term "microbial electrolysis cell" is used to describe a system in which electricity is employed to raise the potential difference between the anode and the cathode to facilitate or accelerate the electrode reactions (MEC). While releasing protons into the solution, electroactive microorganisms in a typical MEC oxidize organic waste substrates to carbon dioxide at the anode. Protons diffuse to the cathode over a proton-exchange membrane located between the two electrode compartments, while electrons travel from the anode to the cathode via an external circuit. A desired product is formed when electrons unite with a soluble electron acceptor at the cathode in the presence of an appropriate (bio) catalyst (Rabaey and Rozendal., 2010). To reduce substrates in wastewater, such as NO_2, NO_3, SO_4, $C_2H_2Cl_4$, and others, electroactive bacteria can also employ a cathode as the electron provider (Clauwaert et al., 2007). It is necessary to supplement the potential created by substrate oxidation at the anode with an external power supply for these processes to be thermodynamically viable at the cathodic electrode. Galvanostatic (based on input current flow) or potentiostatic mode of operation selects the source of energy (based on a fixed potential between two electrodes). While just two electrodes are required for operation in galvanostatic mode, a third is required when it is desired to keep one electrode at a voltage below a chosen value (the working electrode). The counter, or auxiliary electrode, is the one whose potential is determined by the amount of current in the circuit. Crucial to the investigation of the microbial-electrode interaction, this setup permits regulation of the anodic or cathodic processes. In a three-electrode setup, the anode is the working electrode, where organic matter is oxidized when the anode voltage is regulated. For this reason, engineers have devised several devices, including the basic MEC layout, which features a two-chamber structure divided by an ion exchange membrane. For the removal of pollutants from fluids via reduction processes, such as SO_4 from groundwater (Coma et al., 2013) and N from low COD effluents, a fixed cathode potential is used (Tong and He, 2013). Additionally, more manageable methods have been created for use in wastewater treatment facilities. An exemplary MET system uses bioelectrochemistry for denitrification. These setups eliminate the need for a potentiostat by directly connecting to a power source, eliminating the need for a reference electrode and ion exchange membranes. In comparison to traditional heterotrophic denitrification, bioelectrochemical denitrification has several benefits, such as the elimination of organic matter addition and the use of an endless electron supply provided by a cathode (through electric flux). Microbial electrochemical fluidized bed reactors (ME-FBRs) are another device that relies on non-spontaneous reactions and combines a traditional fluidized reactor with a MET (Tejedor-Sanz et al., 2017). The ME-FBR was created to increase the amount of exposed anode surface area available to electroactive microorganisms while also enhancing the kinetics of the catalysis by using a fluid with advantageous mixing qualities. Brewery effluent from these systems has been studied, and it has been shown that the ME-FBR can remove up to 95% of the COD.

7.9 SPONTANEOUS REACTION SYSTEMS

7.9.1 Methods of Producing Electricity

To date, the most researched application of MET is in the generation of electric power via MFC. Microorganisms use the anode as an electron acceptor to oxidize organic compounds. For oxygen reduction, electrons leave the anode and travel through an external circuit to the cathode (Logan et al., 2006). Since the overall reaction, organic oxidation, and oxygen reduction are thermodynamically favourable, electrical energy can be recovered from the external circuit (Modin and Gustavsson, 2014). In most cases, MFCs can be broken down into two distinct types: dual- and single-chambered layouts (Figure 7.1). To facilitate either oxidative or reductive conditions, a dual-chamber setup typically immerses the anode and cathode in a suitable solution. An ion exchange membrane channels the proton current, while an external circuit transports the electrons. In a single container, both the anode and the cathode may be submerged in the same electrolyte, or the cathode may be exposed to air. In this setup, protons pass through a polymer electrolyte membrane en route to the cathode, while electrons travel in an external circuit. The following are descriptions of some MFC developments in wastewater treatment and power generation. A simple-to-install-and-operate floating all-in-one MFC system is proposed based on the principles of generating conditions for dissolved oxygen gradient by exposing the cathode to the atmosphere. Carbon-cloth or carbon-brush anodes are attached to the same floater via a stainless-steel wire, and an external stainless-steel wire and a resistor round out the circuit. Because of its design, this system can be plugged in and used immediately in wastewater pre-treatment facilities (e.g., settling tanks). The annular single-chamber MFC, also known as the spiral microbial electrochemical cell, is the result of research into new electrode designs and electrode material combinations that improve MFC performance (Mardanpour et al., 2012). The anode electrode in this type of MFC reactor is a stainless-steel mesh spiral coated with graphite, and the cathode is either carbon cloth or stainless-steel mesh that has been treated with Pt (0.5 mg cm^2) to improve its electro-conductive qualities. This helical arrangement minimizes the distance between electrodes while maximizing the anodic surface area. These features allowed for impressive wastewater treatment outcomes, with COD removal rates of up to 91% and a maximum power density of 20 W m^3 (anode working volume), and it has the potential to be scaled up due to the relatively inexpensive cost of the applied materials. So-called tubular MFCs are another novel take on the standard MFC layout. One variant of this setup is a vertical flow reactor made from two polypropylene tubes that operate as a double shell, a layer of carbon cloth cathode that is exposed to air, an intermediate layer made of hydrogel, and an ion exchange membrane within. A concentric monolithic-activated carbon anode is put within this reactor, and an external circuit with a resistor (1000 at start-up, 150 under normal operating circumstances) connects the anode to the cathode (Kim et al., 2009). Energy generation up to 1.75 Wh g^1 COD was achieved in studies using two tubular MFC units supplied with synthetic wastewater and at varying organic loading rates. These findings show promise for using this reactor at larger scales in complementing modular systems to refine effluents from treatment units like anaerobic digesters (Kim et al., 2010). Another comparable device, the MFC stack, was created by employing PVC tubing as

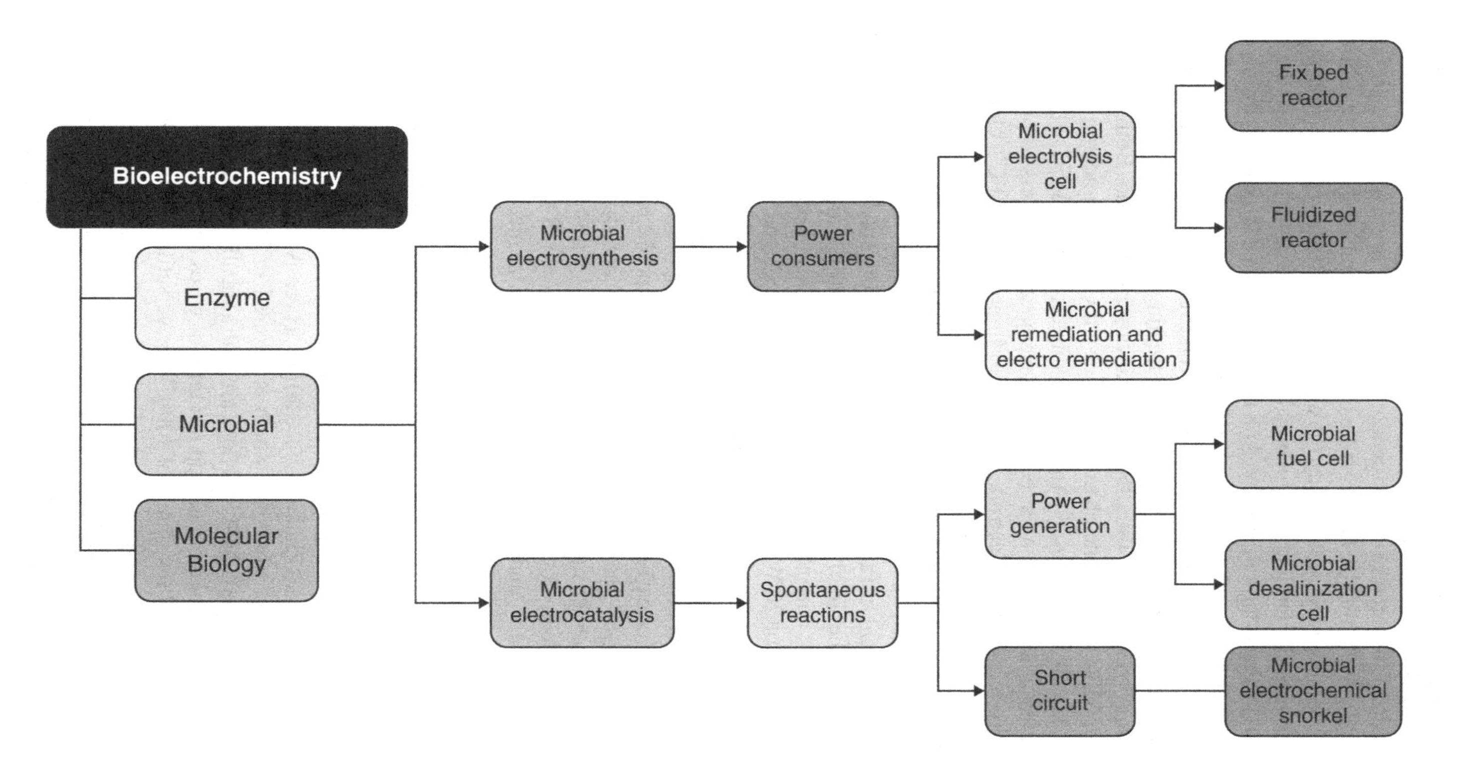

FIGURE 7.1 Microbial electrochemical technologies, including the subdisciplines of biochemistry and their connections to bioelectrochemical systems.

the foundation for a modular tubular air-cathode system that was then subdivided into separate anodic chambers. There is a cation exchange membrane within each anodic chamber, which is covered in MnO_2-containing carbon fibre fabric. The air-cathode MFC parts are hydraulically connected using silicone gel tubes, and the circuit is made up of titanium wires connecting the anode and cathode with an external 1000 resistor. This versatile modular design may be used in either series or parallel circuit operation, providing added flexibility. At a loading rate of 1.2 kg COD m^3 d^1, with a maximum power density of 176 W m^2 at 0.38 V and 43 mA, it has been reported that 84% of the COD and 91% of the NH_4 may be removed. These numbers point to the possibility of generating electricity while also treating wastewater. Circuit Breakers Figure 6. Some typical MET installations for sewage purification. Both the standard two-chamber MFC and the one-chamber floating MFC reactor undergo spontaneous reactions. R stands for "resistor." An annular single-chamber MFC, also known as a spiral microbial electrochemical cell, was developed via efforts to produce novel forms and mixtures of the electrode material to improve an MFC's efficiency. The anode electrode in this MFC reactor is made from a stainless steel mesh that has been coated with graphite, while the cathode is made from either carbon cloth or stainless steel mesh that has been treated with Pt (0.5 mg cm^2) to improve its electro-conductive qualities. This helical arrangement minimizes the distance between electrodes while maximising the anodic surface area. These features allowed for impressive wastewater treatment outcomes, with COD removal rates of up to 91% and a maximum power density of 20 W m^3 (anode working volume), and it has the potential to be scaled up due to the relatively inexpensive cost of the applied materials. So-called tubular MFCs are another novel take on the standard MFC layout. One variant of this setup is a vertical flow reactor made from two polypropylene tubes that operate as a double shell, a layer of carbon cloth cathode that is exposed to air, an intermediate layer made of hydrogel, and an ion exchange membrane within. This reactor has a concentric monolithic-activated carbon anode mounted within its interior, with the anode and cathode being connected through an external circuit with a resistor (1000 during start-up, and 150 under normal operating circumstances) (Kim et al., 2009). Energy generation up to 1.75 Wh g^1 COD was achieved in studies using two tubular MFC units supplied with synthetic wastewater and at varying organic loading rates. These findings show promise for using this reactor at larger scales in complementing modular systems to refine effluents from treatment units like anaerobic digesters (Kim et al., 2010; Shah, 2020). For the same purpose, an MFC stack was designed utilizing PVC pipe as the framework for a modular tubular air-cathode system with separate anodic chambers (Zhuang et al., 2012). There is a cation exchange membrane within each anodic chamber, which is covered in MnO_2-containing carbon fibre fabric. The air-cathode MFC portions are hydraulically connected with silicone gel tubes, and the circuits are made up of titanium wires connecting the anode and cathode with an external resistance of 1000. This versatile modular design may be used in either series or parallel circuit operation, providing added flexibility. At a loading rate of 1.2 Kg COD m^3 d^1; peak power density 176 W m^2 at 0.38 V and 43 mA, it has been claimed that 84% of COD and 91% of NH_4 can be removed using such a system. These numbers point to the possibility of generating electricity while also treating wastewater.

7.9.2 Bioelectrochemical Technology

MFCs and MECs are the two primary categories of BESs (MECs). Creating electricity and cleaning up wastewater at the same time is a promising new use for them. The electrochemical processes in these systems are catalyzed by microorganisms. While MFCs use the chemical energy of organic material in wastewater to create electricity, MECs instead use external power to generate a product (such as hydrogen) at the cathode (Rozendal et al., 2008). Due to their lower sludge formation yield (2.4–26.5 times lower) compared to aerobic-activated sludge processes, BESs have seen extensive usage for the removal of organic pollutants (anodic oxidation) (Cheng, 2009). This is why biological eutrophication systems (BESs) are considered a green option for wastewater purification.

7.9.3 Nitrate Reduction through Bioelectrochemistry

Nitrate may be removed using BESs and a cathodic reduction method. Denitrifying microorganisms are introduced to an otherwise electrochemical nitrate reduction process to boost nitrate removal efficiency.

7.10 MECHANISMS FOR THE CATALYTIC NITRATE REDUCTION

Compared to heterotrophic denitrification, autotrophic denitrification provides greater efficiency in BESs. Hydrogen gas, created at the cathode surface by the electrolysis of water, is used by autotrophic denitrifying bacteria as an electron donor. Autohydrogenotrophs are the microorganisms responsible for denitrification in an autotrophic metabolic process (Ghafari et al., 2008). Nitrate is converted to nitrite (NO_2^-) by hydrogen gas, which then interacts with more hydrogen to produce nitric oxide (NO). This substance is further broken down into nitrous oxide (N_2O) and then nitrogen gas (N_2).

7.11 REMOVAL OF NITRATE CONTROLLING FACTORS

The use of BES to remove nitrate from actual aquaculture effluent is not widely documented in the literature. As a result, we've started using synthetic wastewater as a stand-in for the wastewater produced by aquaculture. Denitrification processes have had several key characteristics studied, including cathodic material, beginning pH, current density, and nitrate loading rate.

7.11.1 Cathodic Substance

Due to their high surface area, graphite granules have lately been implemented as a third bipolar electrode in wastewater treatment (Prosnansky et al., 2002). Traditional BESs typically employ an abiotic electrode material, such as graphite, carbon, platinum, or stainless steel. Standard materials like platinum and stainless steel are pricey and not sustainable when applied to antimicrobial treatments (Clauwaert et al., 2007; Shah, 2021). The creation of a layer of platinum oxide (PtO) on the cathodic electrode surface is another drawback of platinum electrodes (Du et al., 2007). Since they are both affordable and sustainable, biological electrodes have recently gained popularity. Microorganisms that oxidize hydrogen gas and organic substrate or decrease

TAN to nitrogen gas inhabit the anodic and cathodic compartments of traditional BESs. These microorganisms are employed as passive electrodes to take in electrons in a bio-electrode. According to Gregory et al. (2004), these androphilic bacteria can reduce nitrate at room temperature using an inert electrode to draw electrons from graphite electrodes. It was demonstrated by Clauwaert et al. (2007) that a bio-anode oxidizing acetate may be used in tandem with a bio-cathode reducing nitrate to nitrogen gas. Since graphite and carbon are both affordable and conducive to the growth of electrochemically active biofilms, they are widely used as bio-anodes and biocathodes. The high electrical resistance of these materials, however, results in greater electrode ohmic losses when the system is ramped up. Therefore, a conductive metal current collector, such as a stainless-steel mesh, is used to hold up graphite or carbon electrodes (Rozendal et al., 2008).

7.11.2 pH

For hydrogenotrophic denitrification to function well, wastewater pH is the most important aspect to consider (Wan et al., 2010). Denitrification operates best at a certain pH level, although that value varies with the specific cultures and working circumstances. Problems with nitrite build-up can arise at pH values over 8.6, whereas denitrification rates are slowed when pH drops below 7.0 due to the decomposition of carbonate ions. To neutralize the acidity of wastewater, carbon dioxide is often added to the system in the form of H_2CO_3 or HCO_3^- (Prosnansky et al., 2002; Shah, 2021). According to the findings of Clauwaert et al. (2009), the pH of anolyte and catholyte is modified by the addition of NaOH and HCl, respectively. Nitrate removal is just 26.3% without constant pH adjustment and increases to 74% when the pH is kept at 7.2. According to the authors, the pH range of 7–8 is ideal for denitrification. Wan et al. (2010) find that a pH range of 7.05–7.20 is optimal for denitrification, lending further credence to this number as an optimal pH range.

7.11.3 Current Density

Hydrogen generation at the cathode is influenced by the current density, which in turn affects the efficiency of nitrate reduction due to the importance of hydrogen supply in hydrogenotrophic denitrification. Nitrate reduction efficiency is improved from 72% to 74% by increasing the input current density from 21.1 mA/cm^2 to 23.4 mA/cm^2 (Clauwaert et al., 2009). Similar trends may be seen in the results obtained by Sakakibara and Nakayama, who found that an increase in current density input from 0.46 mA/cm^2 to 0.82 mA/cm^2 increased average current-denitrification efficiency from 61% to 70%. To get rid of nitrate in manmade water, Wan et al. (2010) employ a membrane and bioelectrochemical process. After isolating the anodic and cathodic halves of the cell using a proton exchange membrane, the cathodic half is seeded with autotrophic denitrifying bacteria. Since the hydrogen gas produced by water electrolysis is saturated in the cathodic half-cell, the scientists observed that a current input of 30–100 mA does not affect the denitrification rates. Accordingly, the denitrification process is unaffected by the rate at which hydrogen is supplied. When the current is 15 mA, the rate of nitrate removal is 42.5 per cent; however, the hydrogen gas produced is not sufficient for denitrification.

7.11.4 Amount of Nitrate Loading Rate

Clauwaert et al. (2009) conducted an experiment that demonstrated the relationship between the nitrate loading rate and the nitrate removal rate, up to a certain optimal value. The removal rate of nitrate was compared for three different starting nitrate-nitrogen concentrations (22.13, 15.76, and 11.31 mg/L NO_3—N) using a proton exchange membrane electrodialysis cell connected with autotrophic microorganisms (Wan et al., 2010). Since the amount of hydrogen created by the cathode was adequate for nitrate reduction, the findings followed a similar pattern to that described by Clauwaert et al. (2009). The reduction rate of nitrate was indirectly influenced by hydrogen.

7.11.5 TOC Oxidation via Bioelectrochemistry

Microbial oxidation of all organic molecules yields carbon dioxide and protons (Yu et al., 2007). When growth conditions, including temperature and organic carbon supply, are kept essentially consistent, it is considered that microbial growth is stable. It is assumed that the chemical compounds used in synthetic wastewater are similar to those found in real aquaculture wastewater because there are few publications regarding the application of BESs for the removal of TOC in actual wastewater.

7.12 INFLUENCING VARIABLES IN TOTAL ORGANIC CARBON REMOVAL

Electrode material and pH are two major factors that affect bioelectrochemical TOC removal. In the parts that follow, we'll go further into these two aspects.

7.12.1 Electrode Substance

Activation polarization losses impact the performance of BESs; hence, using different electrode materials can provide varying results. Before electrons and ions may flow through an electrode, the interacting species must first overcome an energy barrier known as activation polarization. The materials used to make electrodes should be cheap, conducive to microbial growth, and resistant to corrosion and fouling. Using carbon as both an anode and cathode, as Virdis et al. (2008) noted, can result in the removal of organic matter at a rate of 93.59%.

7.12.2 pH

A decrease in BES efficiency may result from the acidification of the biofilm by the protons generated during the AO of TOC. When microbes break down organic matter, the pH drops, but after the organic matter is gone, the pH stays the same (Picioreanu et al., 2010). Both the oxygen reduction at the cathode in MFCs and the hydrogen production at the cathode in MECs are proton-consuming processes, leading to a rise in pH at the cathodic compartment. Because of this, both the rate at which TOC is oxidized and the total driving force (E.M.F.) of the BESs are diminished. While MEC operation requires higher energy input, MFC has lower electrical energy output (Sakakibara and Nakayama, 2001).

Methane is produced when an electron and hydrogen ion are reduced by bacteria at neutral pH, as discovered by Mohan et al. (2009). Methanogenic bacterial activity is most suppressed between 5.5 and 6.0 pH. While this may seem impossible, Virdis et al. (2008) report that at a pH of 7.0, organic matter removal can be as high as 93.59%. Both cell layout and wastewater recirculation rates play a role in explaining the divergent outcomes. This demonstrates that the removal of total organic carbon (TOC) from aquaculture wastewater is strongly influenced by the interplay between the initial pH of the electrolyte and other factors.

7.12.3 The Simultaneous Nitrate and Organics Removal

The simultaneous removal of nitrate and TOC from aquaculture effluent has gained considerable traction recently due to its economic effectiveness. The electro-Fenton and bioelectrochemical systems are prominent applications for this method. Using the Electro-Fenton technique, Virkutyte and Jegatheesan (2009) have zeroed down on eradicating organics and nitrate at the same time. Hydroxyl radicals eat away at organic contaminants, and nitrate is converted to nitrogen at the cathode. It has been observed that nitrate and organic matter may be removed with 94.8% and 97.3% efficiency, respectively. Virdis et al. (2008) employ MFCs to remove both nitrate and organic debris at the same time, and they find removal rates of 70% and 93%, respectively. The nitrate removal efficiency was found to be 97.3%, and organic matter removal efficiency was reported to be 98.8% in a research by Xie et al. (2011). Due to the development of hazardous intermediates, the use of Fenton's reagent in wastewater treatment is discouraged. Therefore, more study into reactor layouts for bioelectrochemical removal of nitrate and organics is needed. It will be important for future studies to use real-world aquaculture effluent in such setups.

7.13 POSSIBLE SOLUTIONS AND FUTURE DIRECTIONS FOR CW-MFC SYSTEMS

Most of the described systems have been created and operated at laboratory size, and their implementation as an appropriate real-scale system is still in the works, as shown by the studies that combine CW with MFC. There are many of the same obstacles to overcome while expanding the use of CW-MFC technology as there are when using conventional MET to treat wastewater. Electrolyte (ohmic) resistance, kinetic (slow response rates on electrodes), and transport (slow diffusion) resistance all contribute to this limitation (He et al., 2005). As a result of these constraints, achievable power densities are low (in the range of 9–72 mW m^2), and coulombic efficiencies are low (in the range of 0.05–10.48%). Other factors must be taken into account, such as (i) the internal resistance of a CW-MFC, which increases linearly with the size and distance between electrodes; (ii) the over-potential during activation and the insufficient electrical contact between bacteria and anode (Nitisoravut and Regmi, 2017); (iii) competition between EAB and other microorganisms (such as methanogenic bacteria) for electrons or substrates. Also limiting the free flow of ions and electrons in the system is the design of the circuits and the materials used for the membranes and electrodes (Logan et al., 2015). Therefore, work must be put into learning the processes of electron release, transfer, and acceptance, and reducing the losses that occur along the way (Gude, 2016). To design and develop cost-effective electrodes and circuits, more work must be put into determining optimal inoculums, substrate conditions, ionic strength of water, internal and external resistance of the systems, and inventive electrode spacing between them. The reviewed CW-MFC literature reveals a clear bias towards studies that improve power generation as a primary objective. However, the tested systems have not yet achieved competitive energy yields in comparison to other options, like biogas collection from conventional wastewater treatment facilities. Because of this, expanding the CW-MFC infrastructure is difficult. The study and development of systems based on the combination of CW and MET that do not aim to harvest energy but instead rely on the explained bioelectrochemical principles to improve the performance of CW is an approach that seems to be underappreciated as a research field. Combining CW with MEC or MRC, where potentials are set using an external power supply, allows for control of the system's internal conditions, allowing the lack of electron acceptors to be overcome and maximizing treatment efficiency, which is one option worth exploring. Researching CW-MET systems that use electro-conductive materials and function in short-circuit mode rather than electrodes and external circuits is another option. Among the potential benefits of the proposed alternatives is an increase in the efficiency of removal procedures within the CW, which in turn might reduce the amount of land needed to construct one.

Multiple aspects, including changes in the material separators, material electrodes, plant species, reactor architecture, and connections, must be considered to create a well-balanced CW-MFC system with optimal pollutant removal efficiency and parallel power production. So far, most CW-MFC research has centred on the development of prototype or small-scale operational systems. Although a great deal of data has been collected in the lab with synthetic substrates and accompanying microbial culture/consortia development, dealing with real wastewater or effluents has not been tried adequately, and this is where research needs to be prioritized immediately (Gupta et al., 2020). We need to understand how to keep the anode environment thermodynamically and kinetically favourable for EAB while keeping the electrodes at their optimal spacing. To improve bioelectricity generation, research into biofilm formation and microbial community identification, as well as their interaction with plant roots and root exudates, is required. To better exploit renewable energy sources, researchers need to investigate the factors that govern microbial and electrode compatibility, including the biofilm growth process and the electron transport kinetics on the electrode surface, both of which affect internal resistance and energy loss (Wang et al., 2020). Altering the electrodes to generate more power is another area of study that needs further attention. Potentially fruitful studies include the design of biosensors based on CW-MFCs, which would use changes in electrical output to reflect changes in system parameters. Energy generation in CW-MFC might be improved with the use of more economically viable materials, such as granular graphite, activated carbon, or low-cost minerals like coal. Different CW-MFC setups using different electrode and substrate materials require LCA-based research to fully comprehend their effects. The lack of data with a regional emphasis appears to be a drawback of LCA-based CW studies, since doing successful data prediction using LCA research requires the use of a more localized and precise data collection as opposed to a generic data set. The creation of data collection for various ecosystem services associated with CW in a given location would make LCA a more useful tool. Furthermore, most CW-based LCA studies limited their analysis to the building and operating phases, paying little if any mind to the waste products generated after the system's useful life. Although the operating phase is thought to have the greatest impact on the CW system's life cycle due to the technology's vast, low-tech, and low-energy requirements, an end-to-end perspective with well-considered system limits would be preferable for reliable forecasting. To make the technology more long-lasting, it would be beneficial to broaden the system's application to include the final use of system by-products as resources, such as sludge and macrophyte. Resource recovery and lowering environmental effects might be achieved by factoring in the possibility of nutrient recovery from macrophytes, sludge, and the fertilizer replacement value of sludge. It is hoped that CW technology may be more eco-friendly by including process optimization for greater P collection and enhanced denitrification.

7.14 CONCLUSIONS

The use of microbial electrochemical technologies to remove organic matter and other pollutants of interest, as well as to recover potential energy stored in chemical form in wastewater, is a relatively new and innovative approach to wastewater treatment. The CW-MFC is a cutting-edge MET design that combines the novel method of MFC with the tried-and-true efficacy of artificial wetlands in wastewater treatment. The potential for this setup to be used as a novel option for intensified wetland systems that maintain high performance with a smaller footprint makes this combination stand out as an intriguing and promising option among the various options of CW technology. Scaling the technology up, however, still faces several obstacles that must be conquered before it can develop a setup that can compete with other wastewater treatment options in terms of recovery energy, for example, in the form of biogas. Efforts in this direction should centre on learning more about the mechanisms involved in the release, transport, and use of electrons in these systems and developing new materials, internal conditions of the systems, inoculums, and configurations. Systems that make use of bioelectrochemical principles but do not produce electricity at the same time are worth pursuing. One solution would be to devote more resources to studying CW transmissions using MEC or MERC systems, short-circuit systems like snorkel-based METs, or novel MET-CW transmissions like METland systems. In its most efficient form, a CW is a complex wastewater treatment system with several components. A CW incorporating proper design considerations that have minimal life cycle implications enables operation throughout a wide climate range, increasing the likelihood of successful energy collection. Based on the results of the present evaluation, it is clear that in-depth comprehension of the interconnections between the underlying components is crucial to the effectiveness and longevity of any successful CW. This evaluation concludes that CW's temperature-sensitive variations in its pollutant removal capability may be compensated to an acceptable level through the adoption of appropriate operating techniques. It was shown that the average removal efficiency of ammonia-N dropped by 51% during the winter compared to the summer, with reductions of 9% for TSS and COD and 22–24% for BOD and Ortho-P. Heat source and heat preservation method (use of composting heat, geothermal energy source), using suitable cold-adapted microbial species (psychrophilic and psychrotrophic), optimizing loading rate, and pre-treatment strategies (aerobic tanks, sedimentation tanks, septic tanks, anaerobic biodigesters, anaerobic bio tanks, and floating biological beds). And insulating materials have all been used to mitigate the negative impact of low temperatures on CW (Phragmites, Typha, Scirpus, Potamogeton, Lolium perenne, Carex aquatilis, etc.). Significant laboratory-based data have been collected using synthetic substrates, and the current review reveals that CW-MFC can help in increased treatment efficiency (27–50% greater removal for main wastewater contaminants). Research into expanding the technique to actual wastewater, developing cost-effective electrode and substrate material, modifying electrodes, and elucidating microbial interlinkage with plant-specific root exudates are all areas that might improve the economic viability of CW-MFC applications. Studies on the LCA of CWs show that they are more environmentally friendly than traditional wastewater treatment methods like activated sludge, membrane bioreactors, membrane chemical reactors, sequencing batch reactors, high-rate algal ponds, and mechanical sludge. In addition, a more sustainable CW technique might be identified with the use of LCA employing a system- and region-specific data set that takes into account the possible effects of various end-use alternatives to system by-products.

ACKNOWLEDGEMENT

We would like to acknowledge the Department of Agronomy at the Lovely Professional University, Phagwara, Punjab, India, for their consistent moral support and encouragement throughout the writing process.

Conflicts: None

REFERENCES

Arends, J.B.A.; Verstraete, W. 100 years of microbial electricity production: Three concepts for the future, Microb. Biotechnol. 2012, 5, 333–346.

Bonanni, P.S.; Schrott, G.D.; Busalmen, J.P. A long way to the electrode: How do geobacter cells transport their electrons? Biochem. Soc. Trans. 2012, 40, 1274–1279.

Cavalcanti, E.B.; Garcia-Segura, S.; Centellas, F.; Brillas, E. Electrochemical incineration of omeprazole in neutral aqueous medium using a platinum or boron-doped diamond anode: Degradation kinetics and oxidation products, Water Res. 2013, 47, 1803–1815.

Cheng, K.Y, Bioelectrochemical system for energy recovery from wastewater, PhD Thesis, Murdoch University, Australia, 2009.

Clauwaert, P.; Desloover, J.; Shea, C.; Nerenberg, R.; Boon, N.; Verstrate, W. Enchanced nitrogen removal in bio-electrochemical systems by pH control, Biotechnol. Lett. 2009, 31(10), 1537–1543.

Clauwaert, P.; Rabaey, K.; Aelterman, P.; De Schamphelaire, L.; Pham, T.H.; Boeckx, P.; Boon, N.; Verstraete, W. Biological denitrification in microbial fuel cells, Environ. Sci. Technol. 2007, 41, 3354–3360.

Clauwaert, P.; Van der Ha, D.; Boon, N.; Verbeken, K.; Verhaege, M.; Rabaey, K.; Verstraete, W. Open air biocathode enables effective electricity generation with microbial fuel cells, Environ. Sci. Technol. 2007, 41, 7564–7569.

Coma, M.; Puig, S.; Pous, N.; Balaguer, M.D.; Colprim, J. Biocatalysed sulphate removal in a BES cathode, Bioresour. Technol. 2013, 130, 218–223.
Corbella, C.; Guivernau, M.; Viñas, M.; Puigagut, J. Operational, design and microbial aspects related to power production with microbial fuel cells implemented in constructed wetlands, Water Resour. 2015, 84, 232–242.
Dalmau, M.; Atanasova, N.; Gabarrón, S.; Rodriguez-Roda, I.; Comas, J. Comparison of a deterministic and a data driven model to describe MBR fouling, Chem. Eng. J. 2015, 260, 300–308.
Desloover, J.; Arends, J.B.A.; Hennebel, T.; Rabaey, K. Operational and technical considerations for microbial electrosynthesis, Biochem. Soc. Trans. 2012, 40, 1233–1238.
DiMuro, J.L.; Guertin, F.M.; Helling, R.K.; Perkins, J.L.; Romer, S.A. Financial and Environmental Analysis of Constructed Wetlands for Industrial Wastewater Treatment. J Ind Ecol. 2014, 18, 631–640.
Dixon, D.A.; Balch, G.C.; Kedersha, N.; Anderson, P.; Zimmerman, G.A.; Beauchamp, R.D.; Prescott, S.M. Regulation of cyclooxygenase-2 expression by the translational silencer TIA-1. J Exp Med. 2003, 198(3), 475–481.
Du, Z.W.; Li, H.; Gu, T.Y. A state of the art review on microbial fuel cell: A promising technology for wastewater treatment and bioenergy, Biotechnol. Adv. 2007, 25, 464–482.
Erable, B.; Du,teanu, N.; Ghangrekar, M.; Dumas, C.; Scott, K. Application of electro-active biofilms, Biofouling. 2010, 26, 57–71.
Fuchs, V.J.; Mihelcic, J.R.; Gierke, J.S. Life cycle assessment of vertical and horizontal flow constructed wetlands for wastewater treatment considering nitrogen and carbon greenhouse gas emissions, Water Res. 2011, 45, 2073–2081.
Garcia-Segura, S.; Brillas, E.; Cornejo-Ponce, L.; Salazar, R. Effect of the Fe^{3+}/Cu^{2+} ratio on the removal of the recalcitrant oxalic and oxamic acids by electro-fenton and solar photoelectroFenton, Sol. Energy. 2016, 124, 242–253.
Ghafari, S.; Hasan, M.; Aroua, M.K. Bio-electrochemical of nitrate form water and wastewater: A review, bioresour, Technol. 2008, 99, 3965–3974.
Gregory, K.B.; Bond, D.R.; Lovley, D.R. Graphite electrodes as electron donors for anaerobic respiration, Environ. Microbiol. 2004, 6, 596–604.
Guadarrama-P´erez, O.; Guti´errez-Macías, T.; García-Sanchez, ´L.; GuadarramaP´erez, V.H.; Estrada-Arriaga, E.B. Recent advances in constructed wetland-microbial fuel cells for simultaneous bioelectricity production and wastewater treatment: A review, Int. J. Energy Res. 2019, 43, 5106–5127.
Gude, V.G. Wastewater treatment in microbial fuel cells—An overview, J. Clean. Prod. 2016, 122, 287–307.
Gupta, S.; Srivastava, P.; Patil, S.A.; Yadav, A.K. A comprehensive review on emerging constructed wetland coupled microbial fuel cell technology: Potential applications and challenges. Bioresour. Technol. 2020, 320(Pt B), 124376.
He, Z.; Minteer, S.D.; Angenent, L.T. Electricity generation from artificial wastewater using an upflow microbial fuel cell, Environ. Sci. Technol. 2005, 39, 5262–5267.
Heidrich, E.S.; Curtis, T.P.; Dolfing, J. Determination of the internal chemical energy of wastewater, Environ. Sci. Technol. 2011, 45, 827–832.
Hernandez, M.E.; Kappler, A.; Newman, D.K. Phenazines and other redox-active antibiotics promote microbial mineral reduction, Appl. Environ. Microbiol. 2004, 70, 921–928.
Huang, G.; Sun, Y.; Liu, Z.; Sedra, D.; Weinberger, K.Q. Deep networks with stochastic depth. In Computer Vision–ECCV 2016: 14th European Conference, Amsterdam, The Netherlands, 2016, Proceedings, Part IV 14 (pp. 646–661). Springer International Publishing.
Jenssen, P.D.; Mæhlum, T.; Krogstad, T.; Vråle, L. High performance constructed wetlands for cold climates, J. Environ. Sci. Heal—Part A Toxic/Hazard Subst. Environ. Eng. 2005, 40, 1343–1353.

Kadlec, R.; Wallace, S. Treatment Wetlands, 2nd ed.; CRC Press: Boca Raton, FL, USA, 2009; ISBN 9781420012514.

Kataki, S.; Chatterjee, S.; Vairale, M.G.; Dwivedi, S.K.; Gupta, D.K. Constructed wetland, an eco-technology for wastewater treatment: A review on types of wastewater treated and components of the technology (macrophyte, biolfilm and substrate), J. Environ. Manag. 2021, 283, 111986.

Kato, S. Biotechnological aspects of microbial extracellular electron transfer, Microbes Environ. 2015, 30, 133–139.

Kim, J.R.; Premier, G.C.; Hawkes, F.R.; Dinsdale, R.M.; Guwy, A.J. Development of a tubular microbial fuel cell (MFC) employing a membrane electrode assembly cathode, J. Power Sources. 2009, 187, 393–399.

Kim, J.R.; Premier, G.C.; Hawkes, F.R.; Rodríguez, J.; Dinsdale, R.M.; Guwy, A.J. Modular tubular microbial fuel cells for energy recovery during sucrose wastewater treatment at low organic loading rate, Bioresour. Technol. 2010, 101, 1190–1198.

Kotloski, N.J.; Gralnick, J.A. Flavin electron shuttles dominate extracellular electron transfer by Shewanella oneidensis, MBio. 2013, 4, 10–13.

Kracke, F.; Vassilev, I.; Krömer, J.O. Microbial electron transport and energy conservation—The foundation for optimizing bioelectrochemical systems, Front. Microbiol. 2015, 6, 1–18.

Larsen, S.C.; Hendriks, I.A.; Lyon, D., Jensen, L.J.; Nielsen, M.L. Systems-wide analysis of serine ADP-ribosylation reveals widespread occurrence and site-specific overlap with phosphorylation. Cell Rep, 2018, 24(9), 2493–2505.

Liu, F.; Rotaru, A.-E.; Shrestha, P.M.; Malvankar, N.S.; Nevin, K.P.; Lovley, D.R. Promoting direct interspecies electron transfer with activated carbon, Energy Environ. Sci. 2012, 5, 8982–8989.

Logan, B.E.; Hamelers, B.; Rozendal, R.; Schröder, U.; Keller, J.; Freguia, S.; Aelterman, P.; Verstraete, W.; Rabaey, K. Microbial fuel cells: Methodology and technology, Environ. Sci. Technol. 2006, 40, 5181–5192.

Logan, B.E.; Rabaey, K. Conversion of wastes into bioelectricity and chemicals by using microbial electrochemical technologies, Science. 2012, 337, 686–690.

Logan, B.E.; Wallack, M.J.; Kim, K.Y.; He, W.; Feng, Y.; Saikaly, P.E. Assessment of microbial fuel cell configurations and power densities, Environ. Sci. Technol. Lett. 2015, 2, 206–214.

Lovley, D.R. Bug juice: Harvesting electricity with microorganisms, Nat. Rev. Microbiol. 2006, 4, 497–508.

Machado, W.; Santelli, R.E.; Loureiro, D.D.; Oliveira, E.P.; Borges, A.C.; Ma, V.K.; Lacerda, L.D. Mercury accumulation in sediments along an eutrophication gradient in Guanabara Bay, Southeast Brazil. Braz Chem Soc. 2008, 19, 569–575.

Malvankar, N.S.; Lau, J.; Nevin, K.; Franks, A.E.; Tuominen, M.T.; Lovley, D.R. Electrical conductivity in a mixed-species biofilm, Appl. Environ. Microbiol. 2012, 78, 5967–5971.

Mao, L.; Verwoerd, W.S. Selection of organisms for systems biology study of microbial electricity generation: A review, Int. J. Energy Environ. Eng. 2013, 4, 17.

Mardanpour, M.M.; Esfahany, M.N.; Behzad, T.; Sedaqatvand, R. Single chamber microbial fuel cell with spiral anode for dairy wastewater treatment, Biosens. Bioelectron. 2012, 38, 264–269.

Memon, F.A.; Zheng, Z.; Butler, D.; Shirley-Smith, C.; Lui, S.; Makropoulos, C.; Avery, L. Life cycle impact assessment of greywater recycling technologies for new developments. Environ Monit Assess. 2007, 129, 27–35.

Modin, O.; Gustavsson, D.J.I. Opportunities for microbial electrochemistry in municipal wastewater treatment—An overview, Water Sci. Technol. 2014, 69, 1359–1372.

Mohan, S.V.; Srikanth, S.; Raghuvulu, S.V.; Mohanakrishna, G.; Kumar, A.K.; Sarma, P.N. Evaluation of the potential of various aquatic ecosystems in harnessing bioelectricity through benthic fuel cell: Effect of electrode assembly and water characteristics, Bioresour. Technol. 2009, 100, 2240–2246.

Nitisoravut, R.; Regmi, R. Plant microbial fuel cells: A promising biosystems engineering, Renew. Sustain. Energy Rev. 2017, 76, 81–89.

Parde, D.; Patwa, A.; Shukla, A.; Vijay, R.; Killedar, D.J.; Kumar, R. A review of constructed wetland on type, treatment and technology of wastewater. Environ. Technol. Innov. 2021, 21, 101261.

Paredes, D.; Vélez, M.E.; Kuschk, P.; Mueller, R.A. Effects of type of flow, plants and addition of organic carbon in the removal of zinc and chromium in small-scale model wetlands, Water Sci. Technol. 2007, 56, 199–205.

Patil, S.A.; Hägerhäll, C.; Gorton, L. Electron transfer mechanisms between microorganisms and electrodes in bioelectrochemical systems, Bioanal. Rev. 2012, 4, 159–192.

Picioreanu, C.; Loosdrecht, M.C.M.V.; Curtis, T.P.; Scott, K. Model based evaluation of the EFFECT of pH and electrode geometry on microbial fuel cell performance, Bioelectrochemistry. 2010, 78, 8–24.

Prosnansky, M.; Sakakibara, Y.; Kuroda, M. High-rate denitrification and SS rejection by biofilm-electrode reactor (BER) combined with microfiltration, Water Res. 2002, 36, 4801–4810.

Rabaey, K.; Rodríguez, J.; Blackall, L.L.; Keller, J.; Gross, P.; Batstone, D.; Verstraete, W.; Nealson, K.H. Microbial ecology meets electrochemistry: Electricity-driven and driving communities, ISME J. 2007, 1, 9–18.

Rabaey, K.; Rozendal, R.A. Microbial electrosynthesis—Revisiting the electrical route for microbial production, Nat. Rev. Microbiol. 2010, 8, 706–716.

Rahimnejad, M.; Adhami, A.; Darvari, S.; Zirepour, A.; Oh, S.-E. Microbial fuel cell as new technology for bioelectricity generation: A review, Alexandria Eng. J. 2015, 54, 745–756.

Reimers, C.E.; Tender, L.M.; Fertig, S.; Wang, W. Harvesting energy from the marine sediment—Water interface, Environ. Sci. Technol. 2001, 35, 192–195.

Rosenbaum, M.; Agler, M.T.; Fornero, J.J.; Venkataraman, A.; Angenent, L.T. Integrating BES in the wastewater and sludge treatment line. In Bioelectrochemical Systems: From Extracellular Electron Transfer to Biotechnological Application; IWA Publishing: London, UK, 2010; pp. 393–408, ISBN 9781843392330.

Rozendal, R.A.; Hamelers, H.V.M.; Rabaey, K.; Keller, J.; Buisman, C.J.N. Towards practical implementation of bioelechemical wastewater treatment, Trends Biotechnol. 2008, 26, 450–459.

Ruiz, M.A.; Díaz, B.; Crujeiras, J.; García, M.; Soto. Solids hydrolysis and accumulation in a hybrid anaerobic digester-constructed wetlands system, Ecol. Eng. 2010, 36(8), 1007–1016.

Sakakibara, Y.; Nakayama, T. A novel multi-electrode system for electrolytic and biological water treatments: Electric charge transfer and application to denitrification, Water Res. 2001, 35, 768–778.

Schröder, U.; Harnisch, F.; Angenent, L.T. Microbial electrochemistry and technology: Terminology and classification, Energy Environ. Sci. 2015, 8, 513–519.

Scialdone, O.; Galia, A.; Randazzo, S. Oxidation of carboxylic acids in water at IrO_2-Ta_2O_5 and boron doped diamond anodes, Chem. Eng. J. 2011, 174, 266–274.

Shah, M.P. Microbial Bioremediation & Biodegradation; Springer, 2020.

Shah, M.P. Removal of Emerging Contaminants through Microbial Processes; Springer, 2021.

Shah, M.P. Removal of Refractory Pollutants from Wastewater Treatment Plants; CRC Press, 2021.

Subedi, B.; Kannan, K. Occurrence and fate of select psychoactive pharmaceuticals and antihypertensives in two wastewater treatment plants in New York State, USA, Sci. Total Environ. 2015, 514, 273–280.

Tejedor-Sanz, S.; Quejigo, J.R.; Berná, A.; Esteve-Núñez, A. The planktonic relationship between fluid-like electrodes and bacteria: Wiring in motion, ChemSusChem. 2017, 10, 693–700.

Tong, Y.; He, Z. Nitrate removal from groundwater driven by electricity generation and heterotrophic denitrification in a bioelectrochemical system, J. Hazard. Mater. 2013, 262, 614–619.

Uggetti, E.; Ferrer, I.; Llorens, E.; García, J. Sludge treatment wetlands: a review on the state of the art. Bioresour Technol, 2010, 101(9), 2905–2912.

UNESCO. The United Nations World Water Decelopment Report 4. Volume 1: Managing Water Report under Uncertainty and Risk, 2012.

Varma, M.; Gupta, A.K.; Ghosal, P.S.; Majumder, A. A review on performance of constructed wetlands in tropical and cold climate: Insights of mechanism, role of influencing factors, and system modification in low temperature. Sci. Total Environ. 2020, 755, 142540.

Virdis, B.; Rabaey, K.; Yuan, Z.; Keller, J. Microbial fuel cells for simultaneous carbon and nitrogen removal, Water Res. 2008, 42, 3013–3024.

Virkutyte, J.; Jegatheesan, V. Electro-fenton, hydrogenotrophic and Fe^2+ ions mediated TOC and nitrate removal from aquaculture system: Different experimental strategies, bioresour, Technol. 2009, 100, 2189–2197.

Voordeckers, J.W.; Kim, B.C.; Izallalen, M.; Lovley, D.R. Role of geobacter sulfurreducens outer surface c-type cytochromes in reduction of soil humic acid and anthraquinone-2, 6-disulfonate, Appl. Environ. Microbiol. 2010, 76, 2371–2375.

Vymazal, J. Constructed wetlands for wastewater treatment, Water. 2010, 2, 530–549.

Wan, D.J.; Liu, H.J.; Qu, J.H.; Lei, P.J. Bio-electrochemical denitrification by a novel proton-exchange membrane electrodialysis system—A batch mode study, J. Chem. Technol. Biotechnol. 2010, 85, 1540–1546.

Wang, H.; Ren, Z.J. A comprehensive review of microbial electrochemical systems as a platform technology, Biotechnol. Adv. 2013, 31, 1796–1807.

Wang, W.; Zhang, Y.; Li, M.; Wei, X.; Wang, Y.; Liu, L.; Wang, H.; Shen, S. Operation mechanism of constructed wetland-microbial fuel cells for wastewater treatment and electricity generation: A review. Bioresour Technol. 2020, 314, 123808.

Wu, H.; Zhang, J.; Ngo, H.H.; Guo, W.; Hu, Z.; Liang, S.; Fan, J.; Liu, H. Review on the sustainability of constructed wetlands for wastewater treatment: Design and operation, Bioresour. Technol. 2015, 175, 594–601.

Wu, S.; Kuschk, P.; Brix, H.; Vymazal, J.; Dong, R. Development of constructed wetlands in performance intensifications for wastewater treatment: A nitrogen and organic matter targeted review, Water Res. 2014, 57, 40–55.

Xie, S.; Liang, P.; Chen, Y.; Xia, X.; Huang, X. Simultaneous carbon and nitrogen removal using an oxic/anoxic-biocathode microbial fuel cells coupled system, Bioresour. Technol. 2011, 102, 348–354.

Xu, L.; Zhao, Y.; Wang, T.; Liu, R.; Gao, F. Energy capture and nutrients removal enhancement through a stacked constructed wetland incorporated with microbial fuel cell, Water Sci. Technol. 2017, 76, 28–34.

Yadav, A.K.; Dash, P.; Mohanty, A.; Abbassi, R.; Mishra, B.K. Performance assessment of innovative constructed wetland-microbial fuel cell for electricity production and dye removal, Ecol. Eng. 2012, 47, 126–131.

Yazdandoost, F.; Yazdani, S.A. A new integrated portfolio-based water-energy environment nexus in wetland catchments, Water Resour Manag. 2019, 33, 2991–3009.

Yildirim, Ö.; Kiss, A.A.; Hüser, N.; Leßmann, K.; Kenig, E.Y. Reactive absorption in chemical process industry: A review on current activities. Chem Eng J. 2012, 213, 371–391. https://doi.org/10.1016/j.cej.2012.09.121

Yu, E.H.; Cheng, S.; Scott, K.; Logan, B. Microbial fuel cell performance with non-pt cathode catalysts, J. Power Sources. 2007, 171, 275–281.

Zhang, Q.; Hu, J.; Lee, D.-J. Microbial fuel cells as pollutant treatment units: Research updates, Bioresour. Technol. 2016, 217, 121–128.

Zhuang, L.; Zheng, Y.; Zhou, S.; Yuan, Y.; Yuan, H.; Chen, Y. Scalable microbial fuel cell (MFC) stack for continuous real wastewater treatment, Bioresour. Technol. 2012, 106, 82–88.

8 Current Status of Wastewater in India/Other Countries/Regions

Lalit Saini, Prasann Kumar, and Joginder Singh Panwar

8.1 INTRODUCTION

Water, air, and land that are completely clean are increasingly unusual in Asia. Even though the overall impact of climate change on water quantity and quality will be marginal compared to socioeconomic changes, even by 2100, the overall impact of climate change on water quantity and quality will be marginal compared to socioeconomic changes (Hanjra and Qureshi, 2010; Khan and Hanjra, 2009; Park et al., 2010). The Asian economy is quickly urbanizing and industrializing, and this trend will have far-reaching effects on ecosystems everywhere (Angel and Rock, 2009). Many Asian nations are adopting proactive policies and methods to increase environmental regulation (Hanjra et al., 2011), which, when paired with institutional changes, might pave the way for cleaner forms of urbanization and economic growth. There is a connection between insufficient wastewater treatment and management practices and home pollution issues. Nearly 144 km^3 of wastewater1 is produced yearly across Asia, with China producing 37%, South Asia producing 27%, Japan producing 20%, Southeast Asia producing 6%, and Central Asia producing 3% (FAO AQUASTAT, n.d.). South Asia (7% treatment rate) and Southeast Asia (14% treatment rate) have the lowest treatment rates in Asia. Approximately 2.3 billion people (41% of the world's population in 1995) lived in river basins that were water stressed (1,700 m^3 $person^{-1}$ yr^{-1}) and this amount was forecast to climb to 3.5 billion by 2025 (48% of the projected population). Wastewater irrigation will likely be an essential component of the mix in the next few decades as we apply several techniques to deal with water stress. Worldwide, a diverse variety of crops and landscapes are irrigated by wastewater from urban and industrial sources. Increased food availability and a corresponding decrease in hunger in low-income neighbourhoods are examples of the former, along with the fact that farmers rely on a steady water supply and, in certain circumstances, the nutritional content of wastewater, to make a living. These have a domino effect, raising living conditions and societal well-being. There are two primary environmental factors (Anderson, 2003; Hamilton et al., 2007). First, the environmental problems associated with the high abstraction of natural surface waters and groundwater can be reduced when wastewater is used in place of normal irrigation water. Second, by recycling and reusing, wastewater is less likely to be dumped into vulnerable ecosystems like the

 DOI: 10.1201/9781003368472-8

ocean, lakes, and rivers. Over the course of the 21st century, water will become one of the most precious commodities because of the exponential rise in the global population (Day, 1996). By 2015, over 5 billion people worldwide will reside in urban areas. Additionally, there will be 23 cities each with populations exceeding 10 million. Eighteen of these will be in developing countries (Black, 1994). The challenges of delivering municipal services and water sector infrastructure, such as potable water and sewage systems, lie at the heart of the urbanization phenomenon. Engineers, planners, and legislators have significant issues today in meeting fundamental human requirements, including housing, healthcare, social services, and access to essential human needs infrastructure like clean water and the disposal of effluent (Black, 1994; Giles and Brown, 1997). Population increases put a higher demand on resources and raise the risk to natural habitats. There is no sustainability in the current uses of freshwater by either developing or developed nations, according to a report by the Secretary-General of the United Nations Commission on Sustainable Development (UNCSD, 1997). Global water usage has been increasing at a rate greater than three times the rate of global population growth, which has led to widespread public health problems, stifled economic and agricultural development, and had negative effects on a variety of other sectors. With a total size of 3.29 million square kilometres and a population of over one billion, 29% of whom reside in urban areas in a total of 5,162 cities and towns, India ranks seventh among the world's biggest countries. India is home to a rapidly expanding economy and a sizable population of technical and scientific experts. Due to the pollution caused by SSIs, Indian authorities must strike a tough balance between promoting economic growth and protecting the environment. Planning and expanding water and sewage systems have become very challenging and expensive due to the unchecked growth of metropolitan areas (Looker, 1998). Due to their large land area, dense population, and widespread and intense irrigation, Asia and the Pacific generate enormous domestic and agricultural return flows, making them the world region with the biggest annual water withdrawal. This is due in significant part to the fact that around 72% of the world's population does not have access to modern sanitary facilities, the vast majority of whom are located in Asia. It is estimated that two-thirds of all sewage from cities is dumped untreated into waterways across the world, with that number rising to 90% in poorer nations (Corcoran et al., 2010). Water, sanitation, and hygiene-related illnesses and deaths account for 40% of all worldwide deaths, and Asia is no exception (WHO, 2004). Unfortunately, many sewers and wastewater treatment facilities are in a dilapidated state, so only a small percentage of the population has access to them. This is especially true in Southeast and South Asia (Corcoran et al., 2010). Nonpoint source contamination is seen as a major danger to water quality, with more than half of worldwide irrigation taking place in Asia and a high level of pesticide usage.

8.2 WATER POLLUTION STATUS IN INDIA

The 299 cities fall into the "class-1" category, each producing an average of 16,652.5 MLD of wastewater each year. Twenty-three large urban areas account for around 59% of this total. Class-1 cities create around 71% of the world's wastewater, with about a third coming from the state of Maharashtra and the rest from the Ganga river basin. Currently, we only collect 72% of the treated wastewater that is produced. Out of 299 class-1 cities, 160 have sewage systems that cover more than 75% of the population, and 92 cover more than 50%. The majority of people in class-1 cities (about 70%) have access to a sewage system. Sewerage systems can be either open or closed, or they can be piped. However, only 4,037.2 MLD (or 24%) of the wastewater created annually is treated before discharge; the other 12,626.30 MLD is flushed away untreated. Only 47 cities have secondary treatment options, while the remaining 49 offer both primary and secondary care. Allowing wastewater to be released into groundwater, surface water bodies, and/or lands without first being treated can have the following negative effects on the environment: It's possible for a lot of foul-smelling gases to be produced during the breakdown of the organic compounds found in wastewater. Discharging untreated wastewater (sewage) with a high concentration of organic matter into a river or stream may deplete the stream's dissolved oxygen supply to meet the Biochemical Oxygen Demand (BOD) of the wastewater, leading to fish deaths and other unfavourable outcomes. Nutrients in the wastewater can promote the expansion of aquatic vegetation and algal blooms, resulting in the eutrophication of water bodies. Many pathogenic bacteria and hazardous substances, such as those found in the human gastrointestinal system or some industrial waste, are present in untreated wastewater. There is a risk that they will poison the environment wherever sewage is dumped (Figure 8.1).

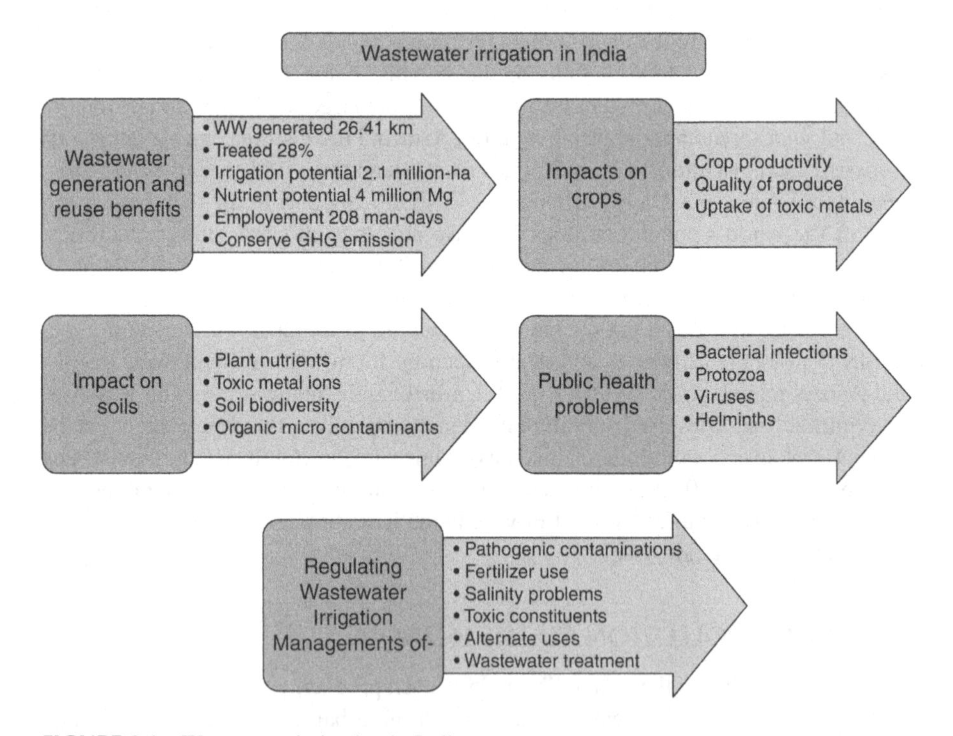

FIGURE 8.1 Wastewater irrigation in India.

8.3 POLLUTION LEVELS IN SOUTH ASIAN WASTEWATER

The Ministry of Environment and Forests (2009) cites a report from the Ministry of Water Resources (MOWR) of the Government of India, which states that the country's surface water resources are contaminated at alarming rates and that the percentage of contaminated groundwater reserves is rising. There has been tremendous growth from the original 18 sites to the current 1,700 in the number of places where routine monitoring of water quality is conducted, with 62 parameters being tracked. The Central Pollution Control Board (CPCB) posts results online annually. According to 2009 data, organic pollutants still pose the greatest threat. Six of the fifty rivers had BOD levels of more than 100 mg/L, and over 36% of the observations had BOD levels higher than the threshold for bathing water (3 mg/L). Only 49% of samples met the minimum total coliform (TC) guideline for swimming pools, while 51% were over the recommended level of 500 MPN/100 mL. Over 30% of locations also have faecal coliform (FC) levels that are higher than this (CPCB, 2010). Since 1995, this has been a minor advancement. Improvements have been made in wastewater collection. There is still no centralized database for monitoring data on Sri Lanka's surface waters. Eutrophication, brought on by an abundance of fertilizer, is a major problem. In addition to larger bodies of water like Kandy Lake and Gregory Lake, several smaller irrigation reservoirs and canals were also discovered to have excessive nutrient levels (Ministry of Environment and Forest, 2001).

8.4 VARIOUS POLLUTION SOURCES

8.4.1 Domestic and Agricultural Waste Contamination

Low-quality wastewater treatment and management contribute to agricultural and household pollution. Nearly 144 km^3 of wastewater is produced yearly across Asia, with China producing 37%, South Asia producing 27%, Japan producing 20%, Southeast Asia producing 6%, and Central Asia producing 3% (FAO AQUASTAT, n.d.). South Asia (7% treatment rate) and Southeast Asia (14% treatment rate) have the lowest treatment rates in Asia. Human health is most negatively impacted by microbial pollution, which can result from inadequate sanitation facilities, inappropriate wastewater disposal, and animal wastes (Corcoran et al., 2010; WWAP, 2009). The rate of disease-related mortality in communities with contaminated water supplies is one indicator of the problem's severity. This number is less than 15 in China, but goes to 100–200 in other regions of Asia (including Afghanistan) and is between 30 and 100 throughout virtually all of South Asia (Corcoran et al., 2010). Many rivers in the area have ten times the OECD's recommended maximum level of FC bacteria (a measure of the amount of human waste in water) and up to three times the global average. Many of Asia's biggest cities rely extensively on septic tanks and latrines since they lack sophisticated waterborne sewage systems. In areas where sewage systems do exist, pollution from leaking pipes is a common problem (Pak-EPA, 2005). Globally, agriculture is responsible for more nonpoint source contamination of surface water and groundwater than any other sector (Chhabra et al., 2010). The widespread use of fertilizers, which are frequently subsidized, results in significant pollution. The potential influence on groundwater quality from the loss of reactive nitrogen from agricultural systems throughout Asia is a major reason for worry, especially in rice-producing areas throughout the Indo-Gangetic Plain and in areas of vegetable agriculture (Chowdary et al., 2005). The use of agricultural chemicals is now low in many regions of Asia compared to more industrialized nations; however, this is expected to change as populations rise and the need for food increases (Biswas and Seetharam, 2008). According to studies on water quality in India, farming is the sector most responsible for degrading water supplies (MoEF, 2009). This is because of the rise in the use of fertilizers, from 70 kg/ha in 1991–92 to 113 kg/ha in 2006–07, and the use of pesticides, which increased by 750% in the second half of the 20th century (CSE, 1999; MoEF, 2009). Increasing water withdrawals and return flow from agriculture, for example in the Amu Darya and Syr Darya, have contributed to the accumulation of salts in Central Asia's rivers and lakes over the past 40 years. Pesticide, herbicide, and fertilizer residues also pose an issue (Ballance and Pant, 2003). Despite a significant decrease in overall pesticide use, the absence of regulations and adequate application, storage, and disposal of leftover pesticides appear to make the use of even relatively modest quantities of uncontrolled imports more dangerous than the past high levels of controlled use (Howarth et al., 2008).

8.4.2 Pollution from Industry

Though just 3% of India's annual water withdrawals are used by industry, it contributes significantly to the country's water pollution problem. Estimates put the daily volume of industrial wastewater production at 55 km^3, with 68.5 million m^3 being discharged directly into water bodies without any treatment whatsoever, and a greater volume receiving just minimum treatment through processes like neutralization and settling. There are 1,532 "grossly polluting" enterprises in the nation, according to the Central and State Pollution Control Boards (MoEF, 2009). Several Asian economies have shifted from being primarily dependent on agricultural exports to relying significantly on industrial revenue. Production in the food and beverage industry, electrical machinery, cement, metals, chemicals, plastics and rubber goods, and textiles has increased by between 20 and 45% in recent years. More than 20% of the gross domestic product is produced by industries in at least 30 Asian and Pacific nations (ESCAP, 2005). Although there has been a surge in environmental consciousness in the manufacturing sector, rules are hard to implement and pollution levels are on the rise since the region is dominated by small and medium-scale manufacturing. The nations of Mongolia, Kyrgyzstan, and Azerbaijan, all located in Central and Northeast Asia, have the greatest levels of industrial pollution in the area, as measured by BOD emissions per US$1,000 of GDP (0.10 kg BOD/US$1,000 GDP), followed by the South Asian countries of Bangladesh and Nepal (8 kg). Mineral extraction is a big component; for instance, Asia is the world's greatest producer of human-made mercury in the atmosphere. Approximately 50% of the world's emissions may be traced back to just one source (Lie et al., 2009). A total of 70 industrial parks have been established in Vietnam, contributing to the daily production of 1 million m^3 of untreated wastewater from these facilities and the country's 1,000 hospitals. It has been estimated that 4,000 businesses are discharging wastewater and need to relocate, close, or implement cleaner technologies and wastewater treatment, as stated by the Ministry of Natural Resources and Environment (MoNRE). The Institute of Tropical Techniques and Environmental Protection (ITTEP) has shown that the levels of toxins in rivers in several of the cities are far higher than the legal limits. Almost 93 tons of rubbish is dumped into the Dong Nai, Thi Vai, and Saigon rivers daily from IPs and EPZs in the Southern Key Economic Zone, as reported by ITTEP (WEPA, 2011).

8.5 WASTEWATER PRODUCTION AND TREATMENT

Similarly, with the growth of urban areas and demands for potable water, the production and treatment of wastewater have also increased rapidly. According to CPHEEO, between 70 and 80% of all home water supplies end up as wastewater. Class-I and class-II cities and towns, home to 72% of India's urban population, produce an estimated 98 lpcd of wastewater per capita daily, while the National Capital Territory of Delhi generates approximately 220 lpcd (discharging 3,663 MLD of wastewater, 61% of which is processed) (CPCB, 1999). The CPCB estimates that there are 498 class-I cities in the country, with a combined wastewater generation of 35,558 MLD, and 410 class-II towns, with a combined wastewater generation of 2,696 MLD. While 26,468 MLD in sewage treatment capacity is needed, only 11,553 and 233 MLD are already installed. The top five states that send in wastewater are Maharashtra, Delhi, Uttar Pradesh, West Bengal, and Gujarat (63%; CPCB, 2007a). Industrial water usage productivity in India (IWP, in billion constant 1995 US$per m^3) is the lowest (i.e., only 3.42), and is around one-thirtieth of that in Japan and the Republic of Korea (Van-Rooijen et al., 2008). By 2050, it is expected that around 48.2 BCM (132 billion litres per day) of wastewater would be created, further worsening this gap (because this wastewater can supply 4.5% of the total agricultural water requirement) (Bhardwaj, 2005). There are 234 wastewater treatment facilities in India (STPs). Most of them were built after 1978–1979 as part of a variety of river action programmes and may be found in (less than) 5% of urban areas that are situated along major river corridors (CPCB, 2005b). Most of the time, activated sludge or an oxidation pond is used to treat wastewater in class-I cities. Technology accounts for 59.5% of all capacity currently in use. Upflow anaerobic sludge blanket technology accounts for the next largest share of installed capacity at 26%. Additionally, 28% of the plants use the Series of Waste Stabilization Ponds technology, albeit its total capacity is just 5.6%. Sewerage infrastructure, including treatment facilities, falls within the purview of state/urban local bodies and their respective agencies for operation and maintenance. State pollution control boards have the authority to take legal action against non-compliant agencies under the Water Act of 1974. The State Governments have disregarded the Water Act of 1974's emphasis on the reuse of treated sewage in irrigation. Effluents from SSI units in clusters may be treated at centralized treatment facilities according to a programme launched by the Indian Government's Ministry of Environment and Forests (MoEF). The capital cost of the project was subsidized by 50% thanks to the Common Effluent Treatment Plant (CETP) financial aid plan, with the Central and State Governments each contributing 25%. More than 10,000 polluting companies have been targeted by the 88 CETPs across India, with a total capacity of 560 MLD (CPCB, 2005b).

8.6 WASTEWATER IRRIGATION AROUND THE WORLD

Israel has pioneered the reuse of wastewater for agricultural purposes. Greater than 60% of Israel's sewage is recycled, with the majority going towards watering crops (Lawhon and Schwartz, 2006). Israel's agriculture used to rely primarily on two huge aquifers, but the country's growing population and rising standards of living have created an unsustainable water deficit (Oron, 1998). As a partial reaction, Israel built the National Water Carrier to transport water from the Sea of Galilee to the country's southern regions (Kinnerret Lake). There are already five large-scale reuse programmes using recovered wastewater for irrigated agriculture, which is the second key strategy (USEPA and USAID, 2004). Dan Region Reclamation Scheme, located not far from Tel Aviv, was the first and is by far the largest of these projects (Kanarek and Michail, 1996). In this plan, treated effluent from the Soreq STP is first deposited into an aquifer and then later retrieved and redistributed to irrigation networks in the southern coastal plain and the northern Negev area. The Kishon project, the second-largest system, is quite a different operation. About 15,000 hectares of non-edible crops, predominantly cotton, are irrigated using water from a 12-million-m^3 irrigation storage reservoir located 30 km east of Haifa. This reservoir is supplied with treated wastewater from the Haifa STP, as well as local wastewater and rainwater. The Israeli method of wastewater irrigation is distinguished by its heavy reliance on drip irrigation (especially sub-surface) to reduce the likelihood of crop contamination and water loss (Oron, 1998). By 2010, Israel aimed to treat and reuse nearly all of its wastewater, a milestone that would establish recycled wastewater as 20% of all freshwater reserves (USEPA and USAID, 2004). Wastewater reuse for irrigation is particularly common in the Near East due to the region's extreme water shortage and high population density. In Iran, around 70 million m^3 y^{-1} of wastewater is utilized for agricultural irrigation, while in Kuwait, roughly a quarter of agriculture, including amenity horticulture, is irrigated with reclaimed wastewater (Radcliffe, 2004). While only approximately half of Oman's wastewater is treated, more than 90% of that is reused, largely in arboriculture (FAO, 2001). When it comes to reusing wastewater for agricultural purposes, few countries can compare it to Tunisia (Shetty, 2007). The country has a very developed sewage system for a developing nation, with around 70% and 20% of the urban and rural populations, respectively, connected to sewers. This has allowed for the adoption of widespread planned reuse (Ministry of Agriculture, 1998). In Sub-Saharan Africa, the extent to which wastewater is used for irrigation is unknown. Formal reuse schemes are difficult to implement in many countries due to a combination of factors, including weak economies, inadequate institutional frameworks, and limited or deteriorating public assets (sewerage and irrigation networks). In Kenya, 34% of farmers surveyed said that they have used raw sewage from trunk sewers on their farms (Hide et al., 2001; Hide and Kimani, 2000). Wastewater reuse is garnering modest and growing interest despite the abundance of water resources throughout most of Central and, especially, South America. There are two main reasons behind this. I would want to start by saying that freshwater may not be as abundant or as easily available as one may assume due to the location and size of many megacities. For example, Brazil is home to around 6% of the world's freshwater, but 80% of it is in the Amazon basin in the north, while 65% of the people live in the southeast, south, and central-western areas (USEPA and USAID, 2004). More than 40 irrigation districts spread across Mexico rely on untreated wastewater for irrigation, irrigating more than 350,000 hectares (ha) of agriculture (Peasey et al., 2000). Although it may not appear so at first glance, roughly half of Europe's countries are in a situation where water availability is becoming a constraint on development and significant investment is required to secure sufficient water supplies (i.e., WSI > 10%), as measured by the water stress index (WSI). The WSI compares a country's total water withdrawal to its total renewable freshwater resources (Bixio et al., 2005a). Large populations paired with relatively high standards of living in many nations contribute to such supply-demand mismatches (Minhas et al., 2022).

8.7 COUNTRIES THAT ARE STILL DEVELOPING

Unfortunately, many emerging nations are seeing a decline in economic activity, which has led to political instability and environmental damage. Water is becoming scarcer and of lower quality in many parts of the world, notably in Africa and South Asia. However, other concerns, such as national or racial security, food supply, and epidemic management, take precedence over water contamination (Ujang and Buckley, 2002). The following is a generalization of the issues facing these emerging countries. The water and wastewater management sectors are seen as less vital than other sectors like military empowerment, road improvement, electricity, mass education, and healthcare facilities due to a lack of environmental understanding among most policymakers and the public. Knowledge gaps exist between desired policies and actual execution due to a lack of experts. Water conservation rules that aren't adequate, such as those that don't make it illegal to cut down stress in watersheds. Rapid urbanization and population growth rates lead to conflicting public expenditures, resulting in insufficient financing for water supply and sanitation services. Scarcity of water, especially in urban and desert areas. Inadequate institutional backing and managerial procedures for supplying water and sanitation infrastructure.

Water and wastewater management, solid and hazardous waste, economic and social factors, and community and governance are just some of the areas that need to be taken into account when trying to solve the water and wastewater management issues plaguing developing countries. This is what "integrated urban water management" means. Integrated water management is an area in which developing nations may make significant strides, and the urban setting provides an excellent testing ground for new ideas (Odendaal, 2000). Catchment-level IUWM, or integrated catchment management, has been implemented in practice. When referring to water and wastewater technology, "appropriate" means "acceptable" and "reliable" for use in poor nations. However, what may be socially acceptable in Thailand may not be in Nepal or Ghana. Several elements, including resources, knowledge, institutions, time, and the environment, all have a role in determining whether or not something is suitable and trustworthy. Malaysia's cities relied mostly on waste stabilization ponds (WSP) 20 years ago to clean their water after human waste.

8.8 COMMUNITIES OF MICROORGANISMS, INCLUDING BIO FLOCS, BIOFILMS, AND GRANULES

In 2014, professionals in the field of environmental engineering and science commemorated 100 years of activated sludge (AS). AS was first reported by Ardern and Lockett in 1914 and has since been used extensively for aerobic wastewater treatment across the world. Wastewater was treated, and the suspended biomass that formed during the aeration phase was recovered and reused. At the end of the aeration phase, the sludge that had been produced and settled out was called "activated." At its core, this microbial colony flocculates out of treated effluent. As a whole, AS flocs have an asymmetrical form and a size of less than 100 m. They have a disorganized microbial composition that is often dominated by filamentous bacteria (Nancharaiah et al., 2006). In biological wastewater treatment, the settling qualities of biomass are just as important as its functional capabilities (contaminant removal). Sludge volume index (SVI) is a measure of the settling qualities that is expressed as the volume (mL) occupied by 1 g of sludge after a 30-minute settling time. Typically, the SVI30 of AS is above 100 ml/g. Traditional AS plants cannot sustain large biomass concentrations (>4 g/L) when treating weak wastewater, such as sewage. This is because AS has a weak microbiological structure and settles more slowly than other particles. The concentration of AS in the bioreactor tanks may be increased, unlike in conventional AS plants, thanks to membrane bioreactors and batch reactor sequencing. The term "biofilm" refers to microbial populations that have become entangled in their own self-produced extracellular bimolecular matrix, which is often made up of carbohydrates, proteins, and extracellular DNA (Flemming and Wingender, 2010). Numerous bacteria have adapted their lifestyles to include biofilm development in a wide range of habitats. Biofilm-forming microbes have uses in bioremediation and pollution biodegradation (Mitra and Mukhopadhyay, 2016). There are two ways to cultivate these biofilms for wastewater treatment: on a static surface or on suspended carriers. The development of biofilms is an efficient method of treating watered-down waste streams and retaining biomass (Nicolella et al., 2000; Shah, 2021). Because of this, biofilm reactors may be used to retain slow-growing microorganisms (such as nitrifiers), keep biomass concentrations high, and treat effluents like sewage and some industrial effluents that have been heavily diluted (Chaali et al., 2018).

8.9 SCOPES AND BIO-ELECTROCHEMICAL TREATMENT OF WASTEWATER

Managing large quantities of chemicals and toxic sludge is a significant challenge in the chemical industry, as it involves handling many chemicals. Companies must adhere to strict regulations and safety protocols when handling these chemicals, as they can be very dangerous and can even cause environmental damage if not properly managed. Companies must also be able to store and transport the chemicals safely and securely. Coagulation/flocculation refers to the processes of coagulation and flocculation. This process is used to separate materials from a mixture, such as solids from liquids, or particles from gas. Coagulation involves adding chemicals that cause particles to stick together, while flocculation involves adding chemicals that cause the particles to form larger clumps that can then be more easily separated from the mixture. Safe sludge disposal has become a pressing issue across the globe, especially in developing nations such as India, Pakistan, and Bangladesh, where the issue of safe sludge disposal in the environment has become a pressing one. Coagulation and flocculation are used as part of the process for treating wastewater and sewage, which can then be more easily disposed of without causing environmental damage. The increased use of these processes in developing nations has helped to reduce the amount of untreated wastewater and sewage, thus reducing the risk of water contamination and disease. The investigation into this issue found that an advanced chemical treatment solution was available for the treatment of effluents from tanneries based on observations that had been made in the industry. The advanced chemical treatment solution was able to reduce the levels of chromium, ammonia, and other contaminants in the wastewater, making it safe to be released into the environment while at the same time reducing the risk of water contamination and the spread of disease. Taking into account these factors, several novel technologies have been developed over the past few years to treat the wastewater generated by tanneries. These solutions use a combination of chemical and physical processes to remove the contaminants from the wastewater, such as adsorption, ion exchange, and oxidation. This helps to reduce the levels of toxic chemicals in the wastewater, while at the same time reducing the cost of wastewater treatment. Some of these methods include electrochemical treatment (Min et al., 2004; Shah, 2020), adsorption, ozonation (Preethi et al., 2009), and advanced oxidation (Dogruel et al., 2006; Schrank et al., 2005). Adsorption works by using an adsorbent material to bind the contaminants to its surface, while ion exchange relies on the exchange of ions between the wastewater and the treatment material. Oxidation breaks down the contaminants, turning them into simpler compounds that can be more easily removed from the wastewater. The electrochemical treatment uses an electric current to create chemical reactions that break down the contaminants. The aim is to enhance the efficacy of the tried-and-true coagulation-flocculation method of water purification to enhance its effectiveness. An electrochemical oxidation process can be accomplished either directly or indirectly with electricity. There is a significant impact on the efficiency of this procedure based on the treatment condition, the wastewater composition, the electrode material type, and the operation mode (batch or continuous). There is no doubt electrochemical treatment is the most appropriate treatment method for tannery effluent due to the high amount of dissolved particles, primarily chlorides, present in the soak yard (Sundarapandiyan et al., 2010; Szpyrkowicz et al., 2001). There is also truth in the fact that different electrode materials with varying electrolytic characteristics can significantly alter the treatment efficacy of a reactor because of different electrode materials. Since electrochemical processes involve the exchange of electrons between electrodes, the exchange of electrons is dependent on the type of electrode material used. Different electrode materials, such as titanium, zinc, and carbon, have different electrolytic characteristics, which can affect the rate of reaction, the efficiency of the process, and the overall performance of the reactor. Based on the kinetics of elimination, it was determined that organic contaminants and nutrients were removed significantly more quickly through chemical treatment compared with biological treatment. This is because different electrode materials have different oxidation and reduction potentials, which can affect the rate at which contaminants are removed from the water. Additionally, chemical treatment is typically more efficient at removing organic contaminants and nutrients because the reaction process occurs more quickly than the natural process of biodegradation. As far as nitrogen, phosphorus, chromium, arsenic, and other heavy toxic metals are concerned, electrochemical treatment is effective for the removal of these pollutants from wastewater; however, there are some limitations to its use in raw tannery effluents. Chemical treatment is more efficient because it can completely break down organic compounds and other pollutants in a shorter period of time. Electrochemical treatment is effective for the removal of nitrogen, phosphorus, chromium, arsenic, and other heavy toxic metals, but it is not as efficient in removing organic compounds. Additionally, electrochemical treatment has some limitations since it cannot be used on raw tannery effluents.

8.10 CONCLUSIONS

Wastewater reuse poses challenges in developing countries such as India due to a lack of wastewater treatment. The difficulty is in finding such low-cost, low-tech, user-friendly technologies that, on the one hand, maintain our large wastewater-dependent livelihoods and, on the other, prevent the destruction of our priceless natural resources. Wastewater irrigation is common not just in arid locations where freshwater is scarce, like the Middle East, but also in wetter areas where negative environmental impacts from wastewater discharge to waterways like the ocean and rivers provide an incentive for its reuse. Because of the proximity of municipal wastewater and food distribution networks, much of the wastewater irrigation occurs on the peri-urban outskirts of major cities. Sustainable development is a paradigm in which social and environmental considerations are integrated into the planning and implementation of development programmers; it is this paradigm that the developing world should adopt. In the past, researchers and practitioners in the field of water and wastewater management in developing nations focused their attention on low-income neighbourhoods. We paid less attention to nations like Malaysia, Thailand, and South Africa, which are developing quickly. Industrial discharges and non-point source pollution are major contributors to the decline of water quality in these nations; hence, tailored approaches and models are required to address the problem.

ACKNOWLEDGEMENT

We would like to acknowledge the Department of Agronomy at the Lovely Professional University, Phagwara, Punjab, India, for their consistent moral support and encouragement throughout the writing process.

Conflicts: None

REFERENCES

Anderson, J. (2003) The environmental benefits of water recycling and reuse. Water Science and Technology: Water Supply, 3(4), 1–10.

Angel, D. & Rock, M. T. (2009) Environmental rationalities and the development state in East Asia: prospects for a sustainability transition, *Technological Forecasting and Social Change*, 76, pp. 229–240.
Ballance, R. & Pant, B. D. (2003) Environment Statistics in Central Asia: Progress and Prospects. ERD Working Paper Series, No. 36 (Economics and Research Department, Asian Development Bank). Available at: http://www.adb.org/Documents/ERD/Working_Papers/wp036.pdf
Bhardwaj, R. M. (2005) Status of Wastewater Generation and Treatment in India, IWG-Env Joint Work Session on Water Statistics, Vienna, 20–22 June 2005.
Biswas, A. K. & Seetharam, K. (2008) Achieving water security for Asia, *International Journal of Water Resources Development*, 24, pp. 145–176.
Bixio, D., De hyer, B., Cikurel, H., Muston, M., Miska, V., Joksimovic, D., Schäfer, A. I., Ravazzini, A., Aharoni, A., Savic, D. & Thoeye, C. (2005a) Municipal wastewater reclamation: where do we stand? An overview of treatment technology and management practice, *Water Science Technology: Water Supply*, 5, pp. 77–85.
Black, M. (1994) Mega-slums: the coming sanitary crisis. Water Aid, London.
Central Pollution Control Board [CPCB]. (1999) Status of Water Supply and Wastewater Collection Treatment & Disposal in Class I Cities-1999, Control of Urban Pollution Series: CUPS/44/1999-2000.
Central Pollution Control Board [CPCB]. (2005b) Performance Status of Common Effluent Treatment Plants in India. Central Pollution Control Board, India.
Central Pollution Control Board [CPCB]. (2007a) Evaluation of Operation and Maintenance of Sewage Treatment Plants in India-2007, Control of Urban Pollution Series: CUPS/68/2007. Central Pollution Control Board, India.
Central Pollution Control Board [CPCB] (2010) Status of Water Quality in India: 2009, Monitoring of Indian Aquatic Resources Series: MINARS 2009-10 (Ministry of Environment and Forests, Government of India).
Centre for Science and Environment [CSE] (1999) Perpetual thirst: faucets of the problem, *Down to Earth*, 7(19), p. 28.
Chaali, M., Naghdi, M., Brar, S. K. & Avalos-Ramirez, A. (2018) A review on the advances in nitrifying biofilm reactors and their removal rates in wastewater treatment, *Journal of Chemical Technology & Biotechnology*, 93(11), pp. 3113–3124.
Chhabra, A., Manjunath, K. R. & Panigrahy, S. (2010) Non-point source pollution in Indian agriculture: estimation of nitrogen losses from rice crop using remote sensing and GIS, *International Journal of Applied Earth Observation and Geoinformation*, 12, pp. 190–200.
Chowdary, V. M., Rao, N. H. & Sarma, P. B. S. (2005) Decision support framework for assessment of non-point-source pollution of groundwater in large irrigation projects, *Agricultural Water Management*, 75, pp. 194–225.
Corcoran, E., Nellemann, C., Baker, E., Bos, R. D. & Osborn, H. S. (eds.) (2010) Sick water? The central role of wastewater management in sustainable development: a rapid response assessment. United Nations Environment Programme, UN-Habitat, GRID-Arendal.
Day, D. (1996) How Australian social policy neglects environments, *Australian Journal of Soil and Water Conservation*, 9, pp. 3–9.
Dogruel, S., Genceli, E. A., Babuna, F. G. & Orhon, D. (2006) An investigation on the optimal location of ozonation within biological treatment for a tannery wastewater, *Journal of Chemical Technology and Biotechnology*, 81, pp. 1877–1885.
Economic and Social Commission for Asia and the Pacific [ESCAP] (2005) State of the environment in Asia and the Pacific. United Nations ESCAP, Bangkok.
FAO (Food and Agriculture Organisation of the United Nations). (2001) Experience of Food and Agriculture Organisation of the United Nations on Wastewater Reuse in the Near East Region. Proc. Regional Workshop on Water Reuse in the Middle East and North Africa. July 2–5, 2001, The World Bank Middle East and North Africa Region. Cairo, Egypt.

FAO AQUASTAT (n.d.) FAO's Information System on Water and Agriculture (Food and Agriculture Organization of the United Nations), Available at: http://www.fao.org/nr/water/aquastat/main/index.stm (accessed 7 June 2011).

Flemming, H. C. & Wingender, J. (2010) The biofilm matrix, *Nature Review. Microbiology*, 8(9), p. 623.

Giles, H. & Brown, B. (1997) And not a drop to drink, *Water and Sanitation Services in the Developing World Geography*, 82, pp. 97–109.

Hamilton, A. J., Stagnitti, F., Xiong, X., Kreidl, S. L., Benke, K. K., & Maher, P. (2007) Wastewater irrigation: the state of play. Vadose zone journal, 6(4), 823–840.

Hanjra, M. A., Blackwell, J., Carr, G., Zhang, F. & Jackson, T. M. (2011) Wastewater irrigation and environmental health: implications for water governance and public policy, *International Journal of Hygiene and Environmental Health*, 215(3), DOI: 10.1016/j.ijheh.2011.10.003.

Hanjra, M. A. & Qureshi, M. E. (2010) Global water crisis and future food security in an era of climate change, *Food Policy*, 35(5), pp. 365–377. doi:10.1016/j.foodpol.2010.05.006.

Hide, J. M. & Kimani, J. (2000) Informal Irrigation in the Peri-urban Zone of Nairobi, Kenya. Findings from an Initial Questionnaire Survey. Report OD/TN98. HR Wallingford Ltd, Wallingford, UK.

Hide, J. M., Hide, C. & Kimani, J. (2001) Informal Irrigation in the Peri-urban Zone of Nairobi, Kenya. An Assessment of Surface Water Quality for Irrigation. Report OD/TN. HR Wallingford Ltd, Wallingford, UK.

Howarth, R. W. (2008) Coastal nitrogen pollution: a review of sources and trends globally and regionally. Harmful algae, 8(1), 14–20.

Kanarek, A. & Michail, M. (1996) Groundwater recharge with municipal effluent: Dan region reclamation project, Israel, *Water Science and Technology*, 34, pp. 227–233.

Khan, S. & Hanjra, M. A. (2009) Footprints of water and energy inputs in food production: global perspectives, *Food Policy*, 34, pp. 130–140.

Lawhon, P. & Schwartz, M. (2006) Linking environmental and economic sustainability in establishing standards for wastewater reuse in Israel, *Water Science and Technology*, 53, pp. 203–212.

Lie, P., Feng, X. B., Qiu, G. L., Shang, L. H. & Li, Z. G. (2009) Mercury pollution in Asia: a review of the contaminated sites, *Journal of Hazardous Materials*, 168, pp. 591–601.

Looker, N. (1998) Municipal Wastewater Management in Latin America and the Caribbean, R.J. Burnside International Limited, Published for Roundtable on Municipal Water for the Canadian Environment Industry Association.

Min, K. S., Yu, J. J., Kim, Y. J. & Yun, Z. (2004) Removal of ammonium from tannery wastewater by electrochemical treatment, *Journal of Environmental Science and Health*, 39, pp. 1867–1879.

Minhas, P. S., Saha, J. K., Dotaniya, M. L., Sarkar, A. & Saha, M. (2022) Wastewater irrigation in India: current status, impacts and response options, *Science of the Total Environment*, 808, p. 152001.

Ministry of Agriculture. 1998. Développement de la stratégie pour promouvoir la réutilisation des eaux usées épurées dans le secteur agricole ou autres. p.138. Direction Générale de Ressources en Eau, Groupement Bechtel International/SCET-Tunisie.

Ministry of Environment and Forest [MoEF] (2009) State of environment report, India 2009. Government of India.

Mitra, A. & Mukhopadhyay, S. (2016) Biofilm mediated decontamination of pollutants from the environment, *AIMS Bioeng*, 3(1), pp. 44–59. doi:10.3934/bioeng.2016.1.44.

Nancharaiah, Y. V., Schwarzenbeck, N., Mohan, T. V., Narasimhan, S. V., Wilderer, P. A. & Venugopalan, V. P. (2006) Biodegradation of nitrilotriacetic acid (NTA) and ferric-NTA complex by aerobic microbial granules, *Water Research*, 40, pp. 1539–1546.

Nicolella, C., van Loosdrecht, M. C. M. & Heijnen, J. J. (2000) Wastewater treatment with particulate biofilm reactors, *Journal of Biotechnology*, 80, pp. 1–33.

Odendaal, P. (2000) Integrated urban water management – a vision for developing countries, *New World Water*, pp. 10–12.

Oron, G. (1998) Water resources management and wastewater reuse for agriculture in Israel. pp. 757–778. *In* T. Asano (ed.) Wastewater reclamation and reuse. CRC Press, Boca Raton, FL.

Pakistan Environmental Protection Agency [Pak-EPA] (2005) State of the Environment Report. Commissioned by Pak-EPA, Ministry of Environment, Government of Pakistan. Available at: http://www.environment.gov pk/Publications.htm%20

Park, J. H., Duan, L., Kim, B., Mitchell, M. J. & Shibata, H. (2010) Potential effects of climate change and variability on watershed biogeochemical processes and water quality in northeast asia, *Environment International*, 36, pp. 212–225.

Peasey, A., Blumenthal, U. J., Mara, D. D. & Ruiz-Palacois, G. (2000) A Review of Policy and Standards for Wastewater Reuse in Agriculture: A Latin American Perspective. Water and Environmental Health at London and Loughborough. London School of Hygiene and Tropical Medicine, UK WEDC, Loughborough, UK.

Preethi, V., Parama, Kalyani, K. S., Iyappan, K., Srinivasakannan, C., Balasubramaniam, N. N. & Vedaraman, N. (2009) Ozonation of tannery effluent for removal of cod and color, *Journal of Hazardous Material*, 166, pp. 150–154.

Radcliffe, J. C. (2004) Water Recycling in Australia. Australian Academy of Technological Sciences and Engineering, Parkville.

Schrank, S. G., José, H. J., Moreira, R. F. P. M. & Schroder, H. F. (2005) Applicability of fenton and H_2O/UV reactions in the treatment of tannery wastewaters, *Chemosphere*, 60, pp. 644–655.

Shah, M. P. (2020) Microbial bioremediation & biodegradation. Springer.

Shah, M. P. (2021) Removal of emerging contaminants through microbial processes. Springer.

Shetty, K. V., Kalifathulla, I., & Srinikethan, G. (2007) Performance of pulsed plate bioreactor for biodegradation of phenol. Journal of Hazardous Materials, 140(1-2), 346–352.

Sundarapandiyan, S., Ramanaiah, B., Chandrasekar, R., & Saravanan, P. (2010) Degradation of phenolic resin by Tremetes versicolor. Journal of Polymers and the Environment, 18, 674–678.

Szpyrkowicz, L., Kelsall, G. H., Kaul, S. N., & De Faveri, M. (2001) Performance of electrochemical reactor for treatment of tannery wastewaters. Chemical engineering science, 56(4), 1579–1586.

Ujang, Z. & Buckley, C. (2002) Water and wastewater in developing countries: present reality and strategy for the future, *Water Science and Technology*, 46(9), pp. 1–9.

United Nations Commission on Sustainable Development [UNCSD] (1997) Comprehensive Assessment of the Fresh Water Resources of the World: A Report of the Secretary-General.

USEPA and USAID. (2004) Guidelines for Water Reuse. Rep. EPA/625/R-04/108. United States Environmental Protection Agency and United States Agency for International Development.

Van-Rooijen, D. J., Turral, H. & Biggs, T. W. (2008) Urban and industrial water use in the Krishna Basin, *Irrigation and Drainage*, DOI: 10.1002/ird.439.

Water Environment Partnership in Asia [WEPA] (2011) State of Water Environmental Issues: Indonesia, Available at: http://www.wepa-db.net/policies/state/indonesia/indonesia.htm (accessed 23 November 2011).

WHO. (2004) Global Health Observatory Data Repository, World Health Organization. Available at: http://appswho.int/ghodata/?vid=10012# (accessed 12 January 2012).

World Water Assessment Programme [WWAP] (2009) The United Nations world water development report 3: water in a changing world. UNESCO, Paris; London: Earthscan).

9 Implementation of Bioelectrochemical Oxidation Systems in Existing Wastewater Treatment Plant

Challenges in Retrofitting

Monika Sharma, Prasann Kumar, and Joginder Singh Panwar

9.1 INTRODUCTION

The three primary areas in which bioelectrochemical systems (BESs) have attracted interest recently are energy production from organic substrates, product development, and the delivery of bull's eye environmental services (Arends and Verstraete, 2012; Chandrasekhar et al., 2015a). They are one-of-a-kind because they use bacteria as catalysts to change chemical energy into electrical energy and vice versa (Bajracharya et al., 2016). These systems may host microorganisms in the form of planktonic cells or biofilms. Over the past 15 years, BESs have evolved into a flexible and promising technology. These include (1) microbial fuel cells (MFCs), which break down organic matter and generate electricity (Capodaglio et al., 2013); (2) microbial electrolysis cells (MECs), which generate valuable hydrogen gas at the cathode (Miller et al., 2019); (3) microbial desalination cells (MDCs), which provide desalinated water from saltwater or brackish water (Brastad and He, 2013); and (4) microbial electrosynthesis (Wang and Ren, 2013).

Hybrid system topologies, including BESs, membrane bioreactors, algal photobioreactors, and capacitive deionization, have been developed to improve energy consumption/production and contaminant removal (Yuan et al., 2017). In situ GW bio-electroremediation, or remediation utilizing BESs, has emerged as a promising niche among BES applications due to its unique properties. Among these are the adaptability to work in a variety of redox conditions (both anode and cathode), the ability to function at a range of potentials, and so on (Modin and Aulenta, 2017; Wang et al., 2020). Additional interesting and useful elimination mechanisms can be developed when anodic and cathodic redox conditions are combined with microbial metabolism. However, the focus of this review will be on various elements of in

 DOI: 10.1201/9781003368472-9

situ bio-electroremediation, rather than the basic process and biological mechanisms of electron-electrode transfer. Critical bio-electroremediation problems, existing research gaps, and prospective future research paths, including energy consumption and scalability, will be highlighted and discussed after a thorough evaluation of various in situ applications of BESs for groundwater remediation (Cheng et al., 2009; Clauwaert et al., 2007; Comninellis et al., 2008; Cordell et al., 2009).

9.2 BIOELECTROCHEMICAL SYSTEMS

BESs utilize electrons created when COD is oxidized by microbes to create energy. Microbial electrosynthesis processes can be carried out in these setups, with energy powering the reduction of carbon dioxide (CO_2) and the reduction or oxidation of other organic feedstocks like wastewater. Each BES has an anode and a cathode separated by a membrane. The anode side is responsible for oxidation processes (such as wastewater or acetate oxidation), whereas the cathode side is responsible for reductive reactions (like O_2 reduction or H_2 evolution). Microorganisms can maintain redox balances without oxidizing substrates or producing reduced by-products by donating or accepting electrons from electrodes (Puyol et al., 2017). Either the cell and the electrode can exchange electrons directly, or soluble molecules that can be both reduced and oxidized can accept electrons from the cell and deliver them to the electrode. Maximizing such electron transfer rates is a primary focus of current research because they are crucial to the efficiency of a larger-scale BES (Logan and Rabaey, 2012).

There are three different ways to use a BES:

- To produce and distribute power in the form of an MFC.
- The anode and cathode of an MEC are directly connected, eliminating the need for a resistor.
- As a Microbial Electrolysis Cell (MEC) into which energy is poured to speed up a reaction and/or make thermodynamically unfavourable reactions possible.

In addition to electricity generation, in theory, three product groups are particularly suited to wastewater resource recovery using a BES that has real benefits over traditional production methods. These categories include:

- Bulk chemicals, which include biofuels, platform chemicals, and plastics.
- Precursors to medications, antimicrobials, and insecticides, all of which are considered to be of high value.
- Inorganic materials, such as nutrients, can be used as fertilizers, etc.

However, the high total costs (particularly for costly metal catalysts and membranes) and the fact that most research is limited to lab-scale applications are the key hurdles preventing widespread BES-based wastewater resource recovery. Water leaks, low power output, influent variations, and unfavourable product forms all continue to plague pilot plant performance outside the laboratory. BESs need to be scaled up to at least cubic metre proportions, with reactor topologies that allow easy integration into current plant designs and infrastructures, to offer a viable alternative to traditional wastewater treatment (Heidrich et al., 2011).

Energy recovery by BES is thought to remain, at best, a niche application in wastewater treatment due to the aforementioned technical constraints and the poor value of power. In terms of BES-based H_2 production, there are significant constraints because of the slow metabolic rates of microorganisms and the very narrow physical and chemical working parameters required (Schröder, 2008). Further, even at moderate temperatures, MECs are unable to compete with methane generation in standard anaerobic digesters (Clauwaert and Verstraete, 2009). Thus, methane recovery via electromethanogenesis from high-strength wastewater is not anticipated to displace anaerobic digestion (Villano et al., 2013). Ultimately, bioelectrochemical pathways are not yet a viable option for WWTP resource recovery.

9.2.1 Hydrogen and Methane Generation via Electrochemical and Biochemical Systems

Energy security, combined with the expansion of a more sustainable energy sector, is a major global concern due to the ongoing stimulation of fossil fuel use and energy demand. It is believed that it may be possible to harness clean energy from fewer renewable sources. Hydrogen is ideal for clean and sustainable energy; however, it is often created through the use of non-renewable resources like natural gas or water. BESs can be used to recover hydrogen from any biodegradable organics on the (bio-) cathode by utilizing the biocatalytic electrolysis (bio-) anode, i.e., MEC.

Oxidation of organic materials by biocatalysts to hydrogen can take place at either the anode or the cathode. Electrons can be generated via the biological oxidation of biodegradable organics at the anode in a microbial electrochemical cell (MEC). To overcome thermodynamic hurdles, a tiny voltage (0.2–1.2 V) is applied across an external circuit (from the anode to the cathode), which causes electrons to flow and ultimately combine with protons to produce hydrogen (Cheng and Logan, 2007). Microorganisms can use H_2 as an electron shuttle, making it possible for them to create methane and other small molecular molecules. MEC, as a bioelectrochemical power-to-gas that can convert renewable excess electricity into hydrogen and methane, has greater H_2-/CH_4-producing efficiency and a wider diversity of substrate usage, making it more favourable, especially for the valorization of low concentration and tortuous organic matter (Logan et al., 2008).

9.2.2 Functional Communities Involved in Bioelectrochemical Systems

Electroactive microorganisms (EAMs) are microorganisms that take electrons from an electrode or contribute to an electrode (in either direction); these organisms are also known as exoelectrogens (electrotrophs). Presently, most environmental exoelectrogens are species of the phyla Proteobacteria and Firmicutes, which are largely facultative anaerobic microbes capable of obtaining growth energy by anaerobic respiration and fermentation metabolism. In the respiratory chain, oxidized iron serves as the last electron acceptor; hence, most exoelectrogens are Fe(III)-reducing bacteria (FRB). Direct interspecies electron transfer (DIET) and mediated interspecies electron transfer (MIET) (Cai et al., 2020) are two methods for moving electrons from the cathode to the anode. The latter is only possible on electrically conductive surfaces if the outer membrane cytochromes and electron transport proteins are in physical contact. Components of the MIET are:

1. Electrons transport molecules that are produced internally and assist in the transfer of electrons from the cells to the anode;
2. Pili that are electrically conductive and can transmit electrons over long distances;
3. Diffusion of electrons between species is facilitated by water-soluble electron transporters (electron-accepting microbes are methanogens).

9.3 WATER RECLAMATION AND REUSE TECHNOLOGIES

Since water accounts for nearly all of the mass in wastewater, its recovery and reuse may be preferable to other methods such as desalination or the transportation of large quantities of freshwater over great distances. As a result of climate change and other factors, freshwater supplies are becoming increasingly scarce worldwide, making water conservation and reusing wastewater from homes an important aspect (Wang et al., 2015). Effluents from secondary wastewater treatment operations still retain trace quantities of organic contaminants such as pharmaceuticals, polychlorinated biphenyls (PCBs), and pesticides, and biological oxygen demand (BOD) is only reduced by 95%. The effluent from secondary wastewater treatment procedures must undergo additional processing on advanced treatment lines to meet the stringent legal regulations for microbe and micropollutant concentrations in reclaimed water (Eslamian, 2016). The three main categories of modern water purification methods are filtration, disinfection, and advanced oxidation.

9.3.1 Membrane Filtration

Reliable advanced treatment is made possible by membrane technologies, which are widely regarded as foundational technology in the development of cutting-edge approaches to wastewater reclamation and reuse. Their benefits include effective microbe retention without inducing resistance or corollary generation, reduced space requirements, and acting as a physical barrier against particle material. Worldwide, membranes are a key component of several well-known, large-scale, advanced treatment designs that are utilized for artificial groundwater recharge, indirect potable reuse, and industrial process-water production. Effluents treated with ultrafiltration (UF) membranes are of high quality since they are free of colloids, proteins, polysaccharides, and even most bacteria and certain viruses. Nanofiltration (NF) and reverse osmosis (RO) are two membrane-based separation processes that can be used to purify water by removing lamentable ions and dissolved particles (Wintgens et al., 2005). This method involves multiple treatment stages and produces large volumes of reclaimed water, which are used to replenish the city's natural drinking water reserves. The initial membrane filtration phase inherent in membrane bioreactors (MBRs) may make them ideal for wastewater reuse. Following primary sedimentation, the NEWater project's pilot application of MBR/RO/UV effectively recovered water of drinkable quality (Lee and Tan, 2016). MBRs use microporous membranes for solid-liquid separation in addition to the activated sludge process (ASP). They suffer from several drawbacks when compared to the CAS method, including higher complexity, less easily dewaterable sludge, and increased vulnerability to shock loads. In addition, MBRs have greater equipment and operational expenses, mostly because of the need for more aeration at higher loading rates and more frequent membrane cleaning (Judd et al., 2008). Membrane technologies are expensive to operate but provide high-quality effluent for water reuse. At large fluxes, wastewater membrane fouling can be a major issue. Low fluxes diminish operational costs, but more membrane units are needed, increasing capital expenses (Pearce, 2008). Extensive pre-treatment of secondary effluents reduces fouling and blockage (Wintgens et al., 2005). The disposing of complex retentate increases the expense of large-scale membrane-technology wastewater reuse (Banjoko and Sridhar, 2016). Membrane filtering requires tremendous pressure. MF/RO systems require 3 kWh per m^3, which may exceed wastewater's chemical energy. Côté et al. (2005) estimated that water reclaimed for municipal wastewater treatment costs US$0.3 per m^3 (Zanetti et al., 2010). Some potential benefits of MBRs include higher mixed liquor-suspended solids concentrations that allow for smaller reactors and independent management of sludge and hydraulic retention times. Costs for retentate discharge and revenues from water valorization were factored into the estimated total cost of €0.8 per m^3 for the CAS procedure followed by UF/RO. When compared to more traditional choices, the high process costs of reclaiming potable water for families and/or industry from wastewater were shown to be uneconomical in the Amsterdam area (van der Hoek et al., 2016). Significant amounts of electricity are always needed for membrane-based filtration processes (Batstone et al., 2015), although reduced water viscosity in warm areas may reduce these needs. However, increasing the consumption of one resource to make another available needs careful consideration in our resource-constrained society (Daigger, 2008).

9.3.2 Activated Carbon Filtration

Effluent filtered using activated carbon (AC) is of higher quality and can be reused, making AC filtration a sophisticated treatment technique. Coal, peat, petroleum coke, and nutshells are the raw ingredients that go into making an air conditioner. When subjected to high temperatures and a combination of physical and chemical agents, these carbonaceous compounds become highly efficient at filtering out chemical oxygen demand (COD), total organic carbon (TOC), chlorine, and a wide variety of hydrophobic organic pollutants like pharmaceuticals (Stefanakis, 2016). Accumulation of soluble contaminants on air conditioner filters can be attributed to two primary forces: (1) the pollutant's solubility in water and (2) the pollutant's affinity for the adsorbent. AC can be used in the form of either a fine powder (PAC) with a grain width of 0.07 mm or less or as granular activated carbon (GAC). It requires a dedicated pressure- or gravity-driven filtration unit; PAC can be applied straight to the activated sludge unit ahead of subsequent stages of filtration. GAC may be regenerated at a low cost on-site, while PAC must be disposed of along with the sludge after use (Trussel, 2012). It has been proven in several studies that AC filtration, combined with other sophisticated treatment stages, successfully removes pollutants from water. Research from 1976 demonstrated that AC combined with ozone oxidation effectively eliminates 90% of pesticides from water supplies. Given that AC can act as a catalyst in the ozonation reaction and that ozone enhances the pore size and active surface area of AC, the combination of AC with ozonation promotes the removal/degradation of many flourish contaminants (Reungoat et al., 2012). Furthermore, the filtration efficacy of membrane filtration systems is greatly enhanced if AC is applied upstream of the membrane filtration units (Sagbo et al., 2008). However, the efficiency of AC as a membrane pre-treatment step is debatable when compared to other recourses. Low molecular weight and highly polar chemicals, such as amines, nitrosamines, glycols, and certain others, may not be absorbed by AC filtering (Çeçen, 2012). It's also important to factor in the cost of replacing or cleaning the filter when contaminants are transferred there from the water (Oller et al., 2011).

9.3.3 Advanced Oxidation Processes

Wastewater treatment is becoming increasingly disquieted with the removal of new pollutants like pharmaceuticals, and this issue must be taken into account during water reclamation. Pharmaceuticals, dyes, and pesticides are not biodegradable; hence, they must be destroyed using advanced oxidation processes (AOPs), which generate hydroxyl radicals (OH) as highly reactive oxidant agents. External energy sources, such as electricity or light, are commonly used to power AOPs. After biological treatment, they are often used as a last cleansing and disinfecting step (Petrovic et al., 2011). This system's setups vary depending on the nature and quantity of contaminants present, as well as the quality of effluent that must be produced. The oxidation rate of organics can be greatly increased through the sequential application of multiple AOPs or the combined application of single AOPs. However, there may be drawbacks to using AOPs, such as the high cost of reagents like ozone and hydrogen peroxide or the required energy source, such as UV radiation (Agustina et al., 2005). The two most common forms of AOP, ozone and UV irradiation, are briefly discussed here.

9.3.3.1 Ozone (O_3)

It is a typical oxidizer made on-site from dry air or pure oxygen. Water reuse removes bacteria, viruses, and protozoa. Pressure, pH, and contact duration improve pollutant decomposition, while temperature inhibits it. Ozonation's energy requirements and short-term stability make it expensive. Bromide in water may convert to bromate during ozonation, forming carcinogenic bromated chemical molecules. This applies mostly to saltwater desalination, drinking water treatment, and wastewater effluent polishing. AC filtration reduces biodegradable chemicals after ozonation (Stefanakis, 2016).

9.3.3.2 Ultraviolet (UV) Irradiation

It is one of the most popular alternatives to chemical disinfection since it is quick, effective, safe, and inexpensive. Ninety ultraviolet light wavelengths pack enough energy to cause unstable pollution molecules to release electrons. In addition to its photolytic effect on dissolved chemicals, UV technology may also be able to degrade other pollutants through the photochemically assisted formation of oxidants such as hydroxyl radicals and photochemically assisted catalytic reactions (Masschelein and Rice, 2002).

DNA reactivation can occur following treatment with UV light because microorganisms have evolved methods to repair their partially denatured DNA. The UV dose used, the persistence of additional disinfectants, the duration of contact, the value of pH and temperature, and the variety and abundance of bacteria in the wastewater all have a role in the risk that could be posed. Furthermore, UV light waves may not reach all microorganisms due to the physiochemical properties of the treated effluent, such as turbidity, hardness, suspended particles, iron, manganese, and humic acid concentration (Brahmi et al., 2010). The treatment alters the composition of the microbial community in terms of the types of bacteria present, but the total number of bacteria in the water can return to pre-treated levels in as little as five days. To enable adequate pollutant removal from wastewater by UV irradiation, the aforementioned parameters must be modified accordingly (Guo et al., 2009). Chlorination is the most common practice for disinfecting water for reuse without introducing harmful bacteria, viruses, or protozoa. Disinfecting wastewater with chlorine is done all over the world using chlorine gas, hypochlorite solution, and solid chlorine. Chlorination may eliminate disease-causing microorganisms, but it isn't without its own set of hazards. Chlorination by-products are formed when normally safe items react with the disinfectant (Jegatheesan et al., 2013). Additionally, studies have demonstrated that some viruses and bacteria may survive being chlorinated. Therefore, for secure water reclamation, it is best to combine this technology with other modern treatment procedures (Shareefdeen et al., 2016). Chlorine dosages typically range from 5–20 mg with contact times between 30 and 60 minutes. Reclaimed water must undergo dichlorination if the concentration of residual chlorine is too high for the water's intended reuse. Because of this, the price of chlorination may rise by as much as 30% (Lazarova et al., 1999).

9.3.4 Microbial Fuel Cell

MFCs are integrated oxic-anoxic dual reactor that converts chemical energy from organic components in wastewater into electrical output via the biocatalytic processes of electrogenic bacteria (Jadhav et al., 2015). The anaerobic anodic compartment is juxtaposed with an aerobic cathodic chamber. In the anodic chamber, bacteria decompose the organics from the wastewater, releasing protons and electrons. An ion-exchange membrane transfers protons to the cathodic reduction site. In addition, the external electric circuit transfers electrons to collect the energy. This means that MFC can generate electricity and clean wastewater effectively (Jadhav et al., 2014).

The efficiency of MFC is affected by a wide range of variables, including those related to its design, electrode qualities, operating parameters, wastewater characteristics, and bacterial conditions. The oxidation and reduction reactions possible on such a platform make this system well-suited for the removal of a wide range of wastewater contaminants, as demonstrated by the work of Chandrasekhar and

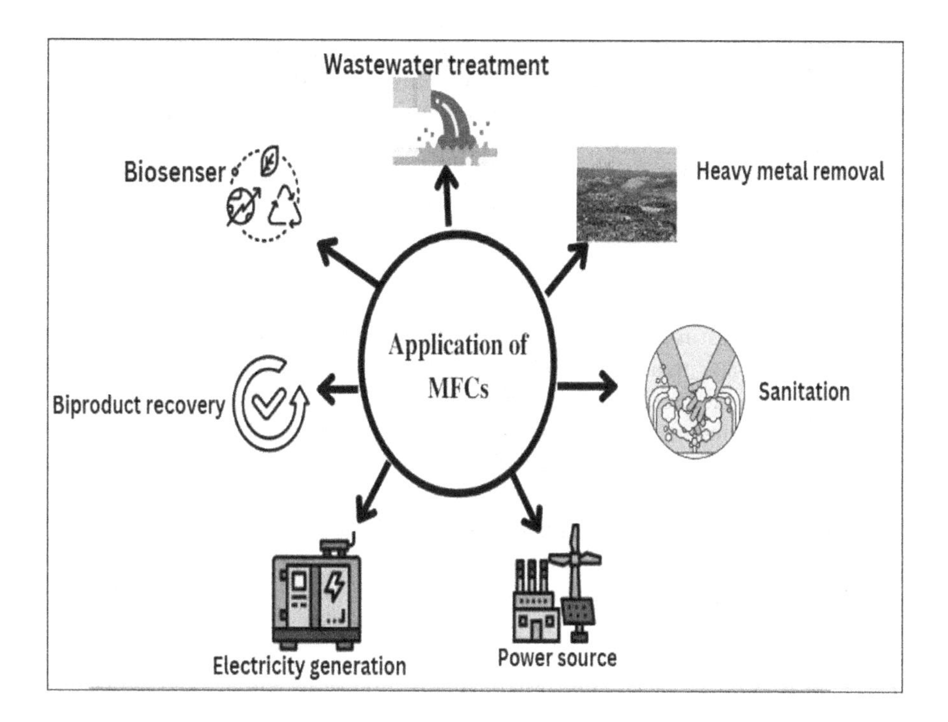

FIGURE 9.1 Various applications of MFC.

Young-Ho (2017). It can also be used to power sensors, LED lamps, and a wide variety of electrical gadgets. Bio-stimulation can make MFC a reliable biosensor for determining heavy metal concentration, COD dosage, pH, and BOD. The potential for recovering valuable by-products during WWT also makes MFC a topic of interest. Electrochemical reactions allow for the recovery of useful products and resources such as struvite from urine, dung, H_2O_2, NaOH, H_2 and methane gas, and other compounds (Figure 9.1).

9.4 WASTEWATER TREATMENT

Human activities and industrialization produce waste, which presents a concern when flushed down the drain. Due to the complexity of its degrading mechanism, wastewater contains a wide variety of complicated contaminants that contribute to a variety of environmental problems. Also, wastewater is a source of numerous bacterial communities that may thrive in harsh environments (Jadhav et al., 2019). Biogas reactors (MFCs) use these microbial populations to produce bioelectricity from treated wastewater effluent. Advantages of MFC include:

a. Lower energy consumption compared to the ASP
b. Not requiring regulated and controlled distribution systems like other fuel cells. For example, MFCs can be used to treat low COD wastewater with volatile fatty acids, both of which are unsuitable for anaerobic digestion.

c. MFCs can generate enough electricity to offset the cost of half the electricity needed for aeration in the ASP (Rittmann, 2008; Watanabe, 2008). Between 50 and 90% less sludge is produced using MFCs (Du et al., 2007). It's the only alternative method that can turn biodegradable trash into usable energy. MFCs have been discussed as a possible wastewater treatment option since 1911 (Habermann and Pommer, 1991).

9.4.1 Wastewater Treatment System

In 2014, authorities in France observed increased methane production in biological mode at wastewater treatment facilities and water treatment facilities; at the same time, laws were modified to allow for the authorization of a well-developed injection network serving the natural gas distribution sector. As an alternative to injecting bio-mode methane into natural gas distribution network locations, SUEZ has demonstrated its idea or technologies (Moustakas et al., 2020; SUEZ report, 2019). The La Wantzenau wastewater treatment station, the fourth largest in France, treats water contaminated by organic and inorganic debris for one million people. BioVALSAN has used two-thirds of the CO_2 emissions from this process. Biofuel synthesis from several stations using combined new optimization methodologies for sludge treatment and biogas recovery has the potential to have the least environmental footprint in France (Parra-Orobioa et al., 2018). This form of energy production can be used as evidence of Strasbourg's progress towards a new type of local, sustainable, carbon-absorbing energy. SUEZ runs the wastewater source treatment facility, and the area natural gas distributors are participating in the cutting-edge BioVALSAN project (SUEZ report, 2019). Biogases from the BioVALSAN project are said to be enough to supply the annual energy needs of 5,000 low-income dwellings. These biogases are produced by injecting bio-mode methane into natural source gas network stations. BioGNVAL programmes have been shown to provide clean fuel with significantly reduced emissions of fine particles, noise, and CO_2 gas compared to diesel fuel combustion (des Arcis and Euzen, 2016). The BioGNVAL project has proven its potential to produce clean fuels with 50% lower noise levels and 90% lower CO_2 emissions compared to diesel engines. The BES process is a cutting-edge technology that has been designed to improve upon the energy generation and waste disposal processes of the past. Analysis of the effects of BES technology on two-stage fermentation processes that facilitate H_2 and CH_4 biosynthesis reveals the advantages of dark fermentation for hydrogen production (Sasaki et al., 2018). When working with carbon sheets, the BES device is used since it is a cheap method that has low reactivity (for cathode or anode). Measurements made with a potentiostat are used to maintain a constant cathodic potential (1.0 V in Ag/AgCl) during these procedures. Operating a BES requires the addition of a glucose solution (concentration 10 g/L) as the primary carbon source, which allows for the generation of a large amount of electric current density with low value throughout (0.3–0.9 A/m^3/electrode) and hydrogen production (0.5–1.5 mM/day) (Zhang et al., 2017; Zhao et al., 2015). The substrate at pH 6.5 BES unit has decreased the quantity of gas mixture synthesis H_2 (52%) and CO_2 (47%), as compared to the non-bioelectrochemical system (NBES). This is reported and observed during hydraulic retention (i.e., 2 days). First, the methane producer fermenter (MF) is used after the BES operations that increase gas output (85% CH_4 and 15% CO_2) (Villano et al., 2017). MF's application came in after the NBES's. When *Ruminococcus* sp. or *Veillonellaceae* sp. flourishes, the BES process speeds improve. However, it has been reported that *Clostridium* or *Thermoanaerobacterium* species can hinder BES's effectiveness.

9.4.2 Sludge Incineration

Complete oxidation of the organic component of sewage sludge occurs during incineration, resulting in the formation of CO_2, water, and inert material (ash), all of which must be disposed of. There are several potential uses for the ash. The energy released from the combustion process can be harnessed and used. Compared to digested sludge, the heating value of raw sewage sludge is 30–40% higher, making it a potentially viable combustion fuel for electricity generation. The treatment system, the procedures utilized for sludge drying, and the type of incineration all have a role in determining whether sludge digestion or incineration is the more energy-efficient option (Bolton et al., 2001). To recover energy from organic matter, several different plant configurations are used on a global scale for biomass combustion, such as dried sewage sludge. Independent biomass combustion facilities typically have electrical efficiency between 25 and 30%. Such plants require cheap fuels, carbon fees, or set tariffs on the power they produce to be profitable. By using fluidized bed technology in power plants, electrical efficiencies can be raised to 40% with reduced costs and increased fuel versatility. Another technology that is frequently used in the European Union is the co-combustion of sludge in coal-fired power plants (Energy Resources Center, 2019). The primary problem with burning sludge is that it contains a lot of water. The water content must be lowered to below 30% to obtain a positive energy balance from combustion, which often demands energy and, thus, creates costs. Because a lot of energy is needed to evaporate the water content of the sludge, the real energy recovery potential of sludge incineration is substantially lower than the energy value of the organic matter in the sludge. Heat exchanger and thermal pump systems provide a viable option for extracting useful heat energy from WWTP wastewater (Tassou, 1988). This inexpensive heat can be distributed to the facility's dewatering and drying systems to enhance waste sludge's heating value (Dong, Y et al., 2015; Durruty et al., 2012; European Commission 2018; Fornero et al., 2008; Gai 2008; Gerrity et al., 2011; Hai et al., 2007; Huang et al., 2008).

9.5 ENERGY RECOVERY TECHNOLOGIES FROM WASTEWATER

The use of fossil fuels is projected to meet over 80% of the world's increased energy demand that is forecast to occur between 2010 and 2040. It follows that we can expect a similar rise in emissions from the use of fossil fuels. As a result of these forecasts, WWTPs will need to drastically cut their energy use through a more efficient process design that emphasizes energy recovery.

9.5.1 Methane

Globally and on a varied scale, biogas production via anaerobic sludge digestion is the most popular kind of energy recovery (Rulkens, 2008). In fully mixed reactors, the biodegradable COD fraction of the sludge can be transformed into usable biogas at a rate of about 80%. Biodegradation productivity and broth methane recovery may be enhanced by using more sophisticated reactor designs (Ma et al., 2015). The recovered methane needs to be pressurized and delivered to customers if it is not to be used locally. In places where CH_4 is abundant and inexpensive, and where it is transported via an extensive pipeline grid, this may be prohibitively expensive (Rabaey and Rozendal, 2010). Digesters have a high heating cost because up to 40% of the generated methane dissolves in the broth at moderate temperatures. Eventually, this dissolved methane could contribute to global warming. Methane leakage can be reduced by strictly regulating anaerobic wastewater treatment and sludge digestion. Biogas recovery can be improved by first capturing as much COD as possible at the plant's inlet and then digesting the primary sludge (Frijns et al., 2013). Applying chemically high-rate activated sludge as an A stage in a WWTP might increase COD concentration. This kind of energy recovery results in a 40% decrease in net energy consumption for most plants. On the other hand, large energy conversion losses of around 60% are implied when employing the produced biogas for combined heat and power recovery. The energy produced by anaerobic digestion and CHP from only 60% of the influent COD is only about half of what is needed for complete COD removal in a CAS process (Wan et al., 2016). Anaerobic membrane bioreactors (AnMBRs) and upflow anaerobic sludge blanket (UASB) reactors are two examples of direct wastewater treatment systems that don't require the addition of oxygen. Even though these procedures can remove carbon with little effort, they leave behind pathogens that must be eliminated in a separate post-treatment process (Batstone et al., 2015). However, municipal wastewater lacks sufficient organic carbon content for direct anaerobic treatment. For this reason, large conventional facilities only use anaerobic digesters for sludge line treatment (Logan and Rabaey, 2012).

9.5.2 Thermal Energy

Thermal energy in municipal wastewater is approximately 2.5 times higher than the maximal chemical energy contained in the COD (assuming a 6°C effluent temperature change). Effluent heat comes primarily from domestic and industrial water heating, with some contribution from microbial reaction heat (Hartley, 2013). Effluent can be used as a reliable heat source that can be recovered with heat pumps due to its comparatively minor seasonal temperature changes compared to ambient temperatures (Kim et al., 2007; Kracke et al., 2015; Lee et al., 2010; Mathuriya 2014; Mehta et al., 2015; Nouri et al., 2006; Oh et al., 2010; Ormad et al., 2008; Oturan and Aaron 2014). Since the influent still contains many impurities that can cause fouling issues in the equipment, it is recommended that the effluent be used as the intake source for heat pumps. In addition, heat exchangers might induce a drop in influent temperature, which could potentially disrupt biological reactions in the treatment process (Chae and Kang, 2013). In most cases, a heat pump can produce three to four times as much thermal energy from low-temperature wastewater as it does from electrical energy (Mo and Zhang, 2013). An alternative on-site application of recovered thermal energy to conventional heating and cooling systems is sludge drying. However, similar to water reuse, thermal energy recovery may hit a snag due to a time-and-location mismatch in supply and demand. Using thermal energy storage facilities, like aquifers, could be a viable option for dealing with this issue (van der Hoek et al., 2016). Although it's a good idea to sell any excess heat to adjacent consumers, there may not be enough demand in the spring and fall when there is less of a need for district heating and cooling (Chae and Kang, 2013). A total of 500 wastewater heat pumps with capacities between 10 and 20 MW were reportedly in use in 2008 (Schmid, 2008). In many places of the world, thermal energy from wastewater has been used to construct large-scale district-heating systems (Mo and Zhang, 2013). Wastewater heating and cooling systems have been demonstrated to significantly cut energy use, especially in Japan. By implementing thermal energy recovery from effluents, the city authority of Osaka, for instance, cut energy use by 20–30%. Sapporo, Japan, uses its effluents every winter to melt massive amounts of snow directly (Shareefdeen et al., 2016).

9.5.3 Hydropower

Recovering electricity through the use of hydropower technologies applied to effluents is a common practice, as it makes use of the WWTPs' steady flow and, depending on the site, a particular hydraulic head. The Archimedes screw, water wheels, and turbines are only a few examples of useful techniques that may be applied to an effluent flow and produce dependable results. However, if these systems are to be used with raw sewage, they must be constructed of corrosion-resistant materials like stainless steel (Berger et al., 2013).

9.5.4 Hydroelectricity's Power Generation Capacity

The hydraulic head and flow rate are crucial to the success of any technological endeavour. Electricity rates, taxes, financial incentives, and the cost of connecting to the power grid are all non-technical elements that affect the economic sustainability of energy recovery routes. When energy prices rise, the system becomes more cost-effective if the recovered electricity is consumed locally. Therefore, economic viability is always location-specific and is influenced by both current and future market conditions in addition to physical factors like the technology chosen (Power et al., 2014). While specific large-scale implementations in Australia, the UK, and Ireland have demonstrated the economic viability of hydropower technology in WWTPs, this aspect is rarely analyzed in depth in scientific case studies. The flow rate of a WWTP effluent stream is the single most critical factor in determining the hydropower potential of that stream, and it is highly variable due to factors such as the seasons, the economy, the infrastructure, and the population. Flow and pressure are often specified during the installation process, and it is recommended that these values be maintained as close to design specifications as feasible to ensure optimal operation (McNabola et al., 2014).

9.5.5 Other Biofuels

Wastewater streams from cities can be exploited for a variety of fuels, not just methane. Feedstock accounts for 40–80% of total production expenses in sugar-based conventional biofuel manufacturing. Therefore, there may be substantial economic potential in converting wastewater COD into biofuels (Chang et al., 2010; Shah, 2021). Even downstream processing and the considerable dilution of the recoverable materials remain critical (Puyol et al., 2017). However, wet sewage sludge can be rapidly gasified to make syngas. It's a method of transforming carbon-based materials into carbon monoxide and hydrogen through heating, often at high pressures of 15–150 bar (Sohi et al., 2009). Syngas produced from sewage sludge must be purified before being utilized as a fuel since it contains contaminants that could harm fuel cells, engines, or turbines if not removed (Manara and Zabaniotou, 2012). (Figure 9.2.)

Extreme water treatment technologies can be used to extract syngas from municipal sewage sludge. Both the temperature and pressure in supercritical water gasification and partial oxidation processes are above the critical point of water (374°C, 221 bar). Under these conditions, biomass can be transformed into syngas at very high rates and with very high energy efficiency. In addition to syngas, solids (metal oxides and salts), as well as a stream of clean water suitable for disposal, are produced as by-products (Goto et al., 1999). The sludge is turned into an energy carrier in substantially shorter residence times, only a few minutes, than with existing sludge-handling technologies. Even though main product chemicals generated in reactors may be predicted using known thermodynamic equilibrium models, not all parameters controlling the final gas composition are perceived. Corrosion of the reactors owing to extreme operating circumstances is a problem that arises during operations. Another issue is a blockage from salt precipitation, which occurs because the solubility of

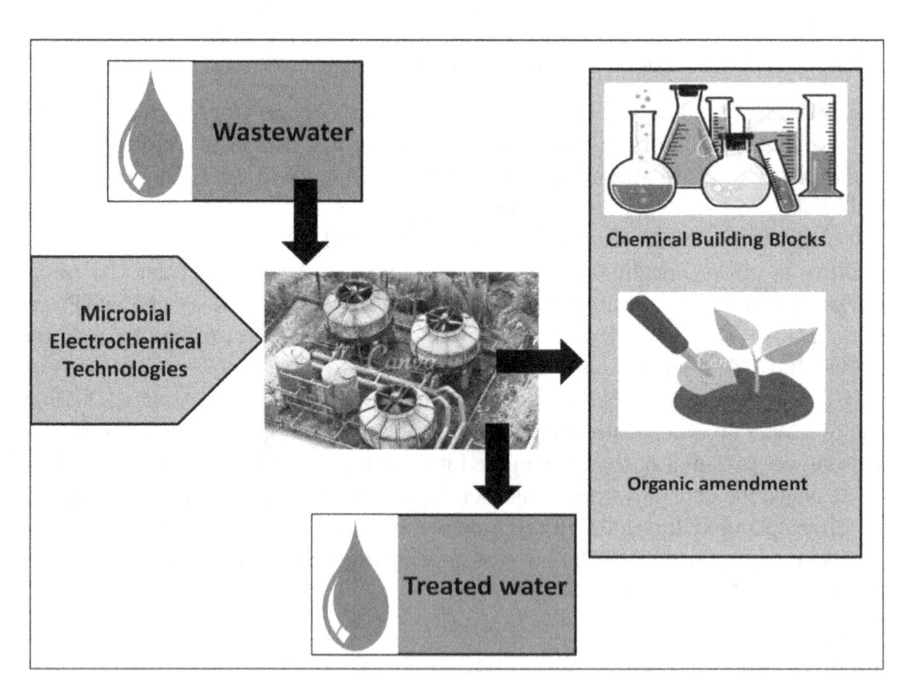

FIGURE 9.2 Bioelectrochemical systems on the recovery of water, carbon, and nitrogen in wastewater treatment plants.

salts rapidly decreases in supercritical water (Yakaboylu et al., 2015). The technique has been partially shown economically viable through several commercial applications. Potential causes for success and failure, COD destruction efficiencies, and research needs about commercial procedures have all been recorded and analyzed elsewhere (Qian et al., 2016). Additionally, phototrophic and/or lithotrophic bacteria can be used in a two-step anaerobic sludge treatment process to hydrolyze and ferment the wastewater into acid. Given that dark fermentation converts only about a third of the COD into hydrogen and the rest into VFA, the latter is often utilized in conjunction with the former to increase overall hydrogen output (Lee et al., 2014). Photo fermentation is commonly used in conjunction with dark fermentation.

One such fuel that can be made from sewage sludge is biodiesel. Anaerobically adapted microbes may ingest and accumulate lipids, which can make up a sizable amount of the organic fraction in urban wastewater. If the surface of wastewater treatment reactors were skimmed for this lipid-rich biomass, it could be used as a feedstock for high-yield biodiesel manufacturing (Muller et al., 2014). In high-rate ponds, photoautotrophic microalgae have been examined as a potential biodiesel production pathway (Muniraj et al., 2015). One key limitation is that in nations with a winter season, the climatic conditions necessary for optimal performance by phototrophic organisms are not always present. However, the price of land, photobioreactors, and algae harvesting all add up to make this method of biodiesel generation expensive (Gao et al., 2014).

9.5.6 Nitrogenous Fuels Can Also Be Recovered from Wastewater

The CANDO process is one way to achieve this goal; it consists of three stages: (i) nitridation of NH_4^+ to NO_2, (ii) partial anoxic reduction of NO_2 to N_2O, and (iii) chemical N_2O conversion to N_2 with energy recovery. Direct recovery of NH_3 from concentrated side streams by methods such as stripping is another option. NH_3 can be burned to produce electricity or used as fuel in vehicles. Sludge digestion can recover methane and turn it into N_2O for burning after nitridation and additional abiotic or biological reduction. However, ammonia fuel recovery processes typically waste more energy than they generate, rendering them untenable from a monetary standpoint. Municipal wastewater often has very low N contents, which is a major hindrance to these methods. Therefore, it appears more prudent to recover ammonia as fertilizer than to use it as a source of energy (Gao et al., 2014).

9.6 FERTILIZER RECOVERY TECHNOLOGIES FROM WASTEWATER TREATMENT

There is a connection between WWTPs and global nutrient cycles because some of the nitrogen and phosphorus fertilizers used in agriculture ultimately make their way into the wastewater stream (Daigger, 2009). According to one global estimate, creating fertilizer uses a significant amount of energy and contributes to anthropogenic greenhouse gas (GHG) emissions. Ammonium fertilizer manufacturing is responsible for more than 90% of these emissions (Sheik et al., 2014). Ammonia fertilizer is produced via the energy-intensive Haber-Bosch process and then destroyed by the energy-intensive biological nitrification and denitrification processes in wastewater treatment plants, which raises questions about the system's overall efficiency. Thus, if ammonia recovery can be completed with less energy usage than industrial production (Daigger, 2009; Shah, 2020), it might result in significant energy savings. Because it is finite and expected to become scarce, P recovery is far more pressing than N recovery. Mining P from rocks has a tremendous environmental impact since it generates by-products like gypsum, which is typically contaminated with radioactive elements and heavy metals and is not disposed of in an environmentally acceptable manner. The normal concentration of P in wastewater is about 6 mg P/L (Xie et al., 2016), and it is introduced in faeces, home detergents, and industrial effluents. Ecological damage can be done by influent P reaching surface water if it is not eliminated during treatment. Many different approaches have been created and researched for nutrient-recovery technology.

9.7 BIOSENSING

The creation of bioelectrochemical-based biosensors for environmental monitoring has recently gained a lot of interest (Capodaglio et al., 2016; Ivars-Barceló et al., 2018). As a result, research has been focused on particular uses for groundwater pollutant detection and monitoring. The system was sensitive to temperature changes but not to changes in salinity or modifications of external resistance (and longer wiring for electrodes' connections). Velasquez-Orta et al. (2017) designed an MFC-based biosensor for such surveillance of faecal and organic pollution in shallow wells. The creation of bioelectrochemical-based biosensors for environmental monitoring has recently gained a lot of interest (Capodaglio et al., 2016; Ivars-Barceló et al., 2018). As a result, research has been focused on particular uses for groundwater pollutant detection and monitoring. The system was sensitive to temperature changes but not to changes in salinity or modifications of external resistance (and longer wiring for electrodes' connections). Velasquez-Orta et al. (2017) designed an MFC-based biosensor for such surveillance of faecal and organic pollution in shallow wells. Despite the large (6 m) separation between the anode and cathode, the electrodes still produced a measurable current, and the Geobacter species was responsible for the electron transfer (Williams et al., 2010). The rise in the concentration of biogenic Fe(II), acting as an indicator, was shown to be linearly associated with the bio-current produced by a bioanode poised at +0.2 V vs. SHE. Fe(II) is a chemical that is frequently employed in GW cleanup, and the system demonstrated its ability to accurately monitor its content (Feng et al., 2013). Using a modified *Shewanella oneidensis* strain, Webster et al. (2014) created a BES-based arsenic biosensor. The sensors have a continuous range of up to 100 M and a 40 M arsenite (AsIII) detection limit. Su et al. (2019) suggested another BES-based biosensor that can track NO_3 in actuality, but it was made to track the wastewater of secondary wastewater treatment plants. Thus, it needs organic matter as its driving energy source. Therefore, this architecture might not be appropriate for GW monitoring. However, with a few setup adjustments, biosensors created to track microscopic activity in anoxic sediments (Wardman et al., 2014) might be used for GW monitoring. Due to their potential for use in off-grid and decentralized applications, competency for in situ and on-site test observing quicker response times, and lower technical skill demands than traditional analysis methods, the development of

BES-based biosensors is of utmost interest to the research community. Ideal BES-based biosensors would provide low-cost, long-term monitoring of subterranean activities, enhancing any remediation procedures. Technologies created by Williams et al. (2010) and Wardman et al. (2014) in particular might effectively support in situ bio-electroremediation. The distance between electrodes would not be a problem since, according to Velasquez-Orta et al. (2017), it has little to no impact on the capacity to sense (Williams et al., 2010).

9.8 WASTEWATER TREATMENT ENERGY DEMAND

The cost of the energy (mainly electric) needed to treat wastewater to acceptable standards is a major factor in the overall price of the urban water cycle. It was calculated that the cost of electricity accounts for 30–35% of the total cost of wastewater treatment facilities in the USA. Energy reduction and recovery represent an important sustainability issue for maintaining required standards as stricter limits for nutrient removal or mandatory removal of currently unregulated contaminants, such as contaminants of emerging concern (CEC) and pharmaceuticals and personal care products (PPCP). The introduction of additional process steps may imply a significant increase in energy demand for treatment facilities. The lack of standardized energy monitoring systems and regular energy audits in these facilities, necessary to detect any possibility for improvement, has thwarted various efforts to adjust conventional processes to make them less energy-intensive. Typically, the BOD or COD, total suspended solids (TSSs), nitrogen (N), and phosphorus (P) in the effluent would not vary significantly from one facility to the next. It is more appropriate to compare energy consumption in terms of units of the pollutant removed, such as kWh/kg removed (Bodik and Kubaská, 2013), or in terms of "electrical energy per order" (EEO), which is defined as the amount of kWh's required to reduce a pollutant's concentration by an order of magnitude (i.e., one log or 90%) in 1 m^3 (Bolton et al., 2001). Thus, EEO values (reported as kWh/m^3 - log) represent real process efficiency rather than absolute pollutant removal. Although they are typically applied to comparing the efficiency of AOPs and other processes that rely on direct electric energy input (Trojanowicz et al., 2017), they could be easily adapted for any process driven directly or indirectly (in this case also accounting for energy conversion efficiency) by an electric power supply. Conventional WWTPs often have significant energy needs because they were built to use aerobic ASPs, which necessitate the pumping of substantial flows of compressed air or, less frequently, oxygen into the biological tank. More than half, and even as much as 70%, of a traditional facility's total energy requirement, may come from one source alone (WSUEP, 2019). While ASP requires nothing in the way of setup or maintenance, it does consume a fair amount of power when in use (Park et al., 2011; Qu et al. 2007; Ranade et al. 2014; Rao 2013; Sedlak 1991; Venkata and Chandrasekhar 2011; Wang and Xu 2012; WWAP UNWWAP 2017; Yong et al. 2014; Yoshizawa et al. 2014).

9.9 CONCLUSIONS

Successful wastewater reclamation and reuse are hindered not only by technology-related bottlenecks but also by more general ones. Since water reuse is rather a new concept in urban planning, current infrastructure seldom considers the distribution of reclaimed water. Consequently, there is little room to install a new separate pipeline network, whilst retrofitting is costly, impractical, and inconvenient. The use of reclaimed water for the irrigation of crops also entails risks, including the uptake by plants of sodium and other ions that can lead to yield losses, alter soil structures, change water infiltration rates, and contaminate soils. Various cases have shown the significant contribution reclaimed water can make to sustainable agricultural production. Describe a variety of successful reuse projects undertaken in cooperation with the agricultural sector. Although complete recovery of all the energy contained in wastewater may be unrealistic due to conversion losses, energy-neutral or even energy-positive WWTPs are increasingly becoming practicable.

ACKNOWLEDGEMENT

We would like to acknowledge the Department of Agronomy at the Lovely Professional University, Phagwara, Punjab, India, for their consistent moral support and encouragement throughout the writing process.

Conflicts: None

REFERENCES

Agustina, T.E., Ang, H.M. and Vareek, V.K. (2005). A review of synergistic effect of photocatalysis and ozonation on wastewater treatment, J. Photochem. Photobiol., C., 6(4), 264–273.

Arends, J.B.A. and Verstraete, W. (2012). Minireview 100 years of microbial electricity production: three concepts for the future, Microbial Biotechnol., 5, 333–346.

Bajracharya, S., Sharma, M., Mohanakrishna, G., Dominguez, X., Strik, D.P.B.T.B. and Sarma, P.M. (2016). An overview on emerging bioelectrochemical systems (BESs): technology for sustainable electricity, waste remediation, resource recovery, chemical production and beyond, Renewable Energy, 2016, 1–18.

Banjoko, B. and Sridhar, C.M.K. (2016).Upgrading Wastewater Treatment Systems for Urban Water Reuse, Urban Water Reuse Handbook.

Batstone, D.J., Hülsen, T., Mehta, C.M. and Keller, J. (2015). Platforms for energy and nutrient recovery from domestic wastewater: a review, Chemosphere, 140, 2–11.

Berger, V., Niemann, A., Frehmann, T. and Brockmann, H. (2013). Advanced energy recovery strategies for wastewater treatment plants and sewer systems using small hydropower, Water Utility Journal, 5, 15–24.

Bodik, I. and Kubaská, M. (2013). Energy and sustainability of operation of a wastewater treatment plant, Environ. Prot. Eng., 39, 15–24.

Bolton, J.R., Bircher, K.G., Tumas, W. and Tolman, C.A. (2001). Figures-of-merit for the technical development and application of advanced oxidation technologies for both electric- and solar-driven systems (IUPAC technical report), Pure Appl. Chem., 73, 627–637.

Brahmi, M., Belhadi, N.H., Hamdi, H. and Hassen, A. (2010). Modeling of secondary treated wastewater disinfection by UV irradiation: effects of suspended solids content, J. Environ. Sci., 22(8), 1218–1224.

Brastad, K.S. and He, Z. (2013). Water softening using microbial desalination cell technology. Desalination, 309, 32–37.

Cai, C., Tihelka, E., Pisani, D. and Donoghue, P.C.J. (2020) Data curation and modeling of compositional heterogeneity in insect phylogenomics: a case study of the phylogeny of Dytiscoidea. Molecular Phylogenetics & Evoltuion, 147, 10672. https://doi.org/10.1016/j.ympev.2020.106782

Capodaglio, A.G., Callegari, A. and Molognoni, D. (2016). Online monitoring of priority and dangerous pollutants in natural and urban waters, Manag. Environ. Qual. An Int. J., 27, 507–536.

Capodaglio, A.G., Molognoni, D., Dallago, E., Liberale, A., Cella, R., Longoni, P. and Pantaleoni, L. (2013). Microbial fuel cells for direct electrical energy recovery from urban wastewaters. T Sci World J. https://doi.org/10.1155/2013/634738

Çeçen, F. (2012). Water and Wastewater Treatment: Historical Perspective of Activated Carbon Adsorption and Its Integration with Biological Processes, in Activated Carbon for Water and Wastewater Treatment: Integration of Adsorption and Biological Treatment.

Chae, K.-J. and Kang, J. (2013). Estimating the energy independence of a municipal wastewater treatment plant incorporating green energy resources, Energy Convers. Manage., 75, 664–672.

Chandrasekhar, K. and Young-Ho, A. (2017). Effectiveness of piggery waste treatment using microbial fuel cells coupled with elutriated-phased acid fermentation, Bioresour. Technol., 244, 650–657.

Chandrasekhar, K., Amulya, K. and Mohan, S.V. (2015a). Solid phase bio-electrofermentation of food waste to harvest value-added products associated with waste remediation, Waste Management, 45, 57–56.

Chang, H.N., Kim, N.-J., Kang, J. and Jeong, C.M. (2010). Biomassderived volatile fatty acid platform for fuels and chemicals, biotechnol, Bioprocess Eng, 15(1), 1–10.

Cheng, S. and Logan, B.E. (2007). Sustainable and efficient biohydrogen production via electrohydrogenesis, Proceedings of the National Academy of Sciences of the United States of America, 104(47), 18871–18873.

Cheng, S., Xing, D., Call, D.F. and Logan, B.E. (2009). Direct biological conversion of electrical current into methane by electromethanogenesis, Environ. Sci. Technol., 43(10), 3953–3958.

Clauwaert, P. and Verstraete, W. (2009). Methanogenesis in membraneless microbial electrolysis cells, Appl. Microbiol. Biotechnol., 82(5), 829–836.

Clauwaert, P., Rabaey, K., Aelterman, P., De Schamphelaire, L., Pham, T.H. and Boeckx, P., et al. (2007). Biological denitrification in microbial fuel cells. Environ. Sci. Technol., 41(9), 3354–3360.

Comninellis, C., Kapalka, A., Malato, S., Parsons, S.A., Poulios, I. and Mantzavinos, D. (2008). Advanced Oxidation Processes Critical Review Environmental Science: Water Research & Technology Open Access Article. Published on 13 January 2020. Downloaded on 12/27/2022 11:39:12 AM. This article is licensed under a Creative Commons Attrfor water treatment: Advances and trends for R&D, J. Chem. Technol. Biotechnol., 83(6), 769–776.

Cordell, D., Drangert, J.-O. and White, S. (2009). The story of phosphorus: global food security and food for thought, Glob. Environ. Change, 19(2), 292–305.

Côté, P., Siverns, S. and Monti, S. (2005). Comparison of membrane based solutions for water reclamation and desalination, Desalination, 182(1), 251–257.

Daigger, G.T. (2008). New approaches and technologies for wastewater management, The Bridge, 382, 38–45.

Daigger, G.T. (2009). Evolving urban water and residuals management paradigms: water reclamation and reuse, decentralization, and resource recovery, Water Environ. Res., 81(8), 809–823.

des Arcis, C. and Euzen, C. (2016). The SIAAP and SUEZ Introduce BioGNVAL: An Unprecendented Solution to Covert Wastewater into Liquid Biofuel. Retrieved from: https://www.suez.com/en/News/Press-Releases/The-SIAAP-and-SUEZ-introduce-BioGNVAL-an unprecentent-solutionto-covert-waste water-into-liquid-biofuel.

Dong, Y., Qu, Y., He, W., Du, Y., Liu, J. and Han, X., et al. (2015). A 90-liter stackable baffled microbial fuel cell for brewery wastewater treatment based on energy self-sufficient mode. Bioresour. Technol., 195, 66–72.

Du, Z., Li, H. and Gu, T. (2007). A state of the art review on microbial fuel cells: a promising technology for wastewater treatment and bioenergy, Biotechnol. Adv., 25(5), 464–482.

Durruty, I., Bonanni, P.S., González, J.F. and Busalmen, J.P. (2012). Evaluation of potatoprocessing wastewater treatment in a microbial fuel cell, Bioresour. Technol., 105, 81–87.

Energy Resources Center, University of Illinois at Chicago. Wastewater Treatment Facilities Program—Reducing Energy Usage in Wastewater Treatment. Available online: http://www.erc.uic.edu/energy-efficiency/illinois-energy-now-programs/waste-water-treatment-facilities-program (accessed on 30 October 2019).

Eslamian, S. (2016). Urban Water Reuse Handbook, CRC Press, Boca Raton, FL, London, New York, NY, p. 1141.

European Commission (2018). Proposal for a Regulation of the European Parliament and of the Council on Minimum Requirements for Water Reuse.

Feng, Y., Kayode, O. and Harper, W.F. (2013). Using microbial fuel cell output metrics and nonlinear modeling techniques for smart biosensing, Sci. Total Environ., 449, 223–228.

Fornero, J.J., Rosenbaum, M., Cotta, M.A. and Angenent, L.T. (2008). Microbial fuel cell performance with a pressurized cathode chamber, Environ. Sci. Technol., 42(22), 8578–8584.

Frijns, J., Hofman, J. and Nederlof, M. (2013). The potential of (waste)water as energy carrier, Energy Convers. Manage., 65, 357–363.

Gai, X.-J. and Kim, H.-S. (2008). The role of powdered activated carbon in enhancing the performance of membrane systems for water treatment, Desalination, 225(1–3), 288–300.

Gao, H., Scherson, Y.D. and Wells, G.F. (2014). Towards energy neutral wastewater treatment: methodology and state of the art, environ, Sci.: Processes Impacts, 16(6), 1223–1246.

Gerrity, D., Gamage, S., Holady, J.C., Mawhinney, D.B., Quiñones, O. and Trenholm, R.A., et al. (2011). Pilot-scale evaluation of ozone and biological activated carbon for trace organic contaminant mitigation and disinfection, Water Res., 45(5), 2155–2165.

Goto, M., Nada, T., Kodama, A. and Hirose, T. (1999). Kinetic analysis for destruction of municipal sewage sludge and alcohol distillery wastewater by supercritical water oxidation, Ind. Eng. Chem. Res., 38(5), 1863–1865.

Guo, M., Hu, H. and Liu, W. (2009). Preliminary investigation on safety of post-UV disinfection of wastewater: bio-stability in laboratory-scale simulated reuse water pipelines, Desalination, 239(1–3), 22–28.

Habermann, W. and Pommer, E.H. (1991). Biological fuel cells with sulphide storage capacity, Appl. Microbiol. Biotechnol., 35(1), 128–133.

Hai, F.I., Yamamoto, K. and Fukushi, K. (2007). Hybrid treatment systems for dye wastewater, Crit. Rev. Environ. Sci. Technol., 37(4), 315–377.

Hartley, K. (2013). Tuning Biological Nutrient Removal Plants, IWA Publ, London, p. 246.

Heidrich, E.S., Curtis, T.P. and Dolfing, J. (2011). Determination of the internal chemical energy of wastewater, Environ. Sci. Technol., 45, 827–832.

Huang, L. and Logan, B.E. (2008). Electricity generation and treatment of paper recycling wastewater using a microbial fuel cell, Appl. Microbiol. Biotechnol., 80, 349–355.

Ivars-Barceló, F., Zuliani, A., Fallah, M., Mashkour, M., Rahimnejad, M. and Luque, R. (2018). Novel applications of microbial fuel cells in sensors and biosensors, Appl. Sci., 8, 1184.

Jadhav, D.A., Ghadge, A.N. and Ghangrekar, M.M. (2015). Enhancing the power generation from microbial fuel cells with effective utilization of goethite recovered from mining mud, Bioresour. Technol., 191, 110–116.

Jadhav, D.A., Ghadge, A.N., Mondal, D. and Ghangrekar, M.M. (2014). Comparison of oxygen and hypochlorite as cathodic electron acceptor in microbial fuel cells, Bioresour. Technol., 154, 330–335.

Jadhav, D.A., Neethu, B. and Ghangrekar, M.M. (2019). Microbial carbon capture cell: advanced bioelectrochemical system for wastewater treatment, electricity generation and algal biomass production, application of microalgae. In Wastewater Treatment: Domestic and Industrial Wastewater Treatment: Biorefinery Approaches of Wastewater Treatment (Vol. 2), Springer Pub, Cham.

Jegatheesan, J., Virkutyte, J., Shu, L., Allen, J., Wang, Y. and Searston, E., et al. (2013). Removal of Lower-Molecular-Weight Substances from Water and Wastewater: Challenges and Solutions, in Wastewater Treatment: Advanced Processes and Technologies.

Judd, S., Kim, B. and Amy, G. (2008). Membrane Bio-reactors. In Biological Wastewater Treatment: Principles, Modelling and Design, IWA Pub, London.

Kim, H.-S., Takizawa, S. and Ohgaki, S. (2007). Application of microfiltration systems coupled with powdered activated carbon to river water treatment, Desalination, 202(1–3), 271–277.

Kracke, F., Vassilev, I. and Krömer, J.O. (2015). Microbial electron transport and energy conservation—the foundation for optimizing bioelectrochemical systems, Front. Microbiol., 6, 575.

Lazarova, V., Savoye, P., Janex, M.L., Blatchley, E.R. and Pommepuy, M. (1999). Advanced wastewater disinfection technologies: state of the art and perspectives, Water Sci. Technol., 40(4–5), 203–213.

Lee, H. and Tan, T.P. (2016). Singapore's experience with reclaimed water: NEWater, Int. J. Water Resour. Dev., 32(4), 611–621.

Lee, H.-S., Vermaas, W.F.J. and Rittmann, B.E. (2010). Biological hydrogen production: prospects and challenges, Trends Biotechnol., 28(5), 262–271.

Lee, W.S., Chua, A.S.M., Yeoh, H.K. and Ngoh, G.C. (2014). A review of the production and applications of waste-derived volatile fatty acids, Chem. Eng. J., 235, 83–99.

Logan, B.E. and Rabaey, K. (2012). Conversion of wastes into bioelectricity and chemicals by using microbial electrochemical technologies, Science, 337(6095), 686–690.

Logan, B.E., Call, D., Cheng, S., Hamelers, H.V., Sleutels, T.H. and Jeremiasse, A.W., et al. (2008). Microbial electrolysis cells for high yield hydrogen gas production from organic matter. Environ. Sci. Technol., 42(23), 8630–8640.

Ma, X., Xue, X., González-Mejía, A., Garland, J. and Cashdollar, J. (2015). Sustainable water systems for the City of tomorrow—a conceptual framework, Sustainability, 7(9), 12071–12105.

Manara, P. and Zabaniotou, A. (2012). Towards sewage sludge based biofuels via thermochemical conversion—a review, Renew. Sustain. Energy Rev., 16(5), 2566–2582.

Masschelein, W.J. and Rice, R.G. (2002). Ultraviolet Light in Water and Wastewater Sanitation, Lewis, Boca Raton, FL, p. 174.

Mathuriya, A.S. (2014). Eco-affectionate face of microbial fuel cells, Crit. Rev. Environ. Sci. Technol., 44(2), 97–153.

McNabola, A., Coughlan, P., Corcoran, L., Power, C., Prysor Williams, A. and Harris, I., et al. (2014). Energy recovery in the water industry using micro-hydropower: an opportunity to improve sustainability, Water Policy, 16(1), 168–183.

Mehta, C.M., Khunjar, W.O., Nguyen, V., Tait, S. and Batstone, D.J. (2015). Technologies to recover nutrients from waste streams: a critical review, Crit. Rev. Environ. Sci. Technol., 45(4), 385–427.

Miller, G.A., Rees, R.M., Griffiths, B.S., Ball, B.C. and Cloy, J.M. (2019). The sensitivity of soil organic carbon pools to land management varies depending on former tillage practices. Soil Tillage Res. 194, 104299.

Mo, W. and Zhang, Q. (2013). Energy–nutrients–water nexus: integrated resource recovery in municipal wastewater treatment plants, J. Environ. Manage., 127, 255–267.

Modin, O., and Aulenta, F. (2017). Three promising applications of microbial electrochemistry for the water sector. Environ Sci: Water Res Technol. 3(3), 391–402.

Moustakas, K., Parmaxidon, P. and Vakalis, S. (2020). Anaerobic digestion for energy production from agricultural biomass waste in Greece: capacity assessment for the region of Thessaly, Energy, 191, 116556.

Muller, E.E., Sheik, A.R. and Wilmes, P. (2014). Lipid-based biofuel production from wastewater, Curr. Opin. Biotechnol., 30, 9–16.

Muniraj, I.K., Uthandi, S.K., Hu, Z., Xiao, L. and Zhan, X. (2015). Microbial lipid production from renewable and waste materials for second-generation biodiesel feedstock, Environ. Technol. Rev., 4(1), 1–16.

Nouri, J., Naddafi, K., Nabizadeh, R. and Jafarinia, M. (2006). Energy recovery from wastewater treatment plant, Pak. J. Biol. Sci., 9, 3–6.

Oh, S.T., Kim, J.R., Premier, G.C., Lee, T.H., Kim, C. and Sloan, W.T. (2010). Sustainable wastewater treatment: how might microbial fuel cells contribute, Biotechnol. Adv., 28(6), 871–881.

Oller, I., Malato, S. and Sánchez-Pérez, J.A. (2011). Combination of advanced oxidation processes and biological treatments for wastewater decontamination—a review, Sci. Total Environ., 409(20), 4141–4166.

Ormad, M.P., Miguel, N., Claver, A., Matesanz, J.M. and Ovelleiro, J.L. (2008). Pesticides removal in the process of drinking water production, Chemosphere, 71(1), 97–106.

Oturan, M.A. and Aaron, J.-J. (2014). Advanced oxidation processes in water/wastewater treatment: principles and applications. A review, Crit. Rev. Environ. Sci. Technol., 44(23), 2577–2641.

Park, J.B.K., Craggs, R.J. and Shilton, A.N. (2011). Wastewater treatment high rate algal ponds for biofuel production, Bioresour. Technol., 102(1), 35–42.

Parra-Orobioa, B.A., Augulo-Mosquera, L.S., Loaiza-Gualtera, J.S. and Torres-Lozada, P. (2018). Inoculum mixture optimization as strategy for to improve the anaerobic digestion of food waste for the methane production, J. Environ. Chem. Eng., 6(1), 1529–1535.

Pearce, G.K. (2008). UF/MF pre-treatment to RO in seawater and wastewater reuse applications: a comparison of energy costs, Desalination, 222(1–3), 66–73.

Petrovic, M., Radjenovic, J. and Barcelo, D. (2011). Advanced oxidation processes (AOPs) applied for wastewater and drinking water treatment. Elimination of pharmaceuticals, Holistic Approach Environ., 1(2), 63–74.

Power, C., McNabola, A. and Coughlan, P. (2014). Development of an evaluation method for hydropower energy recovery in wastewater treatment plants: case studies in Ireland and the UK, Sustain. Energy Technol. Assess., 7, 166–177.

Puyol, D., Batstone, D.J., Hülsen, T., Astals, S., Peces, M. and Krömer, J.O. (2017). Resource recovery from wastewater by biological technologies: opportunities, challenges, and prospects, Front Microbiol., 7, 2106.

Qian, L., Wang, S., Xu, D., Guo, Y., Tang, X. and Wang, L. (2016). Treatment of municipal sewage sludge in supercritical water: a review, Water Res., 89, 118–131.

Qu, X., Zheng, J. and Zhang, Y. (2007). Catalytic ozonation of phenolic wastewater with activated carbon fiber in a fluid bed reactor, J. Colloid. Interface Sci., 309(2), 429–434.

Rabaey, K. and Rozendal, R.A. (2010). Microbial electrosynthesis—revisiting the electrical route for microbial production, Nat. Rev. Microbiol., 8(10), 706–716.

Ranade, V.V. and Bhandari, V.M. (2014). Industrial Wastewater Treatment, Recycling, and Reuse: An Overview, in Industrial Wastewater Treatment, Recycling, and Reuse.

Rao, D.G. (ed.). (2012). Wastewater treatment: advanced processes and technologies. In Environmental Engineering, CRC Press, Boca Raton, FL, p. 365.

Reungoat, J., Escher, B.I., Macova, M., Argaud, F.X., Gernjak, W. and Keller, J. (2012). Ozonation and biological activated carbon filtration of wastewater treatment plant effluents, Water Res., 46(3), 863–872.

Rittmann, B.E. (2008). Opportunities for renewable bioenergy using microorganisms, Biotechnol. Bioeng., 100(2), 203–212.

Rulkens, W. (2008). Sewage sludge as a biomass resource for the production of energy: overview and assessment of the various options, Energy Fuels., 22(1), 9–15.

Sagbo, O., Sun, Y., Hao, A. and Gu, P. (2008). Effect of PAC addition on MBR process for drinking water treatment, Sep. Purif. Technol., 58(3), 320–327.

Sasaki, K., Sasaki, D., Tsuge, Y., Morita, M. and Konda, A. (2018). Changes in microbial consortium during dark hydrogen fermentation in a bioelectrochemical system increases methane production during a two stage process, Biotechnol. Biofuels, 11, 173.
Schmid, F. (2008). Sewage Water: Interesting Source for Heat Pumps, Zürich. Available from: https://heatpumpingtechnologies.org/publications/sewage-waterinteresting-heat-source-forheat-pumps-and-chillers/.
Schröder, U. (2008). From wastewater to hydrogen: biorefineries based on microbial fuel-cell technology, ChemSusChem., 1(4), 281–282.
Sedlak, R. (ed.). (1991). Phosphorus and Nitrogen Removal from Municipal Wastewater: Principles and Practice, 2nd edn, Lewis Publishers, Chelsea, MI, p. 240.
Shah, M.P. (2020). Microbial Bioremediation & Biodegradation, Springer.
Shah, M.P. (2021). Removal of Refractory Pollutants from Wastewater Treatment Plants, CRC Press.
Shareefdeen, Z., Elkamel, A. and Kandhro, S. (2016). Modern Water Reuse Technologies Membrane Bioreactors, in Urban Water Reuse Handbook.
Sheik, A.R., Muller, E.E.L. and Wilmes, P. (2014). A hundred years of activated sludge: time for a rethink, Front Microbiol., 5, 47.
Sohi, S., Loez-Capel, E., Krull, E. and Bol, R. (2009). CSIRO Land and Water Science Report: Biochar's Roles in Soil and Climate Change: A Review of Research Needs.
Stefanakis, A.I. (2016). Modern Water Reuse Technologists: Tertiary Membrane and Activated Carbon Filtration, in Urban Water Reuse Handbook.
Su, S., Cheng, H., Zhu, T., Wang, H. and Wang, A. (2019). A novel bioelectrochemical method for real-time nitrate monitoring, Bioelectrochemistry, 125, 33–37.
SUEZ Report. (2019). Water and Process Treatment Technologies and Solution. Retrieved from: https://www.suez.watertechnologies.com/2019
Tassou, S.A. (1988). Heat recovery from sewage effluents using heat pumps, Heat Recovery Syst. CHP, 8(2), 141–148.
Trojanowicz, M., Bojanowska-Czajka, A. and Capodaglio, A.G. (2017). Can radiation chemistry supply a highly efficient AO(r)P process for organics removal from drinking and waste water? A review, Environ. Sci. Pollut. Res., 24, 20187–20208.
Trussel, R.R. (2012). In Water Reuse: Potential for Expanding the Nation's Water Supply through Reuse of Municipal Wastewater. National Research Council (U.S.), ed. National Research Council (U.S.), National Academies Press, Washington, D.C, p. 262.
van der Hoek, J.P., de Fooij, H. and Struker, A. (2016). Wastewater as a resource: strategies to recover resources from Amsterdam's wastewater, Resour. Conserv. Recycl., 113, 53–64.
Velasquez-Orta, S.B., Werner, D., Varia, J.C. and Mgana, S. (2017). Microbial fuel cells for inexpensive continuous in-situ monitoring of groundwater quality, Water Res., 117, 9–17.
Venkata, M.S. and Chandrasekhar, K. (2011). Solid phase microbial fuel cell (SMFC) for harnessing bioelectricity from composite food waste fermentation: influence of electrode assembly and buffering capacity, Bioresour. Technol., 102, 7077–7708.
Villano, M., Painano, P., Palma, E., Miccheli, A. and Majone, M. (2017). Electrochemically driven fermentation of organic substrates with undefined mixed microbial cultures, Chemsuschem, 10, 3091–3097.
Villano, M., Scardala, S., Aulenta, F. and Majone, M. (2013). Carbon and nitrogen removal and enhanced methane production in a microbial electrolysis cell, Bioresour. Technol., 130, 366–371.
Wan, J., Gu, J., Zhao, Q. and Liu, Y. (2016). COD capture: a feasible option towards energy self-sufficient domestic wastewater treatment, Sci. Rep., 6(1), 1–9.
Wang, H. and Ren, Z.J. (2013). A comprehensive review of microbial electrochemical systems as a platform technology. Biotechnol Adv. 31(8), 1796–1807.
Wang, J.L. and Xu, L.J. (2012). Advanced oxidation processes for wastewater treatment: formation of hydroxyl radical and application, Crit. Rev. Environ. Sci. Technol., 42(3), 251–325.

Wang, X.C., Zhang, C., Ma, X. and Luo, L. (2015). Water Cycle Management: A New Paradigm of Wastewater Reuse and Safety Control, Springer Briefs in Water Science and Technology, Springer, Heidelberg, p. 98.

Wang, Z., Lin, T. and Chen, W. (2020). Occurrence and removal of microplastics in an advanced drinking water treatment plant (ADWTP). Sci Total Environ. 700, 134520.

Wardman, C., Nevin, K.P. and Lovley, D.R. (2014). Real-time monitoring of subsurface microbial metabolism with graphite electrodes, Front. Microbiol., 5, 1–7.

Watanabe, K. (2008). Recent developments in microbial fuel cell technologies for sustainable bioenergy, J. Biosci. Bioeng., 106(6), 528–536.

Webster, D.P., TerAvest, M.A., Doud, D.F.R., Chakravorty, A., Holmes, E.C., Radens, C.M., Sureka, S., Gralnick, J.A. and Angenent, L.T. (2014). An arsenic-specific biosensor with genetically engineered *Shewanella oneidensis* in a bioelectrochemical system, Biosens. Bioelectron., 62, 320–324.

Williams, K.H., Nevin, K.P., Franks, A., Englert, A., Long, P.E. and Lovley, D.R. (2010). Electrode-based approach for monitoring in situ microbial activity during subsurface bioremediation, Environ. Sci. Technol., 44, 47–54.

Wintgens, T., Melin, T., Schäfer, A., Khan, S., Muston, M. and Bixio, D., et al. (2005). The role of membrane processes in municipal wastewater reclamation and reuse, Desalination, 178(1–3), 1–11.

WSUEP. Ultraviolet Disinfection of Water and Wastewater. Water Disinfectant: Ultraviolet vs. Chemical or Ozone. Washington State University Energy Program. Available online: http://e3tnw.org/ItemDetail.aspx? id=13 (accessed on 23 December 2019).

WWAP UNWWAP, (2017). The United Nations World Water Development Report 2017 – Wastewater: The Untapped Resource, UNESCO, Paris.

Xie, M., Shon, H.K., Gray, S.R. and Elimelech, M. (2016). Membrane-based processes for wastewater nutrient recovery: technology, challenges, and future direction, Water Res., 89, 210–221.

Yakaboylu, O., Harinck, J., Smit, K. and de Jong, W. (2015). Supercritical water gasification of biomass: a literature and technology overview, Energies, 8(2), 859–894.

Yong, X.Y., Feng, J., Chen, Y.L., Shi, D.Y., Xu, Y.S. and Zhou, J., et al. (2014). Enhancement of bioelectricity generation by cofactor manipulation in microbial fuel cell. Biosens. Bioelectron., 56, 19–25.

Yoshizawa, T., Miyahara, M., Kouzuma, A. and Watanabe, K. (2014). Conversion of activated sludge reactors to microbial fuel cells for wastewater treatment coupled to electricity generation, J Biosci. Bioengin., 118, 533–539.

Yuan, Y., Chai, L., Yang, Z. and Yang, W. (2017). Simultaneous immobilization of lead, cadmium, and arsenic in combined contaminated soil with iron hydroxyl phosphate. J Soils Sediments. 17, 432–439.

Zanetti, F., De Luca, G. and Sacchetti, R. (2010). Performance of a full-scale membrane bioreactor system in treating municipal wastewater for reuse purposes, Bioresour. Technol., 101(10), 3768–3771.

Zhang, J.Y., Qi, H., He, Z.Z., Yu, X.Y. and Ruan, L.M. (2017). Investigation of light transfer procedure and photobiological hydrogen of microalgae in photobioreactor at different locations of China, Int. J. Hydrogen Energy, 42(31), 19709–19722.

Zhao, Z., Zhang, Y., Wang, L. and Quan, X. (2015). Potential for direct interspecies electron transfer in an electric anaerobic system to increase methane production from sludge digestion, Sci. Rep., 5, 11094.

10 Microbial Desalination Cell

Shipa Rani Dey, Prasann Kumar, Joginder Singh Panwar

10.1 INTRODUCTION

One of the most important requirements for the continued existence of any living thing is the presence of water. Without water, all forms of life on earth would come to an end. Water is an essential component of life. Even though water is easily accessible due to its abundant natural supply, the lack of access to clean water and drinking water is a major problem in many parts of the world. There is an average of 71% of water covering the surface of the earth, but only about 2.5% of that water is freshwater (Sophia and Gohil, 2018). The availability of freshwater is one of the primary concerns of humans; however, freshwater makes up only 3% of the total water resources on the earth. Only 1%, out of the total of 3%, is available for use (Carmalin Sophia et al., 2016). With the world's population growing all the time and the threat of climate change, there will soon not be enough freshwater for everyone. The unprecedented rate of global population growth is the primary driver behind the growing concerns regarding the availability of freshwater and other essential sources of water supply in many regions of the world (Gholizadeh et al., 2017). According to a report by the World Water Assessment Programme, the world will experience a shortage of water of 40% by the year 2030 as a direct result of climate change. In addition, urbanization and rapid industrialization cause a significant increase in the amount of wastewater produced (Jafary et al., 2018). As a consequence of this, it is of the utmost importance to produce clean water by treating wastewater, seawater, brackish water, and industrial wastewater, amongst other types of water, to fulfil the requirements of the global water market (Ghasemi et al., 2016). Since the 1960s, technological advancements in desalination have been playing an important part in this setting. The number of desalination treatment plants has been steadily growing over the past few decades, and this trend is expected to continue (Moruno et al., 2018a). The processes of electrodialysis, nanofiltration, reverse osmosis (RO), multiple-effect evaporation, ion-exchange resins, and others are all used in commercial desalination. On the other hand, these methods are expensive and call for a lot of energy, which in turn causes an increase in the price of water, making them unsuitable for widespread application (Moruno et al., 2018a; Salehmin et al., 2021). The use of renewable energy sources, such as solar, wind, and bioenergy, has recently come to the attention of a significant number of scientists as a potential method for reducing the amount of energy needed for desalination (Zhang et al., 2018). As a result, it is essential to develop technologies that are capable of desalinating water

DOI: 10.1201/9781003368472-10

while also producing electricity at the same time (Anusha et al., 2018). The desalination technologies that are currently available are required to make seawater and brackish water potable, but they are very expensive to operate. For example, to desalinate one cubic metre of seawater, these technologies require an amount of energy in the range of 6–68 kWh (Carmalin Sophia et al., 2016). Conventional desalination technologies, such as multi-stage flash distillation and multi-effect evaporation, have a high energy requirement because they rely on heating temperatures of more than 90 degrees Celsius to complete the desalination process (Khawaji et al., 2008). Conventional desalination technologies that use thermal and membrane technologies have recently seen some improvements (energy efficiencies), but this has not removed these technologies from the brackets of high-energy consumers (National Research Council, 2008). These technologies continue to use a significant amount of energy. As an illustration, desalination through RO requires 3–5 kWh/m^3 of energy (Shannon et al., 2008), and even the most energy-efficient RO desalination systems still require 3–4 kWh/m^3 (Gude et al., 2013; Shah, 2020) to purify seawater. RO systems use a lot of energy because they need to create high pressure (>28 atm) before desalination can happen (Mathioulakis et al., 2006). The removal of ions from an ionic solution into an electrostatic double layer formed at the electrode/solution interface of an assembled oppositely charged porous electrode pair is the process that is referred to as capacitive deionization (CDI). CDI is determined by a difference in the electrical potential that is applied between electrodes that have opposite charges (Zhao et al., 2010). CDI typically requires a voltage of 1.2 volts to function properly. Because of its relatively lower energy consumption, it is preferred over the traditional methods of desalination, which include thermal processes and RO (Ahmed and Tewari, 2018). The electro-deionization (EDI) process is yet another method that can be utilized in the desalination of salty water. ED and the utilization of ion-exchange materials are both necessary steps in this process (Arar et al., 2014). Anion and cation ion exchange membranes are alternately arranged in EDI compartments to create diluted and concentrated chambers. These chambers are separated from one another. According to Arar et al. (2014), the movement of ions through the dilute compartment, which is comprised of ion-exchange resins, is driven by a direct current. The average consumption of energy by EDI falls somewhere between 0.2 and 0.3 kW-hr/m^3 of space (Alvarado and Chen, 2014). Even though both CDI and EDI produce potable water with a relatively lower energy consumption (_0.3 kW-h/m^3), the need for external energy input raises the question of how much it will cost to run the facility.

Microbial desalination cells, also known as MDCs, are a type of technology that is currently being developed to treat wastewater and desalinate salty water in a single reactor (Cao et al., 2009; Qu et al., 2012). It is an addition to the technology known as modified microbial fuel cells (MFCs), in addition to being a bioelectrochemical system (BES). The MFC consists of positively and negatively charged electrodes that are separated by a proton-selective exchange membrane. This membrane is mediated by a peripheral circuit, and its purpose is to keep the oxygen-rich and oxygen-free environments at the respective electrodes. To compare and contrast the fuel cell and the MDC, the former operates with or without a mediator and requires additional microbial input to metabolize the substrate. The latter does not require either of these things to function. MDCs do not need to rely on superfluous bacteria as a mediator source because they rely on the electro-active internal sludge instead. The wastewater with the organic matter in it flows into the anodic side, where bacteria grow and form a biofilm, and electricity is made (Sevda et al., 2015; Shah, 2021). Watery sludge contains bio-pollutants, which are oxidized by the biofilm adhering to the anode surface, kicking off the bio-catalysis process and releasing protons and electrons. An external circuit allows for the anode and cathode to collect the electrons that have been routed to them. Either aerobic or anaerobic conditions can prevail in the positive chamber of the MDC. The difference in potential that exists between the electrode chambers is what causes the MDC to generate bioelectricity. A specific cationic membrane directs the positively charged ions to the cathode, where they combine with electrons and oxygen species to produce pure water. This process results in the transport of the ions. In the case of MDCs, some of the important key points that need to be addressed for practical applications are as follows: (1) information regarding organic loading; (2) electrodes that are cost-effective and have low potential losses; (3) pH gradients; and (4) an MDC design that is scalable and efficient in terms of cost.

The development of the MDC technology came about as a result of efforts to find more cost-effective methods of desalination. The process of using microorganisms to remove salts from water is referred to as MDC technology. According to Cao et al. (2009), one of its additional benefits is the treatment of wastewater, while another is the generation of electricity. If the MDC technology can be made commercially viable, it will be a good technology to use to help reach the Sustainable Development Goals (SDGs) six and seven. This is because MDC technology has the potential to produce drinkable water from the vast quantities of water found in the ocean. This could make it easier for everyone, but especially those who live in dry areas, to get their hands on drinkable water. Because it can treat wastewater, including black water, this technology can also be used to improve environmental sanitation. This ability allows the technology to be used in applications that improve environmental sanitation. Concerning the seventh SDG, a commercialized MDC will have the ability to generate bioelectricity for remote or rural areas, which may not be served by national grid systems due to their lack of economic viability or unfavourable impact on the environment.

It is estimated that approximately 30% of people living in the world are affected by a lack of access to clean water, and 60–70% of the world's population may face severe water shortages as a result of increasing population, depletion of natural resources, climate change, and pollution in the environment (Macedonio et al., 2012). Even though the technology to reclaim water using traditional methods of wastewater treatment and desalination has been developed and is readily available, the operational costs associated with doing so are extremely high. According to the findings of a recent study, the aerobic-activated sludge processes that are used in residential wastewater treatment plants require a minimum of 0.6 kWh per m^3 of wastewater (McCarty et al., 2011). Energy consumption ranges from 3.7 to 650 kWh per m^3 for conventional high-pressure membrane desalination and thermal desalination processes, respectively (Mehanna et al., 2010). The chemical oxygen demand (COD), which can be thought of as a form of stored energy, can range from approximately 17.8–28.7 kilojoules per gramme of wastewater. Wastewater contains a significant amount of organic carbon sources (Heidrich et al., 2011). A positive net energy balance can be achieved through the harvesting of an abundant energy source from high-strength wastewater through the process of anaerobic digestion. This is an attractive proposition (McCarty et al., 2011).

Cao et al. (2009) made the initial suggestion for the MDC in the year 2009; this cutting-edge technology is designed to perform three distinct functions simultaneously: (1) it can desalinate water, (2) it can treat wastewater, and (3) it can generate electricity. A typical MDC consists of a three-chamber system with ion-exchange membranes (IEMs), and the middle chamber is dedicated to the desalination process. In recent years, many distinct MDC system configurations have been conceived of and put into production (Chen et al., 2011; Jingyu et al., 2017; Mehanna et al., 2010; Zuo et al., 2016). The MDC system, upon its inception, was seen to have some limitations, such as high internal resistance, pH imbalance, biofouling, and a few others. These limitations are similar to those that are present in BESs, which also have their limitations. Numerous studies have investigated these limitations, as well as others, to a great extent and addressed them. The presence of exoelectrogens in the anodic chamber is an essential component of these systems as a biological technology. The performance of the MDC system is primarily dependent on the exoelectrogens that are present in the anodic chamber.

Because of rising populations and decreasing water resources, many regions of the world are facing increasingly difficult challenges regarding the provision of drinking water and the treatment of wastewater (Kokabian and Gude, 2013). Many communities are looking into desalination and water reuse as potential ways to diversify their water supply portfolios to better meet the ever-increasing demands. Conventional methods for treating wastewater, such as activated sludge, and desalination technologies for producing freshwater are both expensive and energy-intensive, and they contribute to the emission of greenhouse gases. For example, the treatment of wastewater in the United States is estimated to require 21 billion kW h (kWh) of primary energy each year, which results in nearly 20.2 billion kg of annual carbon dioxide emissions. The energy requirements of water and wastewater treatment processes, the potential for energy generation, and the effects of these processes on the environment. It is extremely difficult to provide clean water without making significant investments in energy, and the same is true for the treatment of wastewater. It takes 0.14–0.24 kWh/m^3 for a pumping head of 120–200 ft to extract potable freshwater from the groundwater source, while it takes 0.36 kWh/m^3 to treat surface waters to a potable quality (Gude et al., 2010). In a related manner, the most cutting-edge RO process has a specific energy consumption of 2.5 kWh/m^3, and that's including the pre-treatment step (Gude, 2011). The premise for the performance of MDC is based on the principles that BESs convert wastewater into treated effluents while simultaneously producing electricity and that the migration of ionic species (i.e., protons) within the system makes it easier to desalinate seawater. Because the two distinct processes can be combined to result in a process that gains energy, this integrated process is particularly appealing for regions that face severe water issues (highly salty groundwater or seawater). This is because the integrated process can result in an energy-gaining process (Forrestal et al., 2012; Luo et al., 2012a).

10.2 HISTORICAL PROGRESS OF MDC RESEARCH

By searching the Scopus database (www.scopus.com) with the keywords "microbial desalination cell," "upscaling of microbial desalination," "challenges of microbial electrochemical desalination," "wastewater treatment using microbial desalination cell," and "resource recovery by microbial desalination cell," more than 190 articles published between 2009 and 2020 were read and analyzed for information about the engineering challenges and the prospects for upscaling MDC technology. The MDC proof of concept was initially published for the first time in 2009, as was mentioned earlier. As shown in the illustration, the investigations carried out by the MDC could be divided into sixteen distinct classes according to the objectives of the study. Following pollution treatment and recovery (16%), anodic biochemical characterization (11%), and operating parameters (10%), the areas of research that have received the second-highest amount of attention are improved reactor configurations (22%) and operating parameters (10%). Few studies have been done on making new generations of electrodes (4%) and membranes (1%), studying how ions move (4%), and combining MDC with systems that recover resources with added value (2%). Another field that receives insufficient attention is mathematical modelling and simulation (4%), which ought to be improved to achieve a deeper comprehension of the dynamic behaviours of MDCs and ensure that large-scale implementations are feasible from a financial standpoint. It is also important to note that only 3% of the studies that were carried out were focused on MDC scaling up and that there is not a single review article in the body of published research that analyzes the difficulties associated with MDC upscaling and the potential solutions to these difficulties. The most significant advancements have been made in the field of MDCs since the field was first established in 2009.

10.3 MDC TECHNOLOGY

The technology known as BESs encompasses a wide variety of desalination methods, including microbial desalination cells (MDC). Cao et al. (2009) devised the MDC to remove salt from the water and generate electricity simultaneously. MFCs were the inspiration behind their development. An anode chamber and a cathode chamber are the two standard components that make up an MFC (Logan, 2008). In contrast to the MFCs that only have two chambers, a typical MDC has a third chamber that is referred to as the desalination chamber. This chamber is formed in the middle of the anode and cathode chambers by placing anion and cation exchange membranes at specified distances (s). The de-salting of NaCl solution, which has been used to represent seawater, has been the extent of the MDC technology's desalination capacity up until this point in the technology's developmental stages. The desalination process at MDC is, in theory, reliant on the electricity generated by bacterial metabolism (exoelectrogens). Exoelectrogens are responsible for the breakdown of organic matter in wastewater, after which the electrons that are produced are transferred to the anode to obtain energy. Because of this, an accumulation of protons takes place in the anode chamber, which in turn draws negative ions (Cl^-) from the desalination chamber into the anode chamber to maintain charge equilibrium (Abubakari et al.,

2019). According to Cao et al. (2009), the movement of cations (Na^+) from the desalination chamber into the cathode chamber is caused by transferred electrons at the anode surface that travel through an external wire and across an external resistor to reduce oxidized species (electron acceptors) on the cathode. The treatment of wastewater, generation of electricity, and desalination of saltwater are all accomplished through the use of these processes.

The MDC technology is one that vividly depicts the concept of the water-energy nexus, and this description can be used to describe the technology. It is possible that the energy produced by microbes during the treatment of wastewater can be used to recover the energy that was lost when pumping seawater and wastewater into the MDC. This suggests that the technology may not require the addition of additional energy from the outside for it to perform its fundamental functions. Other desalination technologies, such as RO, CDI, and EDI systems, can be supported by using surplus energy if any of it is produced by MDCs. MDCs are very similar to MFCs in that they will only generate electricity if the chemical reactions taking place inside the cells are thermodynamically possible. According to Rismani-Yazdi et al. (2008), the equation known as the Nernst equation can be used to give a thermodynamic description of the maximum voltage that can be generated by an MDC.

10.4 MDC DESIGNS

After the initial prototype of the MDC technology was successfully tested, numerous modifications to its design were implemented. Despite the many different designs, the conventional three-chamber MDC that was developed by Cao et al. (2009) is the one that has received the most attention from researchers. This is made up of a cathode chamber as well as an anode chamber and a chamber for desalination. This straightforward configuration has the potential to generate a 3 mA electric current. Jacobson et al. (2011) built a tubular MDC and referred to it as the upflow tubular microbial desalination cell (UMDC). This was done to add to the design catalogue of the MDC technology. The UMDC was made up of two different chambers that were partitioned off from one another using ion exchange membranes. An anion exchange membrane served as the outermost wall of the interior space, which was known as the anode. The anion exchange membrane and the cation exchange membrane were separated by a space that served as the desalination chamber. The cation exchange membrane that was attached to the outer tube served as the cathode. While processing a salt solution with a concentration of 30 gTDS/L, they were able to generate currents of approximately 70 mA with this apparatus. The stack structure microbial electrolysis desalination and chemical-production cell that Chen et al. (2012) built is another intriguing example of a microbial production cell (MDC). The MDC was employed in the manufacture of acid and alkali. It had two different kinds of built-in stack structures and consisted of four chambers: (1) the anode chamber; (2) the acid-production chamber; (3) the desalination chamber; and (4) the cathode chamber. Both a bipolar membrane-anion exchange membrane-cation exchange membrane stack structure and an anion exchange membrane-cation exchange membrane stack structure were the stack structures. It is possible, using this configuration, to achieve a salt removal rate of 33.9±0.02 mg/h (Chen et al., 2012). To create an osmotic microbial desalination cell (OsMDC), Zhang and He (2012) switched out the conventional MDC's anion exchange membrane for forward osmosis (FO) membrane. This resulted in the MDC functioning as an OsMDC. A combination of a cation exchange membrane and an FO membrane was produced as a result of this configuration for the desalination process. The use of FO membranes offered the additional benefit of water abstraction from analytes and facilitated the production of electricity by facilitating proton transfer across membranes. Both of these benefits were made possible through the utilization of the FO membrane. Zhang and He (2012) were able to demonstrate, with the help of the OsMDC, that microbial activities were primarily responsible for the removal of salt. While osmosis was only responsible for a 3.4% reduction in salt concentration, microbial decomposition was responsible for the removal of 57.8% of the salt. Ping and He (2013) developed the spatially decoupling anode and cathode MDC to simplify the assembly process for MDC technology. Ping and He's (2013) desalination cells were designed to accommodate a 4-litre tank housing the anode and cathode units separately. Because of the way this configuration was set up, additional anode and cathode units could be added. By adding one more cathode unit to a parallel electric circuit connection, it is possible to raise the MDC's current density from 72.3 to 116.0 A/m^3, which is a significant increase. The hydraulic-coupled MDC that Qu et al. (2013) built was yet another MDC design that was analyzed and tested. Because of the way that this MDC was designed, the anolyte from the first cell could flow into the cathode chamber of the second cell, and then from there into the anode chamber of the third cell. Additionally, saltwater was circulated in a series through the various desalination units that were assembled. A total of four cells were connected through the coupling process. Qu et al. (2013) were able to increase the amount of COD that was removed from 21% to 60±2% with the help of this MDC. The results of this study demonstrated that the COD removal efficiency of MDCs can be increased by coupling multiple MDCs together in a single system.

10.5 MDC FUNCTIONS

10.5.1 Seawater Desalination

However, the current commercial desalination techniques have a high energy demand (Drioli et al., 2015; Jamaly et al., 2014; Nikonenko et al., 2014), so membrane-based desalination processes have predominately been used for the production of pure water from saltwater. Research at MDC has focused a lot of attention on desalinating seawater because it has the potential to use less energy than other methods. Studies conducted in laboratories found that MDCs are capable of removing up to 99% of salts; the amount of salt removed was dependent on the hydraulic retention time (HRT). Desalinization was performed on the NaCl solution used in most of the MDC studies. The efficiency of desalination was generally lower when compared to that of the pure NaCl solution due to the complex ion composition of both artificial seawater and real seawater, which was also investigated (Chen et al., 2012; Jacobson et al., 2011). An experiment that lasted for eight months and used artificial seawater that contained 100 mM of NaCl and 100 mM of $NaHCO_3$ showed that the MDC was able to remove 66% of the salt at the beginning of the experiment, but after 5000 hours of operation, the desalination efficiency dropped by 27%, and there was a 47% decrease in the current density. This performance degradation was primarily brought on by the accumulation of fouling in the AEM and scaling in the CEM. Another study that desalinated artificial seawater with a UMDC illustrated that the salt reduction reached 74%, but the desalination efficiency was lesser than that with NaCl at the same concentration (Jacobson et al., 2011). This was because the UMDC had to work harder to remove the salt. Some of the compounds found in seawater, like calcium and magnesium, can cause membrane scaling when they precipitate on the ion exchange membranes. This results in an increase in ohmic resistance, which in turn leads to a reduction in the MDC's overall performance. When treating actual seawater with MDC, there is a greater risk of membrane scaling occurring. As a result, the appropriate problems need to be solved before the MDC technology can be utilized for the long term. An osmotic membrane film concentrator (OsMFC) was connected to a microbial desalination cell (MDC) and served as a pre-treatment of both wastewater (organic removal) and seawater (dilution) (Zhang et al., 2016). This helped to improve the efficiency of the desalination process. It was discovered that when the OsMFC was connected to the MDC, up to 85% of the COD and up to 96.60±0.40% of the salts could be removed. It is believed that when desalinating seawater (or high-salinity water), MDCs may act as a pre-desalination process linked to conventional desalination technologies such as RO or ED (Mehanna et al., 2010). This is a result of the slow desalination process, which results from the slow microbial metabolism in the anode. This hypothesis was put to the test through a series of experiments in a 1.90-L UMDC that served as a pre-desalination unit before an ED. It was discovered that using the MDC cut the amount of time needed for desalination by 25% and the amount of energy used by 45.30%, respectively, when compared to using ED desalination on its own. It was also shown that the MDC's bioenergy could be used to power the ED, which desalinates the MDC's wastewater. In a four-chamber microbial electrolysis desalination and chemical-production cell (MEDCC), which is fed with a mixed solution of NaCl, $MgCl_2$, KCl, and $CaCl_2$ as saltwater, valuable by-products such as alkali, acid, and magnesium can also be recovered in addition to desalinated seawater. These by-products include the ability to produce magnesium. A high desalination efficiency of 19% was achieved using artificial seawater thanks to the recovery of magnesium and calcium ions. This was accomplished by reducing membrane scaling. Therefore, the configuration of MEDCC holds a great deal of promise in terms of bringing MDCs into practise. The ion transfer mechanism and membrane stability still need to be better understood before MDCs can be used to treat actual seawater.

10.5.2 Brackish Water Desalination

One potential application area for MDC technology is the desalination of brackish water, which has total dissolved solids (TDS) of less than 10 g/L (usually 1–5 g/L) (Kokabian and Gude, 2013; Morel et al., 2012). This is possible because the desalination efficiency for high-salinity water is relatively low. MDCs are an option that should be considered when there is a high demand for freshwater and there are large quantities of brackish water available as groundwater or in estuaries. When desalinating water with low salinity, one of the issues that can arise is that the water has a low conductivity, which can increase the water's internal resistance. This results in less electricity being generated and a decrease in the efficiency of the desalination process. Ion-exchange resins were added to the desalination chamber of MDCs to improve their conductivity. Compared to the traditional MDC, the desalination rate increased by 1.5–8 times, while the ohmic resistance decreased by 55–272% (Morel et al., 2012). When using ion-exchange resins at a salt concentration that was 0.70 g/L lower, comparable results were also obtained (Zhang et al., 2012). With an initial salt concentration of 0.50 g/L, this type of stacked MDC (SMDC) was scaled up to a total volume of 10 L and achieved a desalination efficiency of 95.80% in batch mode. The removal of salt results in an increase in the salinity of the anolyte, which has the potential to affect the microbial community and, as a consequence, the performance of conventional MDCs. This is one of the most significant issues that plagues conventional MDCs. The process of desalinating brackish water has inspired the development of several different configurations, the primary focus of which is on preventing ion movement from entering the anode chamber. Salts are adsorbed on specially designed cation exchange membrane assemblies in a microbial capacitive desalination cell, also known as an MCDC (Forrestal et al., 2012). These assemblies also contain layers of activated carbon cloth. The two cation exchange membranes in the MCDC let protons move freely, which makes pH changes less likely to happen. When compared to a conventional MDC, it was discovered that an MCDC could achieve desalination efficiency that was anywhere from 7 to 25 times higher. Utilizing ACC as electrodes for ion adsorption was a further development of the MCDC concept that was being worked on. The development of a capacitive microbial desalination cell (cMDC) was made possible by the incorporation of electrochemical adsorption. cMDC was used, and the results showed that it was able to remove up to 69.40% of the salt from the desalination chamber during each batch cycle. On the other hand, no salt diffused into the anodic or cathodic chambers. However, the initial salt concentration in those systems was limited below 10 g/L NaCl, which suggests that the capacitive process may be more efficient for the desalination of low-salinity water. Both MCDC and cMDC appear to have potential, but the former is more likely to succeed (e.g., brackish water).

10.5.3 Water Softening

The removal of hardness from groundwater is typically accomplished through the utilization of lime soda softening or ion exchange softening methods. The process of lime soda softening uses up a significant quantity of lime, caustic soda, soda ash, and acids, and it generates a significant amount of sludge that needs to be treated after it has been produced (Arugula et al., 2012). Ion exchange processes are frequently utilized in the context of residential water softening. Because salt-saturated resin beds are used in ion exchange water softening, a greater amount of salts is reported to be present in the water that has been softened. Other techniques for water softening, such as nanofiltration, RO, carbon nanotube, electrodistillation (ED), and distillation, require a significant amount of energy. An enzymatic-style MDC was developed to make the process of water softening a process that consumes less energy. In that apparatus, glucose oxidation was catalyzed by a dehydrogenase enzyme to generate electricity, which resulted in the removal of 46% and 76% of hardness, respectively, depending on the corresponding feed $CaCO_3$ concentration. The high cost of capital and the relatively short lifetime of enzymatic fuel cells restrict their use in applications. As a consequence of this, conventional MDCs have been utilized for water softening. More than 90% of the hardness was removed from the water using a bench-scale MDC that was tested to treat the hard water that was collected from several locations across the United States (Brastad and He, 2013). Moreover, the MDC was successful in effectively removing several heavy metals, such as arsenic (89%), copper (97%), mercury (99%), and nickel (95%). This is the first study to show that MDCs can be used to soften water while treating wastewater at the same time. However, additional research and development are required to improve the volumetric ratio between hard water and wastewater. By exchanging the AEM membrane for a FO membrane, the concept of an OsMDC was successfully demonstrated (Zhang and He, 2012). The FO membrane is utilized to remove water from the anodic chamber, and the OsMDC's desalination chamber is utilized to reduce the concentration of salt in the original water. When open and closed circuit modes were compared, it was discovered that low-salt concentrations required current generation as the most important factor to successfully desalinate the water. Nevertheless, when there was a higher concentration of salt (20 g/L), the water flow was the primary factor that led to the decrease in conductivity. It was discovered that a greater number of Na^+ ions were eliminated compared to Cl^- ions. This was because Na^+ ions were easily transported through CEM, whereas Cl^- ions were prevented from passing through the FO membrane. An OsMDC is capable of simultaneously achieving the goals of minimizing wastewater and treating it, extracting freshwater, and desalinating salt water. Because of this, the technology has a lot of potential and could be used in the future when there is more emphasis on water reuse. According to the findings of these studies, MDCs are anticipated to emerge as a sustainable alternative to lime softening as a pre-treatment of hard water before the membrane process. Even so, the process of softening water using MDCs has several drawbacks, including a lengthy operation period, a high initial investment cost, membrane fouling, and scaling. Before MDCs can be used effectively for water softening, these constraints need to be addressed appropriately.

10.5.4 Microorganisms Used in MDC

The majority of the MDCs' anodes were inoculated with pre-acclimatized inoculum taken from previously established MFCs. The predominant microorganism seeds come from either acclimatized microbes obtained from acetate-fed MFCs or inoculum obtained from a local wastewater treatment plant. The improvement in overall system performance can be attributed to the movement of anions and cations from the desalination chamber to the anode chamber and the cathode chamber, respectively (Luo et al., 2012a). This caused an increase in electrolyte conductivity. On the other side, the number of anions like Cl^- grew over time, which stopped exoelectrogenic microorganisms from working. The exoelectrogenic activities of the microbial community were irretrievably lost at a higher salinity of 46 g/L TDS (Kim and Logan, 2013), but the microbial community performed well up to a salinity of 41 g/L TDS. Moreover, by acclimating microbes for a longer period, it is possible to increase their ability to tolerate higher salinities. When compared to a dual chamber MFC, the microbial community found in an MDC that had been running for four months and was using domestic wastewater with a COD of 2744 16 mg/L as the anodic substrate was found to be more diverse. In this particular research project, the MDC demonstrated a 131% increase in coulombic efficiency and a 52% higher COD removal compared to the MFC in a single batch cycle that lasted for 200 hours. In a desalination chamber, the first microbial community analysis in MDC was carried out with two distinct salt concentrations (NaCl) of 5 g/L and 20 g/L. Both of these salt concentrations were used in the study. The clone portion of the Pelobacter propionicus was found to be 63% and 21% with 20 g/L NaCl and 5 g/L NaCl, respectively (Mehanna et al., 2010). The clone portion of the Geobacter sulfurreducens was found to be 29% and 12% with 5 g/L NaCl. When high saline wastewater (250 g/L NaCl) was used in MFC, the researchers found that Halanaerobium predominated on the anode surface 85.7% of the time. It is necessary to conduct studies on microbial communities using varying modes of operation and increasing concentrations of salt to gain a better understanding of the factors that can limit MDCs.

10.5.5 Oxidants Used in Cathodes of MDCs

Potassium ferricyanide is the chemical oxidant that is used most frequently in the cathodes of MDCs. According to Gude et al. (2013), it possesses a high cathodic potential in addition to good electron acceptability and faster reduction kinetics. On the other hand, due to concerns regarding the toxicity of the chemicals and the cost, their use on a large scale is not recommended. The use of this chemical has been deemed acceptable at this point in the development stage of the MDC technology because relatively small quantities of the chemical are required in experiments. It is important to find alternatives that are not only more financially feasible but also better for the environment. Oxygen is an alternative that can be used, and it is easily accessible because it is found in the air (Gude et al., 2013). It is the basis for air and biocathodes, but its oxidation–reduction kinetics are more sluggish under uncontrolled conditions, necessitating the use of an expensive catalyst. In a typical air cathode, for example, the outside of the cathode is revealed to the air in the atmosphere for aeration. For reduction reactions to proceed without a hitch, a catalyst is necessary (Shehab et al., 2014). Biocathodes are referred to as the cathodes of MDCs that are constructed using living organisms that are capable of producing oxygen for reduction reactions. Photosynthesis serves as the basis for biocathode production (Hui et al., 2019). For example, Kokabian and Gude (2015) conducted research to determine the significance of the role played by the photosynthetic activity of microalgae in the operation of photosynthetic microbial desalination cells (PMDCs). They concluded that the cathode and photosynthetic reactions were limiting factors in the PMDCs' ability to function properly. In contrast to the potential reduction that takes place when chemical oxidants such as potassium ferricyanide are used as terminal electron acceptors, microbially catalyzed cathodes produce a voltage that is stable for longer periods (Wen et al., 2012).

Nevertheless, the interruption of photosynthesis during periods of darkness results in a decrease in the voltage of PMDCs. This is because photosynthesis cannot take place in the dark. Despite this, there should not be any encouragement to maintain a constant supply of light. According to Kokabian and Gude (2015), a constant supply of light can change the natural growth pattern of algae by causing alterations in the production and consumption of oxygen by the algae. This is the reason why this occurs. The utilization of microalgae has comprised a significant portion of the focus of recent developments in biocathodes. On the other hand, Abubakari et al. (2019) showed for the first time that the aquatic plant Ceratophyllum demersum could sustain cathodic reactions in a plant MDC. They referred to it as the plant microbial desalination cell, or MDC for short (PMDC). The Ceratophyllum demersum was able to maintain an oxygen concentration of at least 6.3 1.1 mg/L on average; however, when the aquatic plant began to die off, the oxygen concentration was able to drop to 3.8 0.5 mg/L, which caused a decrease in the amount of voltage that was produced. In subsequent works on biocathodes research, the construction of sustainable biocathodes may take into consideration the use of other aquatic plants.

10.5.6 Microbial Desalination Cell Configurations

The many different configurations of MDC reactors and their associated benefits and drawbacks are outlined in this article. In the sections that follow, more in-depth discussions about each particular configuration will be presented.

10.5.7 The Basic Configuration of MDC (Three-Chamber Cubic Design)

When the idea of MDC was initially presented, the cube-shaped reactor that was used was constructed out of polycarbonate material. The MDC was made up of three different chambers: (1) the anode chamber, (2) the cathode chamber, and (3) the desalination chamber. By inserting AEM and CEM in the space between the electrodes, the desalination chamber was created. Both the anode and cathode electrodes were because of the cathode and anode electrode monitors, respectively. The anode chamber, the desalination chamber, and the cathode chamber each had a volume of 27, 3, and 27 mL when they were empty. The total net volume of each chamber was brought down to 11 millilitres by using carbon felt as electrodes in the anode and cathode chambers, respectively. To establish an external electrical connection, the electrodes were linked together using graphite rods with a diameter of 5 millimetres. A 1.6 g/L solution of sodium acetate and a 1.6 g/L solution of potassium ferricyanide were introduced into the anode and cathode chambers, respectively. The desalination chamber was supplied with NaCl at three different concentrations: (1) twenty, (2) thirty-five, and (3) five grams per litre, respectively. The MDC had an operating external resistance of 200Ω when it was in use. The desalination process was driven by the migration of anions (such as Cl^{-1}) and cations (such as Na^{+1}) from the middle chamber to the anode chamber and the cathode chamber, respectively. This migration was made possible by the potential gradient that was developed across the electrodes. The MDC system was able to generate a maximum power output of 31 W/m^3, all while simultaneously removing 90% of the salt from the solution in a single cycle of desalination. Towards the end of the cycle, it was observed that the ohmic resistance of the MDC had increased from 25 to 970Ω, which went hand in hand with the efficient removal of salt. The application of potassium ferricyanide in the cathode chamber in the role of electron acceptor made it possible to achieve a high cathode potential and accelerate reduction kinetics. Moreover, the risks associated with its toxic properties, the high cost of using it, and its application in large-scale processes have been restricted. A cathode chamber that is supplied with ferricyanide solution cannot be regarded as an example of environmentally responsible technology, even though it generates a high power density. This fundamental configuration has been adapted for use in sustainable applications to realize increased desalination and power-generation capacities. This recently developed technology offered the significant benefits of desalination and power recovery by making use of domestic sewage as well as municipal solid and liquid wastes (Al-Mamun et al., 2017b; Baawain et al., 2017).

10.5.8 Air Cathode MDC

Many restrictions were applied to the utilization of the chemical catholyte, such as ferricyanide solution. On the other hand, oxygen was discovered to be an excellent and workable terminal electron acceptor due to the high reduction potential, low cost, and widespread availability it possesses. Mehanna et al. (2010) were the first people to attempt the concept of air cathode MDC using Platinum/Carbon (Pt/C) as the catalyst. This air cathode MDC operation resulted in a decrease of 43±6% in the salt-water conductivity while simultaneously producing a maximum power density of 480 mWm^{-2} and a coulombic efficiency of 68±11%, respectively. According to Park et al. (2004), hydrogen peroxide (H_2O_2), which was initially obtained as an intermediate product in the redox reaction of oxygen, was also utilized as a cathode oxidant. Due to the negligible toxic effects of atmospheric oxygen and the harmlessness of the end products, air cathode MDC processes with oxygen were utilized in a variety of studies (Alvarez-Gallego et al., 2012). Moreover, this air cathode MDC had several drawbacks, including slow redox kinetics at ambient conditions and a high power input for the mechanical equipment that was used to maintain a particular level of dissolved oxygen in the catholyte. Both of these issues were problematic for the device. Because of the first disadvantage, it was necessary to make use of pricey metals like platinum to reduce the activation potential that was superseding the oxygen reduction potential. The disadvantage of the latera was addressed in several studies, which investigated a variety of strategies for reducing the amount of energy required. In earlier designs of the MDC, the necessity of cathodic aeration caused an offset in the amount of power generated by the amount of power necessary for aeration (Werner et al., 2013). As a consequence of an investigation into several drawbacks associated with air cathode MFC that were also applicable to air cathode MDC, Logan (2010) proposed a variety of strategies to improve its performance. These strategies were developed in response to the findings of the investigation. The upgrades that were proposed to mitigate these issues included exposing the air cathode to the atmosphere, making use of passive methods for optimal oxygen transfer in the cathode, and making use of activated carbon with ultra-high surface area for achieving the desired levels of oxygen reduction without the requirement of an expensive catalyst (Biffinger et al., 2007; Freguia et al., 2008). The Pt-catalyst that was on one side of the air cathode was not good for the economy or the environment because it was expensive, always needed to be replaced, and made the compounds in the bacterial solution less pure. In contrast, research conducted on other metals as potential substitutes for platinum has shown promising results with pyrolyzed iron (II), phthalocyanine (FePc), and cobalt tetramethyl phenyl porphyrin (CoTMPP). To investigate the mechanism of electron transfer that takes place during the oxygen reduction reaction, Park and Zeikus (2003) impregnated the cathode electrode with Fe(III) compounds. In the redox reaction caused by oxygen, it was observed that the element iron went through a series of reductions and oxidations, going from Fe(III) to Fe(II) and then from Fe(II) to Fe(III). During the process of transferring electrons from the cathode to the terminal electron acceptor, the iron compounds performed the role of mediators. According to Park and Zeikus (2003), Fe(III)-cathode demonstrated improved power output, which was superior to the results obtained with woven graphite cathode. Another study found that applying a CoTMPP catalyst to the cathode surface resulted in a performance that was comparable to that of a cathode that had been impregnated with platinum. But none of these materials could be used as cathode catalysts sustainably because they needed to be replaced all the time, and using them on a large scale was expensive (Clauwaert et al., 2007b).

10.5.9 Biocathode MDC

Studies relating to biocathode have garnered a significant amount of attention recently. This can be directly attributed to the difficulties that have been associated with air cathode systems. Biocathodes were innovative and sustainable electrodes that promoted electrochemical reduction reactions in the cathode chamber by using microbes as a catalyst. These reactions were brought about by biocathodes. Biocathodes were effective microbial catalytic electrodes because of their lower construction and operational costs, as well as their flexibility in the production of valuable chemicals. They did not require costly catalysts, which made them very attractive (He and Angenent, 2006). As a result, biocathodes seemed to be effective replacements because of their capacity for self-regeneration, their ease of scaling up, and their ability to remain sustainable. It demonstrated how biocathode MDC works in a general sense. The microbial community at the biocathode catalyzed the reduction reactions that took place either directly at the electrode surface or within the catholyte (Croese et al., 2011). As a result of the electro-active bacteria that were present in the cathode chamber, the oxidative-reduction reactions were facilitated, which led to an increase in coulombic efficiency as well as an improvement in water desalination (Wen et al., 2012). Consequently, the objective of the biocathode was to facilitate the reduction reaction of an oxidant in a manner that was either direct or indirect through the utilization of microorganisms as biocatalysts (Huang et al., 2011). The terms "aerobic cathodes" and "anaerobic cathodes" refer to two distinct types of biocathodes that can be distinguished from one another by the kind of terminal electron acceptor that is used in the cathode chamber. Because of its low cost, wide availability, and high redox potential, oxygen was utilized as an oxidant in aerobic biocathodes. Bergel et al. (2005) used a biocathode covered with a biofilm made of seawater in a proton exchange membrane (PEM) fuel cell. This made oxygen reduction happen quickly and efficiently. This air-saturated biocathode produced results with a power density of 0.32 W/m^2 and a current density of 1.34 A/m^2, respectively (Bergel et al., 2005). Desalination, power generation, and anodic sludge stabilization were all accomplished concurrently through the construction of a biocathode MDC by Meng et al. (2014). This resulted in the desired parametric conditions being met, specifically an anodic pH value that was kept between 6.6 and 7.6 and high stability over a considerable period during operation. When compared to initial NaCl concentrations of 5 and 10 g/L, the desalination rates that were achieved were 46.37±1.14% and 40.74±0.89%, respectively. The open circuit voltage of this biocathode MDC configuration was 1.118 volts, which resulted in a maximum power output of 3.178 watts per m^3. In addition to this, a reduction in the start-up period of three days was observed. Wen et al. (2012) created a biocathode membrane direct current (MDC) with an aerobic cathode consisting of carbon felt as the cathode and aerobic consortia as the biocatalyst. It was discovered that the maximum voltage that could be produced with biocathode MDC was 136 mV, which was a higher value than that which could be produced with air cathode MDC. The coulombic efficiency, salinity removal, and total desalination rate (TDR) that were determined to be obtained were found to be 96.2 3.8%, 92%, and 2.83 mgh^{-1}, respectively (Wen et al., 2012). Therefore, the biocathode MDC process gave off the impression of being a potentially fruitful strategy for effective desalination and current generation. The performance of an air cathode MDC and a photosynthetic MDC with a microalgae-catalyzed biocathode was analyzed and compared by Kokabian and Gude (2013) in terms of the production of electricity and the removal of salt and COD. According to Kokabian and Gude (2013), the PMDC operation resulted in a maximum power density of 84 mWm^{-3} and a desalination rate of 40%. Despite the significant amount of attention that has been focused on aerobic biocathodes over the past few years, the primary drawback associated with this method was the level of the supply of dissolved oxygen that was required. This was the case due to the possibility that the system's performance could be hindered if there was insufficient oxygen available. An extensive amount of research effort was focused on developing anaerobic biocathodes to circumvent the limitations of aerobic biocathodes and their use in particular applications. Even so, the start-up of anaerobic biocathodes and the growth of biocathode biofilm was a rather difficult and time-consuming method (Morita et al., 2011). This was because anaerobic biocathodes cannot function in the presence of oxygen. The microorganisms that are responsible for either the direct or indirect acceptance of electrons from the cathode are referred to as electrotrophs, which is a more formal designation. There are a wide variety of terminal electron acceptors that could be utilized. Some examples of these include sulphate, nitrate, iron, manganese, fumarate, arsenate, and carbon dioxide. Microbial electrosynthesis was one of the most revolutionary applications of electrography, as it turned carbon dioxide and water into valuable multi-carbon organic compounds. This application was one of the most transformative applications of electrography. Various strategies have been reported in the literature as potential ways to bring about anaerobic conditions at the biocathode. It was one of the few methods developed for establishing anaerobic biocathodes that showed great potential (Jeremiasse et al., 2012; Villano et al., 2011). One of these methods was the addition of hydrogen and organic compounds in the cathodes.

10.5.10 Stacked MDC

The performance of the desalination process was improved by the addition of multiple IEM pairs, which were inserted between the anode and the cathode during the construction of the stacked MDC (SMDC). The flow of ions through the membrane pairs was improved as a result of this insertion, which increased the charge transfer efficiency (CTE) (Gude et al., 2013). The system was made up of a sequence of cells that alternated between concentrating on the dilution and the concentration of the substance being processed. The migration of an ion pair in each chamber across the membranes was caused by the transfer of a single electron across the electrodes. This resulted in an increased CTE and TDR (Kim and Logan, 2013a). When compared to other MDC configurations, the SMDC proved to be a more cost-efficient system for the recovery of additional energy. Chen et al. (2011) established the very first prototype of SMDC to increase the rate of desalination. They did this by creating two chambers that concentrated on desalination and one chamber that concentrated on concentration, using two pairs of CEM and AEM. According to Chen et al. (2011), a significant increase in TDR was obtained, which was 1.4 times higher than what was obtained using a single desalination chambered MDC. It is possible that variations in design and operating parameters, like the assembly of SMDC electrodes connection and the hydraulic flow process in parallel or series mode, could affect the desalination process. This is because MDC configurations are quite versatile. Choi and Ahn (2013) investigated the effect that changing the electrode connection and hydraulic flow mode had on the desalination performance of air cathode MFC. By connecting the electrodes in series and parallel, the hydraulic flow was able to achieve a higher power density of 420 mWm^{-2} while also removing 44% of the COD (Choi and Ahn, 2013). The performance of the SMDC was significantly improved by Kim and Logan (2011), who also operated four SMDCs in series, with each one containing five desalination chambers. By minimizing the energy losses in ED that were caused by high internal resistance, the intermembrane distances could be restricted. This made it possible for the SMDC to generate power that was comparable to that produced by an MDC with a single desalination chamber. The CTE was 430%, the TDR was 77 mg/h, and the salinity was reduced by 44% with this SMDC configuration (Kim and Logan, 2011). The MDC configuration that was discussed earlier had a few flaws, like an imbalance in pH, higher water losses in dilute chambers, and an over potential. The flaws decreased the efficiency of both the desalination process and the generation of electricity. Because the rate of proton release in the anode chamber was so much higher in response to the exoelectrogenic oxidation of organic matter, the anolyte became more acidic. This was in contrast to the rate of proton diffusion in the cathode chamber. Because of the presence of AEMs in the stack of membranes inserted between the anode and the cathode, proton transportation was impeded, which led to an accumulation of protons in the anode chamber and a subsequent reduction in the anolyte's pH. The respiration of anodic bacteria was hampered when the pH of the anolyte was lowered below neutral, which resulted in a reduction in the activity and growth of microorganisms. Due to the flow of organic medium through a series of MDCs in the SMDC structure, the pH changes were less noticeable than in a single-cell MDC. This made the pH changes go away (Qu et al., 2013). If you were to position the CEM in such a way that it faced the cathode, the pH level in the cathode chamber would rise as a result.

10.5.11 Recirculation MDC (rMDC)

An MDC reactor's anolyte acidification could be traced back to the strategic placement of its various chambers and the insertion of a pair of IEMs in the space between the electrodes (decrease of pH). Because of the bioelectrochemical oxidation and electrochemical reduction reactions that took place at the anode and air cathode, respectively, during the desalination process, protons and hydroxyl ions were produced. This was previously explained. The potential gradient that was caused by electrochemical reactions and the subsequent flow of electrons through the external circuit developed the major flux through IEM that consisted of salt ions rather than protons and hydroxyl ions. This was because the electrochemical reactions caused the potential gradient. Because of this circumstance, there is a possibility that the accumulation of protons and hydroxyl ions in the anode and cathode chambers will be facilitated, which will result in a change in pH within the cell. Because of the transition in pH, there has been less microbial growth and activity in the anode chamber, which has led to the efficiency of the anode being negatively impacted more than that of the cathode. The pH imbalance in the cathode chamber, on the other hand, could result in potential losses (Zhao et al., 2006). The optimal pH range for the majority of enzymes and microbes to function properly was between 6 and 8. He et al. (2008) looked into the performance of an air-cathode MFC at different electrolyte pH levels to observe how it affected the amount of current that was generated. Even when operating with a solution having a pH of 10, it was observed that the performance of the MFC did not noticeably deviate in any significant way. The open circuit potential that was measured showed that the anodic reaction performed at its best when the pH was neutral, while the cathodic reaction performed best when the pH was increasing (He et al., 2008). To improve the desalination performance and the density of the power generated, it was, therefore, important to reduce the pH variations in the MDC. There have been several studies done to correct the pH imbalance that occurs in MDC chambers, and these studies have been reported in the scientific literature. Some of these studies include the utilization of different buffer solutions and the addition of excess anolyte volumes.

10.5.12 Half-Cell Coupled MDCs

Many different applications can be accomplished by utilizing BESs due to their adaptability in terms of reactor design and operation. By using a variety of different combinations of functions that complement one another, the system's efficiency and adaptability can be increased. The development of MDC was plagued by several difficulties, including the positioning of membranes, an increase in internal cell resistance over time that caused a corresponding drop in voltage, a decrease in osmotic pressure, substantial contamination, and the loss of valuable commodities (Cheng et al., 2009). Several configurations of an integrated multi-chambered MDC were proposed to improve the desalination rate while simultaneously producing useful by-products and potentially treating wastewater. The coupling of MDC with microbial electrolysis cells (MECs) would result in the production of valuable chemicals in addition to desalination and the concurrent generation of electricity; the resultant reactor was referred to as a microbial electrolysis desalination cell (MEDC). The authors, Mehanna et al. (2010) and Luo et al. (2011), conducted preliminary research that validated the MEDC concept.

10.5.13 Capacitive MDC (cMDC)

The accumulation of positive (Na^+) and negative (Cl^-) ions within the cathode and anode chambers, respectively, was a major cause for concern during the operation of MDC. The build-up of ions will change the pH of the anolyte and catholyte, which will stop microbial growth, meaning that the anolyte and catholyte need to be replaced more often and make it harder to control TDS for water reuse (Xu et al., 2008). To solve these problems, an improved version of the microbial electrochemical desalination (MED) system that utilizes the CDI method was developed. Yuan et al. (2012) created a combined technology for the treatment of low-concentration brackish water that included both a CDI and an MFC. In this integrated system, the electricity that was generated by the MFC was placed to power the CDI module, which was then used to desalinate the water. When looking into the process of desorption of ions that were adsorbed on electrodes, researchers looked into two distinct modes: (1) discharging and (2) short-circuiting. In the discharging mode, we were able to achieve a higher desalination rate of 200.6 mg/(L/h) than in the short-circuiting mode, where we were only able to achieve 135.7 mg/(L/h) (Yuan et al., 2012). Liang et al. conducted research to determine how the removal of salt was affected by varying configurations and operational conditions of the MFC-CDI system. When it came to MFCs with high internal resistance, the most effective method for salt removal was to connect them in parallel. On the other hand, connecting MFCs with low internal resistance to one another in series led to efficient removal of salt. It was found that the electrical characteristics of certain MFCs and CDIs, in addition to the operating conditions of those devices, were important in determining the MFC-CDI circuit arrangement that worked best (Liang et al., 2015). Wen et al. developed an MDC-MCDI system for the complete desalination of MDC effluent. They did this by capitalizing on the low energy consumption of MCDI, which is driven by a power source made up of multiple assemblies of MDCs connected in parallel or series. With the parallel connected MDCs configuration, a significantly high desalination rate of 3.7 mg/h was observed (Wen et al., 2014).

10.5.14 Osmotic MDC

Due to its ability to function under conditions of low or no hydraulic pressure, the idea of FO has garnered a lot of attention in the field of wastewater treatment over the course of the last few decades. According to McGinnis and Elimelech (2007) and Zou et al. (2017), FO is defined as the movement of water across a semipermeable membrane from a feed solution to a draw solution under the influence of an established osmotic pressure gradient. Due to its low energy input, high rejection of a wide range of contaminants, lower fouling tendency, easy fouling removal, and high water recovery, the FO has several potential advantages over other pressure-driven processes of water treatment (RO, nanofiltration, ultrafiltration, etc.). These potential advantages include high water recovery, low fouling tendency, high rejection of a wide range of contaminants, and easy fouling removal. There have been several research studies that have been reported in the literature that are associated with the incorporation of the FO process in MDCs and MFCs reactors by exchanging the IEM for FO membranes. OsMDC is the name given to this particular embedded form of FO-BES approach, which aims to increase water recovery (OsMDC). The FO membrane simultaneously slows the movement of salt ions across the desalination chamber while allowing water to pass from the anode to the chamber where it is desalinated. Because of this water flux, proton transport from the anode chamber to the desalination/middle chamber is facilitated, resulting in a pH value that is lower in the anode chamber than in the MDC. Only a very small amount of research has been done on the topic of integrating FO membranes into various BES to treat wastewater, extract clean water, and produce bioelectricity all at the same time. Zhang et al. (2016) found that treating wastewater in a BES with an FO integrated produced more electricity and water than a BES with a CEM placed. Desalination of seawater and the reuse of water with seawater serving as the draw solution were two of the potential applications that were envisioned for this constructed reactor after it was completed. If you used seawater as the catholyte, you could eliminate the need to recycle it because it would be diluted with the water that was extracted from the FO membrane and transported. Due to its capacity for extensive wastewater treatment, maximized bioelectricity generation, and the reduction of required pressure in osmotic reactors, it was suggested that this osmotic reactor be linked with an MDC in terms of a combined operation (Zhang et al., 2011). In their study, Zhang et al. (2012) compared the effectiveness of OsMDC to that of FO technology and MDC concerning the treatment of wastewater, the desalination of water, and the recovery of water. It was discovered that the OsMDC reactor was superior to the FO process because it was able to separate a greater quantity of salt ions and generate a greater amount of electricity as a result of the oxidation of organic compounds. When compared to MDC, the water that was recovered from the wastewater using OsMDC had a higher quality, and due to the dilution process, it also had a lower salinity. With a production of an average current of 4.6 mA, a significant desalination rate of 57.8% was accomplished. A high difference in osmotic pressure, which developed as a result of concentration gradients, was observed to be the cause of an increase in water flux that was observed to occur in conjunction with an increase in salinity (Zhang and He, 2012). Pardeshi and Mungray (2014) researched the performance of FO membranes consolidated in MFC using glucose as the substrate. They found that the power density of the FO membranes was 27.38 W/m^3. The operation of the reactor resulted in a COD removal efficiency of 92%, a TDS removal efficiency of 80%, a power density of 48.52 mW/m^2, a current density of 136.30 mA/m^2, and a power yield of 7.46 W/kg, respectively. Zhang et al. (2016) developed a two-membrane-based bioelectrochemical reactor that consisted of a hydraulically connected osmotic microbial fuel cell (OsMFC) and an MDC. This reactor was based on the concept of MFCs. By maximizing the dilution and desalination capabilities of OsMFC and MDC, respectively, the combined system demonstrated significantly improved desalination performance of 95.9% and energy production of 0.160 kWh/m^3.

10.6 ACHIEVEMENTS MADE WITH THE MDC TECHNOLOGY

10.6.1 Desalination Efficiencies of MDCs

According to Shannon et al. (2008), desalinated seawater can be an additional source of drinkable water for our planet, and several technologies are currently available for accomplishing this goal. MDC is a developing technology in this regard, and Cao et al. (2009) were the first people to demonstrate that MDCs are capable of desalinating saltwater using their device. They were able to remove 93±3% of the salt from a solution that contained 35 g/L of salt, which is an impressive result. The extremely encouraging results obtained by Cao et al. (2009) were the impetus for an explosion in the number of studies conducted on MDC technology. These studies were primarily conducted to raise the percentage of desalination from 93±3% to 100%. In this way, Girme (2014) was able to use a "photo MDC" to make 100% desalination. Using a microbial electrolysis and desalination cell, Luo et al. (2011) also achieved an improved 99% desalination from the 93.3% desalination that was achieved by Cao et al. (2009). However, they achieved this at a lower salt concentration of 10 g/L NaCl compared to Cao et al. (2009) (MEDC). MDC desalination efficiencies can be improved via several different methods, and many of these have been reported. The use of membranes with increased ion-exchange capacities and the application of ion-exchange resins are two examples of these various methods. For example, Mehanna et al. (2010) said that they could enhance the ion exchange capacities of the membranes used to enhance the desalination efficiency of an MDC they looked at from 50% to 63%. The movement of Na^+ and Cl^- ions out of the desalination chamber is facilitated by membranes that have high ion-exchange capacities, which in turn makes the desalination process more efficient. On the other side, Shehab et al. (2014) improved the desalination efficiency of an SMDC from 43% to 72% in a reduced time of 80 hours by using ion-exchange resins in the SMDC (earlier HRT was 110 h). Ion-exchange resins can increase the percentage of MDCs that are desalinated because they have a low internal resistance, according to an analysis of the research works that have been done on ion-exchange resins. This occurs as the process of desalination progresses. Morel et al. (2012) also showed that ion-exchange resins could make an MDC more effective at desalinating water. At a saltwater flow rate of 0.063 mL/min, the ion-exchange resin-supported MDC that they investigated was able to produce a 58% conductivity reduction. In contrast, a classical (three-chamber) MDC that they compared the resin-modified MDC to recorded only a 45% conductivity reduction when they tested it. In a solution that had an electrical conductivity of 4.72 mS/cm, Zuo et al. (2013) were able to achieve a desalination efficiency of 99% by utilizing multiple ions and ion-exchange resins in their experiment. They also mentioned the stabilizing effect that ion-exchange resins have on MDCs and acknowledged the fact that the efficiency of MDCs to remove ions is compromised when there are multiple ions present, which is typically the case with seawater. When applied to the process of desalinating real seawater, the MDC technology encounters a challenge involving the competitive migration of ions, which requires additional research to fully comprehend and find a solution. HRT is another factor that has been discovered to affect the desalination efficiencies of MDCs (HRT). Morel et al. (2012) researched the effect of HRT on desalination and concluded that, while a longer HRT might lead to back diffusion of ions due to concentration gradients between the desalination chamber and adjacent chambers, a shorter HRT does not allow for effective desalination because of microbial activities in the anode that drive desalination peaks gradually. This was demonstrated by the fact that a shorter HRT does not allow for effective deals. An examination of the published literature on MDCs revealed that the HRT for desalination ranged anywhere from a single day to some months in the majority of the studies. For example, Jacobson et al. (2011) used an up-flow multidimensional chromatography and worked with a four-day HRT to achieve a salt removal of above 99% from a 30 g (TDS)/L salt solution. Also, Ping et al. (2016) got percentages of desalination between 25.4% and 79.2%. They thought that the differences in performance could be due to the length of the HRT (longer HRT made desalination better). For optimal desalination performance to be attained, it is necessary, regardless of the operating conditions, to strike a balance between the short and long HRT (Morel et al., 2012). Other researchers have focused their attention on modifying the electrodes of MDCs to improve their desalination performance. Forrestal et al. (2012) used electrode adsorption to remove an average of 69.4% of the salt from the desalination chamber using a capacitive adsorption MDC in a batch cycle. Through the process of electrode adsorption, dissociated salt ions can be prevented from contaminating anolytes and catholytes by being adsorbed on the surfaces of the electrodes. This method is exceptionally beneficial for the commercial utilization of chemical catholytes, in particular. This is because the majority of chemical catholytes, also known as potassium ferricyanide, are quite pricey, and it would be uneconomical to continuously replace them due to the accumulation of Na^+ in the cathode chamber.

Kokabian and Gude (2013) developed the algae biocathodes with oxygen as the electron acceptor and recorded a desalination efficiency of 40%. This is in contrast to the majority of MDCs, which rely on pricey chemical electron acceptors such as potassium ferricyanide. In a later review (Kokabian and Gude, 2015), the same authors found that the desalination efficiency could be improved to 64.21±0.5% for 500 mg/L COD anolyte solution and to 63.47±0.1% for 1000 mg/L COD anolyte solution. According to a review of the relevant published research, more than half of the studies conducted on MDCs have reported a percentage of desalination of at least 50%, which is an encouraging sign for the potential expansion of this technology.

10.6.2 COD Removal from Wastewater Using MDCs

The treatment of wastewater, in particular the reduction of organic load and COD, is one objective of the MDC technology. The decrease in COD is caused by microorganisms' consumption of organic matter in wastewater to fuel their metabolic processes. The reduction in COD is an acceptable standard for its estimation because the reduction in organic matter cannot be measured directly. In response to this, several different kinds of MDC and different operating conditions have been researched to remove COD from wastewater. This includes the UMDC that Jacobson and colleagues researched (2011). Using the UMDC, Jacobson et al. (2011) were able to successfully remove a COD concentration of 92.0±0.4% from anolytes, while the COD loading rate was 6.78±0.36 g COD L^{-1} daily. This UMDC was among the few up-scaled MDCs, and it was interesting to note that its COD reduction efficiency was reproducible regardless of whether a salt solution or artificial seawater was used in the desalination chamber. This was because the COD reduction efficiency was the same regardless of which medium was used. In the anode chamber of MDCs, microbial decomposition of organic matter in wastewater is the primary mechanism by which COD is removed. This is an important fact to keep in mind. After using a submerged desalination-denitrification cell (SMDDC), Zhang and Angelidaki (2013) were able to reduce the COD in wastewater by 87.7%, starting with an initial COD concentration of 800 mg-COD/L. Exoelectrogens in the anode chamber of the SMDDC degraded the organic matter in the contaminated water, which resulted in the COD being removed from the water. This was accomplished by submerging the SMDDC into a simulated aquifer. After starting with a COD concentration of 800 mg L^{-1}, the microbial capacitive desalination cell (MCDC) that was constructed by Forrestal et al. (2014) was capable of achieving a COD reduction rate of 170 mg COD per litre per hour. The origin of the wastewater was a significant aspect of the research that Forrestal et al. (2014) conducted. That is, water that was produced from shale gas, and even with this particular kind of wastewater, the desalination chamber was able to remove 85% of the COD in just about four hours. This particular type of MDC will be of great assistance to the oil and gas industry, which faces significant difficulties in the treatment of wastewater and occasional spills. In a different study, Zhang and He (2015) utilized a scaled-up MDC to achieve a high COD reduction of more than 96% from wastewater that had an initial COD concentration of 3000 mg COD. The results of this study were published in the journal Environmental Science and Technology. Moreover, according to Zhang and He's (2015) research, the post-aerobic run set-up that was used in the experiments was likely responsible for making COD removal easier. Before the wastewater was treated in an anaerobic environment, it may have gone through a post-aerobic run, which caused some of the substrates in the wastewater to be used by microbes that require oxygen.

According to the findings of a few studies, the amount of time that hydraulic retention takes does not have a significant impact on the amount of COD that is reduced in MDCs (HRT). Studies conducted by Luo et al. (2012a) and Luo et al. (2012b) offered some evidence to support this assertion. In the first study conducted by Luo et al. (2012a), a COD removal percentage of 52% was achieved in approximately eight days from wastewater with a COD concentration of 2744 16 mg/L. However, in the second study conducted by Luo et al. (2012b), the percentage of COD removal increased by only 3% (from 52 to 55%) even after eight months of operation. In their investigation of multiple MDCs, Qu et al. (2013) also reported an observation that was very similar to this one. The difference in COD reduction between the two different HRTs that were investigated in that study was very minimal, coming in at only 1%. The difference in COD reduction between the two different HRTs that were investigated in that study was very minimal, coming in at only 1%. The effect of COD concentration on the percentage of COD reduction in MDCs was investigated, and the results showed that there was a positive correlation between the two variables. In support of this idea, Mehanna et al. (2010) found that the percentage of removal of COD could be increased to 82 6% when the acetate concentration was 2 g L^{-1}, but it dropped to 77 3% when the COD concentration was reduced to 1 g L^{-1}. This wasn't the only thing they found. Kokabian and Gude (2015), who worked on synthetic wastewater, also showed that the amount of COD removed was related to the amount of COD in the wastewater. Kokabian and Gude (2015) were able to produce a COD reduction in a photosynthetic MDC of 76.06 1.21% using synthetic wastewater with a concentration of 500 mg/L COD, while a 1000 mg/L COD concentration resulted in an 82.17 1.27% COD removal. Both of these results were based on the use of synthetic wastewater. According to Yuan et al. (2015), the ability of bacteria to degrade excess COD may be responsible for the increasing COD reduction efficiency that occurs along with an increase in the concentration of COD in wastewater. Nevertheless, this correlation is valid only in situations in which the substrates are easily hydrolyzable, and the HRTs are optimized for the task at hand (which can vary). Because of this, increasing the COD concentration of wastewater by adding complex hydrocarbons (for instance, high-molecular-weight benzene derivatives) and working within a short HRT will not produce a high COD reduction efficiency in an MDC. This is because increasing the COD concentration of wastewater can be accomplished by adding complex hydrocarbons. In this particular study, decreases in COD were found to be attributable to the actions of attached FO units as well as microbial degradation. In their more recent study, which achieved a COD removal of 70.6%, the researchers came to a similar conclusion (Yuan et al., 2016).

10.7 CONCLUDING REMARKS AND PROSPECTS

This review demonstrates that several different research works on MDC technology have been carried out over the years. On the other hand, relatively few research studies have gone beyond measuring in litres. This is because the issues that were brought up in this review have not been satisfactorily resolved. Moving forward, the collection and utilization of methane gas in MDCs will be the focus of research interests. This does not mean that methanogenesis is a desirable process in MDC operations particularly given the fact that it lowers coulombic efficiency. If the methanogenesis process can't be stopped in the anode chamber, it's better to trap the gases so that more voltage can be made by burning methane gas to turn it into electricity. This will be of particular benefit to the class of MDCs that, to function effectively, require an external voltage supply. This category of MDCs includes both the MDC used to produce hydrogen gas and the MDC that relies on the recirculation of electrolytes to maintain a stable pH. Moreover, with the support of voltages from the conversion of methane gas to electricity, MDCs can be used for fluoride removal. The contamination of the world's water supplies with fluoride is widespread, and researchers are still looking for an efficient method to remove it from water supplies. The MDC technology has the potential to provide a solution that is more environmentally friendly for the removal of fluoride from our water sources. The traditional three-chamber MDC can be utilized for the removal of fluoride from surface waters, whereas the submersible MDC can be utilized for the treatment of fluoride-contaminated groundwater sources. At the moment, our research team is in the process of constructing new MDCs to investigate the potential of the MDC technology for the removal of fluoride. It is also essential that, in future research works, microbial succession in MDCs be taken into consideration. That is, looking into the different kinds of microbes and how many there are right from the beginning of an MDC operation until the very end of an experiment. This is essential information for comprehending how microbial communities develop within an operational MDC. In subsequent research, this will serve as a reference point for the design of electrodes and the selection of substrates. It will be interesting to look into the water-energy nexus and how it applies to the MDC technology, particularly concerning the technology's life cycle, as this will be an area of research that needs to be done.

ACKNOWLEDGEMENT

We would like to acknowledge the Department of Agronomy at the Lovely Professional University, Phagwara, Punjab, India, for their consistent moral support and encouragement throughout the writing process.

Conflicts: None

REFERENCES

Abubakari, Z.I., Mensah, M., Buamah, R., Abaidoo, R.C., 2019. Assessment of the electricity generation, desalination and wastewater treatment capacity of a plant microbial desalination cell (PMDC). Int. J. Energy Water Resour. 3, 213–218.

Ahmed, M.A., Tewari, S., 2018. Capacitive deionization: Processes, materials and state of the technology. J. Electroanal. Chem. 812, 178–192.

Al-Mamun, A., Baawain, M.S., Egger, F., Al-Muhtaseb, A.H., Ng, H.Y., 2017b. Optimization of a baffled-reactor microbial fuel cell using autotrophic denitrifying bio-cathode for removing nitrogen and recovering electrical energy. Biochem. Eng. J. 120, 93–102.

Alvarado, L., Chen, A., 2014. Electrodeionization: Principles, strategies and applications. Electrochim. Acta. 132, 583–597.

Alvarez-Gallego, Y., Dominguez-Benetton, X., Pant, D., Diels, L., Vanbroekhoven, K., Genné, I., Vermeiren, P., 2012. Development of gas diffusion electrodes for cogeneration of chemicals and electricity. Electrochimica Acta. 82, 415–426.

Anusha, G., Noori, M.T., Ghangrekar, M., 2018. Application of silver-tin dioxide composite cathode catalyst for enhancing performance of microbial desalination cell. Mater. Sci. Energy Technol. 1, 188–195.

Arar, Ö, Yüksel, Ü, Kabay, N., Yüksel, M., 2014. Various applications of electrodeionization (EDI) method for water treatment—A short review. Desalination. 342, 16–22.

Arugula, M.A., Brastad, K.S., Minteer, S.D., He, Z., 2012. Enzyme catalyzed electricity-driven water softening system. Enzym. Microb. Technol. 51, 396–401.

Baawain, M.S., Al-Mamun, A., Omidvarborna, H., Al-Amri, W., 2017. Ultimate composition analysis of municipal solid waste in Muscat. J. Clean. Prod. 148, 355–362.

Bergel, A., Féron, D., Mollica, A., 2005. Catalysis of oxygen reduction in PEM fuel cell by seawater biofilm. Electrochem. Commun. 7, 900–904.

Biffinger, J.C., Pietron, J., Ray, R., Little, B., Ringeisen, B.R., 2007. A biofilm enhanced miniature microbial fuel cell using *Shewanella oneidensis* DSP10 and oxygen reduction cathodes. Biosens. Bioelectron. 22, 1672–1679.

Brastad, K.S., He, Z., 2013. Water softening usingmicrobial desalination cell technology. Desalination. 309, 32–37.

Cao, X., Huang, X., Liang, P., Xiao, K., Zhou, Y., Zhang, X., Logan, B.E., 2009. A new method for water desalination using microbial desalination cells. Environ. Sci. Technol. 43, 7148–7152.

Carmalin Sophia, A., Bhalambaal, V.M., Lima, E.C., Thirunavoukkarasu, M., 2016. Microbial desalination cell technology: Contribution to sustainable waste water treatment process, current status and future applications. J. Environ. Chem. Eng. 4, 3468–3478.

Chen, S., Liu, G., Zhang, R., Qin, B., Luo, Y., 2012. Development of the microbial electrolysis desalination and chemical-production cell for desalination as well as acid and alkali productions. Environ. Sci. Technol. 46, 2467–2472.

Chen, S., Liu, G., Zhang, R., Qin, B., Luo, Y., Hou, Y., 2012. Improved performance of the microbial electrolysis desalination and chemical-production cell using the stack structure. Bioresour. Technol. 116, 507–511.

Chen, X., Xia, X., Liang, P., Cao, X., Sun, H., Huang, X., 2011. Stacked microbial desalination cells to enhance water desalination efficiency. Environ. Sci. Technol. 45, 2465–2470.

Cheng, S., Xing, D., Call, D.F., Logan, B.E., 2009. Direct biological conversion of electrical current into methane by electromethanogenesis. Environ. Sci. Technol. 43, 3953–3958.

Choi, J., Ahn, Y., 2013. Continuous electricity generation in stacked air cathode microbial fuel cell treating domestic wastewater. J. Environ. Manag. 130, 146–152.

Clauwaert, P., van der Ha, D., Boon, N., Verbeken, K., Verhaege, M., Rabaey, K., Verstraete, W., 2007b. Open air biocathode enables effective electricity generation with microbial fuel cells. Environ. Sci. Technol. 41, 7564–7569.

Croese, E., Pereira, M.A., Euverink, G.-J.W., Stams, A.J.M., Geelhoed, J.S., 2011. Analysis of the microbial community of the biocathode of a hydrogen-producing microbial electrolysis cell. Appl. Microbiol. Biotechnol. 92, 1083–1093.

Drioli, E., Ali, A., Macedonio, F., 2015. Membrane distillation: Recent developments and perspectives. Desalination. 356, 56–84.

Forrestal, C., Stoll, Z., Xu, P., Ren, Z.J., 2014. Microbial capacitive desalination for integrated organic matter and salt removal and energy production from unconventional natural gas produced water. Environ. Sci.: Water Res. Technol. 1 (1), 47–55.

Forrestal, C., Xu, P., Ren, Z., 2012. Sustainable desalination using a microbial capacitive desalination cell, Energy Environ. Sci. 5 (5), 7161–7167.

Freguia, S., Rabaey, K., Yuan, Z., Keller, J., 2008. Sequential anode–cathode configuration improves cathodic oxygen reduction and effluent quality of microbial fuel cells. Water Res. 42, 1387–1396.

Ghasemi, M., Daud, W.R.W., Alam, J., Ilbeygi, H., Sedighi, M., Ismail, A.F., Yazdi, M.H., Aljlil, S.A., 2016. Treatment of two different water resources in desalination and microbial fuel cell processes by poly sulfone/sulfonated poly ether ether ketone hybrid membrane. Energy. 96, 303–313.

Gholizadeh, A., Ebrahimi, A.A., Salmani, M.H., Ehrampoush, M.H., 2017. Ozone-cathode microbial desalination cell; An innovative option to bioelectricity generation and water desalination. Chemosphere. 188, 470–477.

Girme, G.M., 2014. Algae Powered Microbial Desalination Cells. MSc thesis. The Ohio State University, Ohio, USA.

Gude, V.G., 2011. Energy consumption and recovery in reverse osmosis. Desalin. Water Treat. 36 (1e3), 239–260.

Gude, V.G., Kokabian, B., Gadhamshetty, V., 2013. Beneficial bioelectrochemical systems for energy, water, and biomass production. J. Microb. Biochem. Technol. 6 (5), 1–14.

Gude, V.G., Nirmalakhandan, N., Deng, S., 2010. Renewable and sustainable approaches for desalination. Renew. Sustain. Energy Rev. 14 (9), 2641–2654.

He, Z., Angenent, L.T., 2006. Application of bacterial biocathodes in microbial fuel cells. Electroanalysis. 18, 2009–2015.

He, Z., Huang, Y., Manohar, A.K., Mansfeld, F., 2008. Effect of electrolyte pH on the rate of the anodic and cathodic reactions in an air-cathode microbial fuel cell. Bioelectrochemistry. 74, 78–82.

Heidrich, E., Curtis, T., Dolfing, J. 2011. Determination of the internal chemical energy of wastewater. Environ. Sci. Technol. 45: 827–832.

Huang, L., Regan, J.M., Quan, X., 2011. Electron transfer mechanisms, new applications, and performance of biocathode microbial fuel cells. Bioresour. Technol. 102, 316–323.

Hui, W.J., Ewusi-Mensah, D., Jingyu, H., 2019. Using *C. vulgaris* assisted microbial desalination cell as a green technology in landfill leacheate pre-treatment: A factor performance relation study. J. Water Reuse Desal. 10 (1), 1–16.

Jacobson, K.S., Drew, D.M., He, Z., 2011. Use of liter-scale microbial desalination cell as platform to study bioelectrochemical desalination with salt solution or artificial seawater. Environ. Sci. Technol. 45 (10), 4652–4657.

Jafary, T., Daud, W.R.W., Aljlil, S.A., Ismail, A.F., Al-Mamun, A., Baawain, M.S., Ghasemi, M., 2018. Simultaneous organics, sulphate and salt removal in a microbial desalination cell with an insight into microbial communities. Desalination. 445, 204–212.

Jamaly, S., Darwish, N.N., Ahmed, I., Hasan, S.W., 2014. A short review on reverse osmosis pretreatment technologies. Desalination. 354, 30–38.

Jeremiasse, A.W., Hamelers, H.V., Croese, E., Buisman, C.J., 2012. Acetate enhances startup of a H_2-producing microbial biocathode. Biotechnol. Bioengin. 109, 657–664.

Jingyu, H., Ewusi-Mensah, D., Norgbey, E., 2017. Microbial desalination cells technology: A review of the factors affecting the process, performance and effciency. Desalin. Water Treat. 87, 140–159. (CrossRef)

Khawaji, A.D., Kutubkhanah, I.K., Wie, J.M., 2008. Advances in seawater desalination technologies. Desalination. 221, 47–69.

Kim, Y., Logan, B.E., 2011. Series assembly of microbial desalination cells containing stacked electrodialysis cells for partial or complete seawater desalination. Environ. Sci. Technol. 45, 5840–5845.

Kim, Y., Logan, B.E., 2013. Simultaneous removal of organic matter and salt ions from saline wastewater in bioelectrochemical systems. Desalination. 308, 115–121.

Kim, Y., Logan, B.E., 2013a. Microbial desalination cells for energy production and desalination. Desalination. 308, 122–130.

Kokabian, B., Gude, V.G., 2013. Photosynthetic microbial desalination cells (PMDCs) for clean energy, water and biomass production. Environ. Sci.: Process. Impacts. 15 (12), 2178–2185.

Kokabian, B., Gude, V.G., 2015. Sustainable photosynthetic biocathode in microbial desalination cells. Chem. Eng. J. 262, 958–965.

Liang, P., Yuan, L., Yang, X., Huang, X., 2015. Influence of circuit arrangement on the performance of a microbial fuel cell driven capacitive deionization (MFC-CDI) system. Desalination. 369, 68–74.

Logan, B.E., 2008. Microbial Fuel Cells, John Wiley and Sons Inc, New Jersey, USA.

Logan, B.E., 2010. Scaling up microbial fuel cells and other bioelectrochemical systems. Appl. Microbiol. Biotechnol. 85, 1665–1671.

Luo, H., Jenkins, P.E., Ren, Z., 2011. Concurrent desalination and hydrogen generation using microbial electrolysis and desalination cells. Environ. Sci. Technol. 45 (1), 340–344.

Luo, H., Xu, P., Ren, Z., 2012b. Long-term performance and characterization of microbial desalination cells in treating domestic wastewater. Bioresour. Technol. 120, 187–193.

Luo, H., Xu, P., Roane, T.M., Jenkins, P.E., Ren, Z., 2012a. Microbial desalination cells for improved performance in wastewater treatment, electricity production, and desalination. Bioresour. Technol. 105, 60–66.

Macedonio, F., Drioli, E., Gusev, A., Bardow, A., Semiat, R., Kurihara, M.J.C.E., et al. 2012. Efficient technologies for worldwide clean water supply. Chem. Eng. Process. 51, 2–17.

Mathioulakis, E., Belessiotis, V., Delyannis, E., 2006. Desalination by using alternative energy: A review and state-of-the-art. Desalination. 203, 346–365.

McCarty, P.L., Bae, J., Kim, J., 2011. Domestic wastewater treatment as a net energy producer–can this be achieved? Environ. Sci. Technol. 45, 7100–7006.

McGinnis, R.L., Elimelech, M., 2007. Energy requirements of ammonia–carbon dioxide forward osmosis desalination. Desalination. 207, 370–382.

Mehanna, M., Kiely, P.D., Call, D.F., Logan, B.E., 2010. Microbial electrodialysis cell for simultaneous water desalination and hydrogen gas production. Environ. Sci. Technol. 44, 9578–9583.

Mehanna, M., Saito, T., Yan, J., Hickner, M., Cao, X., Huang, X., Logan, B.E., 2010. Using microbial desalination cells to reduce water salinity prior to reverse osmosis. Energy Environ. Sci. 3, 1114–1120.

Meng, F., Jiang, J., Zhao, Q., Wang, K., Zhang, G., Fan, Q., Wei, L., Ding, J., Zheng, Z., 2014. Bioelectrochemical desalination and electricity generation in microbial desalination cell with dewatered sludge as fuel. Bioresour. Technol. 157, 120–126.

Morel, A., Zuo, K., Xia, X., Wei, J., Lou, X., Liang, P., Huang, X., 2012. Microbial desalination cells packed with ion-exchange resin to enhance water desalination rate. Bioresour. Technol. 118, 43–48.

Morita, M., Malvankar, N.S., Franks, A.E., Summers, Z.M., Giloteaux, L., Rotaru, A.E., Rotaru, C., Lovley, D.R., 2011. Potential for direct interspecies electron transfer in methanogenic wastewater digester aggregates. MBio. 2(4), 10–1128.

Moruno, F.L., Rubio, J.E., Atanassov, P., Cerrato, J.M., Arges, C.G., Santoro, C., 2018a. Microbial desalination cell with sulfonated sodium poly (ether ether ketone) as cation exchange membranes for enhancing power generation and salt reduction. Bioelectrochemistry. 121, 176–184.

National Research Council, 2008. Desalination: A National Perspective, The National Academic Press, Washington DC.

Nikonenko, V.V., Kovalenko, A.V., Urtenov, M.K., Pismenskaya, N.D., Han, J., Sistat, P., et al., 2014. Desalination at overlimiting currents: State-of-the-art and perspectives, Desalination. 342, 85–106.

Pardeshi, P., Mungray, A., 2014. High flux layer by layer polyelectrolyte FO membrane: Toward enhanced performance for osmotic microbial fuel cell. Int. J. Polym. Mater. 63, 595–601.

Park, D., Park, Y.K., So, C., 2004. Application of single-compartment bacterial fuel cell (SCBFC) using modified electrodes with metal ions to wastewater treatment reactor. J. Microbiol. Biotechnol. 14, 1120–1128.

Park, D.H., Zeikus, J.G., 2003. Improved fuel cell and electrode designs for producing electricity from microbial degradation. Biotechnol. Bioengin. 81 (3), 348–355.

Ping, Q., He, Z., 2013. Improving the flexibility of microbial desalination cells through spatially decoupling anode and cathode. Bioresour. Technol. 144, 304–310.

Ping, Q., Porat, O., Dosoretz, C.G., He, Z., 2016. Bioelectricity inhibits back diffusion from the anolyte into the desalinated stream in microbial desalination cells. Water Res. 88, 266–273.

Qu, Y., Feng, Y., Liu, J., He, W., Shi, X., Yang, Q., Lv, J., Logan, B.E., 2013. Salt removal using multiple microbial desalination cells under continuous flow conditions. Desalination. 317, 17–22.

Qu, Y., Feng, Y., Wang, X., Liu, J., Lv, J., He, W., Logan, B.E., 2012. Simultaneous water desalination and electricity generation in a microbial desalination cell with electrolyte recirculation for pH control. Bioresour. Technol. 106, 89–94.

Rismani-Yazdi, H., Carver, S.M., Christy, A.D., Tuovinen, O.H., 2008. Cathodic limitations in microbial fuel cells: An overview. J. Power Sources. 180, 683–694.

Salehmin, M.N.I., Lim, S.S., Satar, I., Daud, W.R.W., 2021. Pushing microbial desalination cells towards field application: Prevailing challenges, potential mitigation strategies, and future prospects. Sci. Total Environ. 759, 143485.

Sevda, S., Yuan, H., He, Z., Abu-Reesh, I.M., 2015. Microbial desalination cells as a versatile technology: Functions, optimization and prospective. Desalination. 371, 9–17.

Shah, M.P., 2020. Microbial Bioremediation & Biodegradation, Springer.

Shah, M.P., 2021. Removal of Emerging Contaminants through Microbial Processes, Springer.

Shannon, M.A., Bohn, P.W., Elimelech, M., Georgiadis, J.G., Marinas, B.J., Mayes, A.M., 2008. Science and technology for water purification in the coming decades. Nature. 452, 301–310.

Shehab, N.A., Amy, G.L., Logan, B.E., Saikaly, P.E., 2014. Enhanced water desalination efficiency in an air-cathode stacked microbial electrodeionization cell. J. Membr. Sci. 469, 364–370.

Sophia, C., Gohil, J.M., 2018. Chapter 19 - microbial desalination cell technology: functions and future prospects, in: P.P. Kundu, K. Dutta, (Eds.). Progress and Recent Trends in Microbial Fuel Cells, Elsevier, pp. 399–422.

Villano, M., De Bonis, L., Rossetti, S., Aulenta, F., Majone, M., 2011. Bioelectrochemical hydrogen production with hydrogenophilic dechlorinating bacteria as electrocatalytic agents. Bioresour. Technol. 102, 3193–3199.

Wen, Q., Zhang, H., Chen, Z., Li, Y., Nan, J., Feng, Y., 2012. Using bacterial catalyst in the cathode of microbial desalination cell to improve wastewater treatment and desalination. Bioresour. Technol. 125, 108–113.

Wen, Q., Zhang, H., Yang, H., Chen, Z., Nan, J., Feng, Y., 2014. Improving desalination by coupling membrane capacitive deionization with microbial desalination cell. Desalination. 354, 23–29.

Werner, C.M., Logan, B.E., Saikaly, P.E., Amy, G.L., 2013. Wastewater treatment, energy recovery and desalination using a forward osmosis membrane in an air-cathode microbial osmotic fuel cell. J. Membr. Sci. 428, 116–122.

Xu, P., Drewes, J.E., Heil, D., 2008. Beneficial use of co-produced water through membrane treatment: Technical-economic assessment. Desalination. 225, 139–155.

Yuan, H., Abu-Reesh, I.M., He, Z., 2015. Enhancing desalination and wastewater treatment by coupling microbial desalination cells with forward osmosis. Chem. Eng. J. 270, 437–443.

Yuan, H., Abu-Reesh, I.M., He, Z., 2016. Mathematical modeling assisted investigation of forward osmosis as pretreatment for microbial desalination cells to achieve continuous water desalination and wastewater treatment. J. Membr. Sci. 502, 116–123.

Yuan, L., Yang, X., Liang, P., Wang, L., Huang, Z.-H., Wei, J., Huang, X., 2012. Capacitive deionization coupled with microbial fuel cells to desalinate low-concentration salt water. Bioresour. Technol. 110, 735–738.

Zhang, B., He, Z., 2012. Integrated salinity reduction and water recovery in an osmotic microbial desalination cell. RSC Adv. 2, 3265–3269.

Zhang, B., He, Z., 2013. Improving water desalination by hydraulically coupling an osmotic microbial fuel cell with a microbial desalination cell. J. Membr. Sci. 441, 18–24.

Zhang, F., He, Z., 2015. Scaling up microbial desalination cell system with a post-aerobic process for simultaneous wastewater treatment and seawater desalination. Desalination, 360, 28–34.

Zhang, F., Brastad, K.S., He, Z., 2011. Integrating forward osmosis into microbial fuel cells for wastewater treatment, water extraction and bioelectricity generation. Environ. Sci. Technol. 45, 6690–6696.

Zhang, F., Chen, M., Zhang, Y., Zeng, R.J., 2012. Microbial desalination cells with ion exchange resin packed to enhance desalination at low salt concentration. J. Membr. Sci. 417–418, 28–33.

Zhang, F., Wang, X., Xionghui, J., Ma, L., 2016. Efficient arsenate removal by magnetite-modified water hyacinth biochar. Environmental Pollution, 216, 575–583. https://doi.org/10.1016/j.envpol.2016.06.013

Zhang, J., Yuan, H., Deng, Y., Zha, Y., Abu-Reesh, I.M., He, Z., Yuan, C., 2018. Life cycle assessment of a microbial desalination cell for sustainable wastewater treatment and saline water desalination. J. Clean. Prod. 200, 900–910.

Zhang, Y., Angelidaki, I., 2013. A new method for in situ nitrate removal from groundwater using submerged microbial desalination e denitrification cell (SMDDC). Water Res. 47 (5), 1827–1836.

Zhao, F., Harnisch, F., Schröder, U., Scholz, F., Bogdanoff, P., Herrmann, I., 2006. Challenges and constraints of using oxygen cathodes in microbial fuel cells. Environ. Sci. Technol. 40, 5193–5199.

Zhao, R., Biesheuvel, P.M., Miedema, H., Bruning, H., van der Wal, A., 2010. Charge efficiency: A functional tool to probe the double-layer structure inside of porous electrodes and application in the modeling of capacitive deionization. J. Phys. Chem. Lett. 1, 205–210.

Zou, S., Qin, M., Moreau, Y., He, Z., 2017. Nutrient-energy-water recovery from synthetic sidestream centrate using a microbial electrolysis cell – forward osmosis hybrid system. J. Clean. Prod. 154, 16–25.

Zuo, K., Liu, F., Ren, S., Zhang, X., Liang, P., Huang, X., 2016. A novel multi-stage microbial desalination cell for simultaneous desalination and enhanced organics and nitrogen removal from domestic wastewater. Environ. Sci. Water Res. Technol. 2, 832–837.

Zuo, K., Yuan, L., Wei, J., Liang, P., Huang, X., 2013. Competitive migration behaviors of multiple ions and their impacts on ion-exchange resin packed microbial desalination cell. Bioresour. Technol. 146, 637–642.

11 Treatment of Wastewater and Energy Recovery by Bioelectrochemical Oxidation Systems

M. Sai Achuth, R. Saravanathamizhan

11.1 INTRODUCTION

In the recent century, wastewater treatment has become an utmost concern and has been prioritized by environmentalists worldwide with technological advancements. Pollutants and toxic contaminants present in the wastewater create an inevitable need for treatment before they let into the environment. From a treatment perspective, wastewater treatment technologies of the current day are authentic, although they contribute to greenhouse gas emissions and other emerging concerns about water quality. There are several conventional methodologies for wastewater treatment including sedimentation, precipitation,[1] filtrations, adsorption, membrane separations,[2] aerobic and anaerobic digestion,[3] advanced oxidation process, etc.

Wastewater provides a number of opportunities for achieving a pathway towards sustainability, in spite of being a concern and a challenge. Currently, well-established technologies and emerging process configurations, when examined preliminarily, bring forth the anaerobic digestion (AD), fermentation and dark fermentation (DF) processes and bioelectrochemical systems (BESs), such as microbial fuel cells (MFCs), microbial electrolysis cells (MECs), and other biorefinery configurations, as technologies that are promising for developing sustainable wastewater treatment systems that are chemical-, energy-, and cost-efficient techniques.[4,5]

BESs are generally producing electrical energy directly from treating wastewater. Bioelectrochemical oxidation occurs due to transfer of electrons and oxidation of substrate by catalytic microbes instead of inorganic catalysts that drive the anodic and/or cathodic reactions to generate electricity and the effluent chemical oxygen demand (COD) is reduced.[6,7] In this chapter, wastewater treatment and energy recovery by these bioelectrochemical oxidation systems are discussed.

11.1.1 POLLUTANTS IN WASTEWATER

Anthropogenic sources consisting of household and agricultural waste and industrial processes pollute water resources. Due to heavy wastewater pollution, public concern over the environment has increased. Wastewater is broadly classified as

DOI: 10.1201/9781003368472-11

municipal wastewater and industrial wastewater. Industrial wastewaters are called effluents in the textile, pharma,[8] leather, and food industries. Wastewater contains high amount of colour, COD, BOD, and TOC. This is caused due to organics and inorganics present in the wastewater. Allergies, cancers, etc. in humans and adverse effects on aquatic life are caused by industrial effluents when high organic effluents are released into water bodies. Hence, the treatment of wastewater is mandatory.[9]

11.2 WASTEWATER TREATMENT TECHNIQUES

Conventional wastewater treatment techniques, including filtration, sedimentation,[10] precipitation, and biological AD upon application, were supposed to eradicate pollution, but there are still disadvantages, specifically that of increased operation costs. Modern techniques for wastewater treatment include adsorption, membrane separation, electrophoresis, advanced oxidation processes, and bioelectrochemical oxidation. Every treatment technique has its own advantages and disadvantages. The important treatment technique, apart from BES is AOP and the classification is shown in Table 11.1.[11,12]

11.3 BIOELECTROCHEMICAL OXIDATION SYSTEMS

BESs are generally producing electrical energy directly from treating wastewater. Bioelectrohemical system has two types: MFC and MEC. Various categories of BESs are listed in Table 11.2. MFCs are electrochemical devices that use catalysts that are microbes as alternatives to inorganic catalysts to drive the anodic and/or cathodic reactions to generate electricity.

Less than half of the voltage or electrical energy needed for conventional water electrolysis was utilized by microbial electrolyzers or MECs to produce hydrogen. Chemical energy enables the reduced electrical input that comes from organic or reduced inorganic substrates that serves as the feedstock for hydrogen production

TABLE 11.1
Various Types of Advanced Oxidation Processes and Their Classification

Advanced Oxidation Process				
Homogeneous AOP (with Energy)				
Ultraviolet Radiation	Ultrasound	Electrical Energy	Homogeneous AOP (without Energy)	Heterogeneous AOP
• O_3/UV • H_2O_2/UV • O_3/H_2O_2/UV • Photo-Fenton • Fe^{2+}/H_2O_2/UV	• O_3/US • H_2O_2/US • Fe^{2+}/US	• Electrochemical oxidation • Electro-Fenton	• O_3 in alkaline • O_3/H_2O • H_2O_2/catalyst	• Catalytic ozonation • Photocatalytic ozonation • Heterogeneous photocatalysis

TABLE 11.2
Various Types of Bioelectrochemical Oxidation Processes

Bioelectrochemical Oxidation Processes				
Microbial Fuel Cell	**Microbial Electrolysis Cell**	**Microbial Solar Cell**	**Microbial Desalination Cell**	**Microbial Electrosynthesis**
• Wastewater treatment • Generate electricity	• Wastewater treatment • Generate electricity	• Wastewater treatment • Generate electricity	• Generate electricity	• Generate electricity • H_2 production • CH_4 production

in MECs. Substrates produce electrons, which are converted into hydrogen at the cathode, operating at ambient temperature.[13,14]

Despite the fact that fuel cells that are microbial (MFCs) have been researched progressively, the most significant advancement was in 1999 when it was discovered that mediators, which are chemicals, were unnecessary for the system to produce electricity. Before air cathodes, BES was aerated to increase the treatment efficacy, as shown in Figure 11.1. After the discovery of air cathodes, applications for wastewater treatment practically were therefore visualized to be viable, possibly permitting both treatment of wastewater and production of electrical power. Although air cathodes and mediator-less MFCs have been discussed for more than ten years, there are currently no commercial uses for the technology. The price of the electrodes has restricted the scaling up of laboratory-scale procedures. With improvements in low-cost anodes, separators, and cathodes based on activated carbon catalysts, this now appears to be feasible. Reduced power at bigger sizes is another aspect that can restrict the development of larger-scale MFCs.[15]

BESs are particularly promising technologies due to application of newer concepts, as well as the development of aliter materials for electrodes, separators, and catalysts, as well as one-of-a-kind designs. BES energy production is being seriously considered as a viable alternative energy/power source. Energy generation from "negative-value" waste streams can assist in fulfilling the world's energy needs while also lowering pollution and expenses associated with water and wastewater treatment.[16]

MFC is frequently regarded as a viable wastewater treatment technology that also generates electricity biologically at a lower operational cost. MFC systems are primarily comprised of electrodes and membranes. The efficiency of MFC systems is found by the anodic and cathodic oxidation and reduction rates. Because of this, critical components of MFC systems are electrodes.[17] Anodic and cathodic chambers are often separated by a PEM in an MFC. Anodic and cathodic chambers in a standard two-compartment MFC are connected by a PEM or occasionally a bridge of salt to allow travel of protons from the anode to the cathode while prohibiting oxygen from diffusing into the anode. Divisions are designed in a variety of useful ways. Upflow mode MFCs are suited for wastewater treatment since scaling them up is very simple.[18]

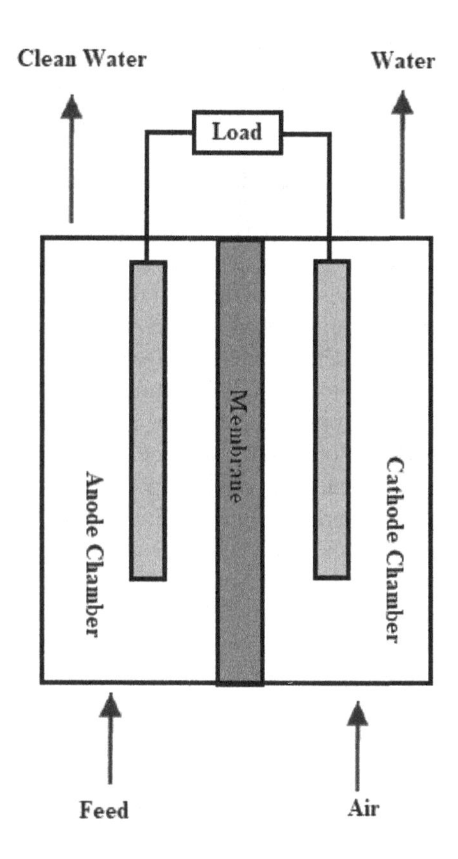

FIGURE 11.1 Bioelectrochemical oxidation system.

11.3.1 Bioelectrochemical Oxidation System for Wastewater Treatment

Many different categories of wastewater from industrial and domestic sources have been analyzed and evaluated in various forms of BES. The viability of using organic compounds in wastewater is determined by the characteristics of the components present and their biological degradability. Kondaveeti et al.[19] studied petroleum industry wastewater treatment by hybrid type BES using the organics present in wastewaters to generate bioenergy and products that add value to waste management. The author reported that a hybrid type BES achieved 96.5% COD, removal, which outperformed one- and two-chamber BES performances.

Dye decolourization was reported by Ling et al.[20] using electro-Fenton at BES cathode powered by anodic Fungus-bacterium. The bioelectro-Fenton system has been used to degrade azo dyes such as Orange II, Rhodamine B. Carbon felt as cathode, composite Fe_2O_3 used as anode. It was observed that COD removal efficiency was ranging from 73.9% to 86.7%. It suggested that bioelectro-Fenton could be full of promising environmentally friendly AOP methods for Azo-dye wastewater treatment. Chen et al.[21] demonstrated that MFC was used to degrade sulfamethoxazole, norfloxacin, and oxytetracycline with a COD removal rate of more than 92.53%,

65.5%, and 99.00%, respectively. Evidently, MFCs have a high antibiotic clearance rate. Most target antibiotics were efficiently eliminated in MECs, up to 86.1%, demonstrating MECs' excellent performance and promise in lowering antibiotics.

11.3.2 Energy Generation Using Bioelectrochemical Oxidation System

According to Zhang et al.,[22] the energy-generating bacteria might develop a layer of biofilm on the anode, which could decide the MFC technology's power generation performance. *Shewanella* species, one of the most researched exoelectrogens, exhibit all known bacterial extracellular electron transfer mechanisms in MFCs, including indirect electron transfer via self-secreted electron mediators and direct electron transfer via outer membrane cytochrome c and nanowire. *Shewanella*, as facultative exoelectrogens, can grow and produce electricity in both anoxic and aerobic environments. Al-Sahari et al.[23] observed that depending on the anodic degradation process of substrates and the external power supply, BES is commonly employed to produce valuable hydrogen and reduce low potential metals at the cathode chamber. In the electrosynthesis cells (MES) system, the substrates and high redox potential metals are degraded in the cathode chamber.

Verma & Mishra [24] demonstrated that the dried banana peel powder was fed into two MFCs, one without and one with *Saccharomyces cerevisiae* inoculant. Dried banana peel powder alone produced no power; however, dried banana peel powder with *S. cerevisiae* had the highest power density. The banana peel slurry was fed into two separate MFCs, one with *S. cerevisiae* and the other without. Maximum power output was observed using banana slurry and *S. cerevisiae*. Without inoculation, banana peel slurry produced the lowest power production. Evidently, microbial community study revealed that the presence of native microbiota was responsible for the higher power output produced from banana slurry-based MFCs. MFCs with banana slurry removed up to 70–88% of COD; while MFCs with dried banana peel powder removed 18–44% of COD. According to S. Y. Liu et al.,[25] the BES can be used as an auxiliary technology in AD to improve organic waste treatment and biogas generation. In thermophilic circumstances, the efficiency of a one- and two-chamber BES-anaerobic digestor system with a cation-exchange membrane demonstrated that an active glucose-fed temperature-tolerant anaerobic sludge could easily boost production of biogas in both reactor configurations by inserting a carbon electrode upstream. However, the rate of production of biogas from the BES reactor dropped due to the formation of volatile fatty acids. Only the two-chamber arrangement could provide biogas methane enrichment.

11.4 FACTORS AFFECTING BIOELECTROCHEMICAL OXIDATION SYSTEMS

Various factors affecting the efficiency of the MFC include pH, temperature, substrate concentration, and electrode material. These factors directly influence the efficiency of the system. Proton transfer affects pH of BES, whereas temperature affects kinetics and thermodynamics of the system. Changes in substrate concentration and compatibility of electrode material used to bring about changes in efficiency of BES.

11.4.1 pH

During the operation of MFC, protons generated at anode are moved to cathode where they, along with electrons, form water upon reaction with oxygen. The rate of transport of proton from anode to cathode affects the pH and the efficiency of operation of MFC. The slow rate of transport of proton results in proton accumulation at anode causing acidification of anode. At cathode, discharge of proton in oxygen reduction and minor rate of proton transfer to cathode raise the pH at the cathode. This pH increase at cathode is associated with the decrease in the reduction of oxygen leading to a reduction in power generation.[26–28]

The performance of MFCs and MECs is significantly influenced by the wastewater pH in the anode compartment. The pH between 5 and 6 results in more hydrogen generation, and the pH between 6.5 and 7.5 produces greater power. If pH falls below 3, the reactor performance suffers because the majority of electrogenic bacteria can no longer thrive in the acidic environment. In the optimum range of pH, both the reactors give COD removal three times greater compared with extremely low or high pH. [29] At a pH below ideal, microbial activity in MFCs and MECs is less active than it is at optimum pH. At high pH (>10), the concentration gradient of proton across the proton exchange membrane is reduced, because of the lower values for both power generation in MFCs and hydrogen production in MECs.

11.4.2 Temperature of Anode and Cathode Chambers

Temperature poses a substantial impact on efficiency of MFC, as various kinetic and thermodynamic characteristics are governed by this parameter. Temperature has been found to affect directly the efficiency of MFC for production of bioelectricity. The proliferation of microorganisms and the formation of electroactive microbial biofilms have been significantly influenced by the temperature. The anode chamber's anaerobic culture is sensitive to temperature changes, which have an impact on both the proliferation of bacteria and their digestion of organic materials. The efficiency of MFCs and MECs is often severely hampered by temperatures that are either too high (>45°C) or too low (<15°C), but temperatures of 30–40°C tend to support the performance of these reactors. More & Gupta [30] studied BES for the removal of chromium from the cathode chamber, and the innoculam is in the anode chamber. With varying reactor temperatures from 25 to 45°C, the author observed that BES resulted in maximum chromium removal efficiency of 95% at 40°C. The cause behind better efficiency at 40°C was supporting growth conditions for microorganisms with the superior intake of the substrate in the anode chamber.

M. Behera's periodic increment in temperature from 20 to 40°C resulted in increment in COD removal efficiency from 62 to 84%. Further increment in temperature beyond 40°C, the COD removal efficiency reduced, and it was 58.8% at 55°C. The optimum temperature of 35–40°C was reported. At higher temperature (>40°C), the inhibition of the growth of mesophilic anaerobic bacteria reduces the COD removal, and at 40°C a power density of 34.38 mW/m^2 was reported.[31] Liu et al. (2011) gave an account in a study that microorganisms have shown optimal activity in a temperature range of 30–45°C and hence give increased electric power generation in MFC. Variation in temperature from recommended optimum temperature may result in poor biofilm formation and thereby reduce MFC potential for generation of bioelectricity.[32]

11.4.3 Substrate Concentration

The substrate concentration is a significant factor when assessing efficiency of the MFC. Due to the potential for the treatment of wastewater and recovery of energy with MFCs, several wastewater types have been used as substrates. Mohan et al. [6] evaluated the feasibility of composite vegetable waste as substrate in a single-chambered MFC. The MFC utilized in that study resulted in increased power density of 57 mW/m^2 with effective COD removal of 62%. Liu et al.[25] reported higher power densities of 305 and 506 mW/m^2 with butyrate- and acetate-fed MFCs, respectively. Efficiencies of energy conversion have also been evaluated with the often-used fermentable (glucose) and non-fermentable (acetate) types of substrates. In several studies, various wastewaters used for the treatment and energy production, as listed in Table 11.3.

11.4.4 Electrode Material

Generally, electrode materials should ideally be biocompatible, conductive, porous, easily made at low cost, recyclable, and scalable. Also, they should possess high specific surface area, corrosion resistance, and high mechanical strength. When the primary aim is to treat wastewater, for example, electrode material costs need to be economical to make them economically viable for this application. Various electrode materials employed include carbon-based materials such as carbon, graphite and its composites, activated carbon, carbon nanotubes, stainless steel, and mild steel. [44,45]

The following electrode material qualities are critical for improving MFC efficiency: (i) Surface area and porosity: This is an important attribute of any electrode material, and the surface area of the electrodes has a large effect on power generation. Increasing the surface area increases the number of reaction sites and the mobility of electrons. Greater porosity, on the other hand, results in reduced electrical conductivity. (ii) Electrical conductivity: By using an electrode material with high electrical conductivity, electrons may travel from the anode to the cathode more easily and quickly across the electrical circuit. (iii) Stability and durability: Because of the oxidation and reduction processes that occur at the anode and cathode, the electrode material should be robust and stable in both acidic and alkaline environments.[46,47]

Several carbon-based compounds for use as anodes in MFC systems have been developed and demonstrated. Carbon paper, graphite plates or sheets, graphite rods, and carbon fabric are examples (Wei et al., 2011). Because of their superior biocompatibility, chemical stability, high conductivity, and low cost, carbonaceous materials are the most often utilized materials for MFC anodes. Carbon-based electrodes are classified into three types based on their configuration: flat, packed, and brush. The accessible surface area is a crucial component that influences the performance of various anode materials.[48]

TABLE 11.3
Various Wastewaters Used for the Treatment and the Energy Production

Type of Wastewater	Concentration (mg/L)	Type of MFC	COD Reduction (%)	Power Density Generated (mW/m²)	Reference
Brewery wastewater	2240	A single-chamber, air-cathode MFC	87	205	[33]
Beer brewery wastewater	626	One-chamber air-cathode MFC with	42.8	64	[34]
bagasse-based paper mill wastewater	8000	Dual-chambered microbial fuel cell	85	53	[35]
dairy industry wastewater	1600	A dual-chambered Microbial Fuel Cell	91	192	[36]
dairy wastewater	3700	MFC with single-chamber configuration	90.63	6.71	[37]
paper mill effluent	4100	Single-chamber BET	55	40.2	[38]
distillery wastewater	82200	Single-chamber MFC	72.84	124.35	[39]
starch processing wastewater	4852	Single-chamber MFC	98.0	239.4	[40]
Dye industry wastewater	1000	Single-chamber air cathode microbial fuel cell	85	515	[41]
tannery effluent	COD 4720	Dual-chamber MFC	60	500	[42]
Meat processing Wastewater	1420	Single-chamber	86	80	[43]

Cropping waste is always available with a high yield after having organic-rich substances that can substitute commercial carbon-made components in the MFC, such as electrodes or catalysts. To meet the demands of waste-to-energy practises in order to meet the objectives of sustainable development goals, the selection of agricultural wastes to generate biochar in various methods, as well as the electrochemical performance of products in MFC, is crucial. The context of negative CO_2 emission technologies is investigated after the inclusion of biochar as sustainable electrode materials for MFC and bioenergy with carbon capture, storage, and use. The utilization of circular economy for waste-derived biochar as electrodes in MFC using agricultural waste would help to achieve sustainable development goals by addressing both the water-food-energy and waste-to-energy concerns concurrently.[49]

11.5 DIFFERENT TYPES OF REACTOR CONFIGURATION

11.5.1 Single-Chamber MFC (SCMFC)

According to the number of chambers, MFCs may be divided into two configurations: (i) single-chamber microbial fuel cells (SCMFC) and (ii) double-chamber microbial fuel cells (DCMFC). By removing the cathode chamber and maintaining the cathode in direct contact with air, SCMFCs may be created, as shown in Figure 11.2.

MFCs provide simpler designs as well as cost reductions. They normally simply have an anodic chamber and do not require aeration in a cathodic chamber. The SCMFC arrangement is more cost-effective, especially in the absence of an air sparger, which requires a lot of energy to deliver air to the cathode. The initial SCMFC consisted of a rectangular carbon anode chamber connected to a porous carbon cloth air-cathode that was directly open to air, with protons travelling from the anolyte solution to the cathode. Catholyte is not necessary in single-chambered designs. Unlike DCMFC, this technology significantly reduces energy usage owing to the aeration stage.[50] The perspective of SCMFCs over DCMFCs are cheap cost, simple design, cathode chamber aeration, and efficient power output. The downside of this type of MFC is the back diffusion of oxygen from the cathode to the anode via the proton exchange membrane.[51] Park and Zeikus created a one-chamber MFC

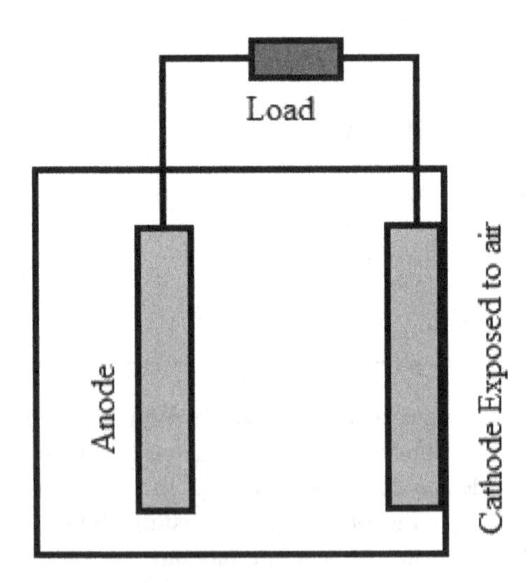

FIGURE 11.2 Single-chamber MFC.

that consists of an anode in a rectangular anode chamber connected with a porous air-cathode that is directly exposed to air. The anolyte solution transfers protons to the porous air-cathode.[52]

Logrono and co-workers designed an air-exposed SCMFC using microalgal bio-cathodes. The reactors were tested for the simultaneous biodegradation of real dye textile wastewater and the generation of bioelectricity. The system was suitable for the treatment of dye textile wastewater and the removal of COD, colour, and heavy metals. High removal efficiencies were observed in the SCMFCs for Zn (98%) and COD (92–98%), but the removal efficiencies were considerably lower for Cr (54–80%). It was observed that this SCMFC simplifies a double-chamber system.[53] An air-cathode MFC with graphite rods to investigate household wastewater treatment resulted in a total COD reduction of roughly 50%. The power density increases by passing the wastewater via the carbon cloth anode, which reduces the total COD by 40–50%.[54]

11.5.2 Dual-Chamber MFC (DCMFC)

SCMFC design is straightforward and cost-effective since it eliminates the need for a membrane. The fundamental issue with a membrane-less MFC, however, is the substrate being consumed by the oxygen diffused through the cathode. DCMFC, which may reduce oxygen diffusion without reducing the power density or raising cell internal resistance, may be able to tackle these issues. Anaerobic (anode) and aerobic (cathode) chambers are separated by a cation exchange membrane and linked by a bridge in the traditional design of DCMFCs (CEM), as shown in Figure 11.3. Both fed-batch and batch modes are frequently used to execute this concept. The basic goal of CEM is to physically separate the liquids in each chamber while allowing protons to pass from anode to cathode.[55]

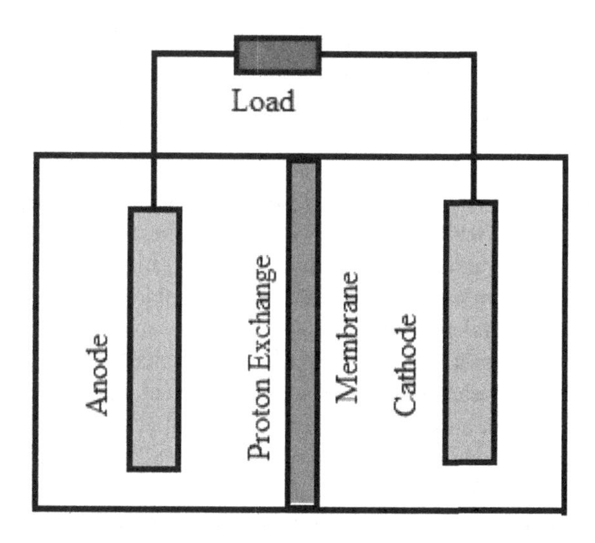

FIGURE 11.3 Dual-chamber MFC.

Lee et al.[56] established a lab-scale BES to reduce Cr to this goal (VI). The system consisted of two compartments separated by a wall. The wall was composed of a vertically aligned carbon nanotube (VACNT) composite that could physically partition but electrically connect the two chambers. The viability of Cr(VI) reduction was examined utilizing this BES, and it showed a Cr(VI) removal rate of 99.6% at 100 mg/L, which is the highest reduction rate achieved using *Pseudomonas aeruginosa.* Ali et al.[57] investigated on the removal of Pb^{2+} using a SCMFC. It was observed that the maximum power density was 41.58 mW/m^2 with a 95% removal efficiency of Pb^{2+}.

11.5.3 Upflow MFC

MFC, which has upflow, is designed in such a way that a cylinder is divided into two sections by glass wool and glass bead layers. These two sections were used as anodic and cathodic chambers. The disc-shaped graphite felt anode and cathode were placed at the reactor's bottom and top, respectively. The feed stream is supplied to the anode's bottom, and the effluent continuously passes through the cathodic chamber and exits at the top. There is no such thing as an anolyte or a catholyte. Furthermore, the diffusion barriers between the anode and cathode provide a DO gradient for proper MFC operation.[18]

The chocolaterie wastewater was used as a substrate in an upflow anaerobic microbial fuel cell (UAMFC) for waste treatment and power generation. Subha et al.[58] noticed the effect of hydraulic retention time (HRT) and organic loading rate (OLR) on organic removal and power generation. The predominant bacterial strains obtained from anode during biofilm analysis were Achromobacter insuavis strain BT1, Achromobacter insuavis strain B3, Bacillus encimensis strain B4, and Kocuria flava strain B5. These bacteria play an important role in substrate degradation and power generation. It was observed that at an HRT of 15 hours, the maximum power density and COD removal were 98 mW/m^2 and 70%, respectively.

Tamilarasan et al.[59] demonstrated the performance of a continuous fed upflow anaerobic MFC operated with surgical cotton industry wastewater at different OLRs to determine the potency of power generation, COD, and TSS removal efficiency. It was observed that at an optimum OLR of 1.9 g COD/L d^{-1}, the highest TCOD and SCOD removal rates of 78.8% and 69%, respectively, were achieved. With an initial TSS concentration of 970±70 mg/L, a TSS removal efficiency of 62% was obtained. While treating surgical cotton industry wastewater, a maximum power density of 116.03 mW/m^2 (2.2 W/m^3) and corresponding coulombic efficiency of 17.8% were achieved at an OLR of 1.9 gCOD/L/d.

Tanneries that process a large amount of leather produce a large amount of saline organic content wastewater. Ghorab et al.[60] reported the treatment of tannery industrial wastewater in an upflow microbial fuel cell (UMFC) under saline (4%). Total chemical oxygen demand (TCOD) removal was 87.12%, 91.2%, and 93.8% at 0.6, 1.2, and 1.8 g COD/L, respectively. Jayashree et al.[61] demonstrated a tubular UMFC utilizing seafood processing wastewater for wastewater treatment efficiency and power generation. It was observed that the MFC removed 83% and 95% of total and soluble COD at an OLR of 0.6 g d^{-1}. At an OLR of 2.57 g d^{-1}, a maximum power density of 105 mW m^2 (2.21 W m^3). The anode biofilm's dominant bacterial communities were identified as *RB1A (LC035455), RB1B (LC035456), RB1C (LC035457),* and *RB1E (LC035458).* All four strains belonged to the *Stenotrophomonas* genus. Clearly, the wastewater from the seafood processing can be treated in a UMFC for simultaneous power generation and wastewater treatment.

11.5.4 Oxygen Exposed Bioelectrode MFC

Y. Wang et al.[62] employed BES with graphene oxide/carbon brush electrode, and the COD removal efficiency was observed to be 91.1±0.1%–97.6±0.4%. Oxygen was considered to be electron acceptor competitor; however, small levels of oxygen aid in the biotransformation or conversion of refractory organics, such as aromatic amines, by micro-aerophilic bacteria to non-ring organic compounds. Meanwhile, supplying oxygen to the bioanode (obioanode) in BES might facilitate energy extraction or recovery from aromatic amine elimination.

Bacteria-encapsulating bioanode was manufactured using coaxial electrospinning technique. A novelly built MFC consists of two distinct chambers joined by a bridge (K-salt bridge/Na-salt bridge), and it was observed that it produced current of about 5.23 mA and COD removal efficiency was 94–97%. In this concept, the first container can be wastewater/microbial-rich water, and the second container can be aerated/salt/aerated wastewater. The wastewater chamber (second one) can be kept under anaerobic conditions (that is, without O_2), and the lids can be airtight with an opening for methane and other gases (that were generated by applied microbial system). The following container (chamber one) was constantly kept aerobic by creating O_2 for aerated/aerated wastewater.[63]

X. Wang et al.[64] used graphite brush cathode and pure graphite granular anode for redox couple Fe(III)/Fe(II) pair on biocathode with Fe(II)-oxidizing bacteria like *Acidithiobacillus ferrooxidans* and *Leptospirillum ferrooxidans.* It was observed that the power density produced was 77.15±1.62 Wm^{-3}, and COD removal efficiency was 31.4–38.9%. Because of the low efficiency of isolation and enrichment, the isolated Fe(II)-oxidizing strain to enrich the biocathode is not practicable in field applications.

11.5.5 Stacked MFC

Stacked MFCs (SMFCs), in which many electrodes are stacked in a single unit, are practical and efficient devices. An SMFC is, as shown in Figure 11.4, for the investigation of performances of several MFCs connected in series and in parallel.[65] Connecting many MFCs in series or parallel might result in increased voltage or current output. There was no discernible negative influence on the maximum power output per MFC unit. Stack was created to increase MFC power generation. Multiple-unit MFCs are coupled in series or parallel in this approach to produce additional power. Protons are released into the anolyte during the oxidation of organic materials in the anode compartment under anaerobic conditions. Stacking single-chambered MFC with double-chambered MFC solves the problem of proton build-up in the anode chamber, which has a detrimental influence on

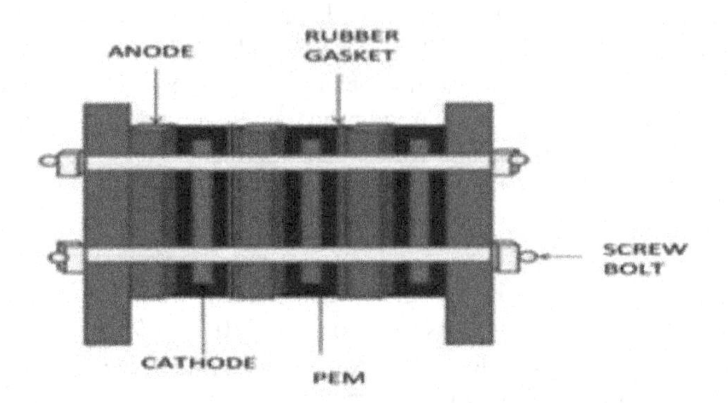

FIGURE 11.4 Stacked MFC.

MFC performance in terms of long-duration operation.[66] This model not only increases total power production, but the increase in power generation has no effect on the power efficiency of each cell. It has been discovered that the power output provided by a parallel connection is greater. The use of SMFCs to handle barley shochu waste and generate energy was investigated. The greatest COD removal efficiency and power density were 36.7% and 15.7 mW/m^2, respectively.[67] MFC stack is made up of a common base and many pluggable units that may be linked in either series or parallel to generate power during waste treatment in septic tanks. When three units are linked in parallel, a power density of 142 6.71 mW m^2 is attained in the lab.[68]

11.5.6 Tubular MFC

A tubular separate (constant cross-section) structure in an MFC allows for near-optimal spatial distribution of the anode, cathode, and (if present) ion exchange membrane, as shown in Figure 11.5. In the reactor, this type of tubular MFC can be operated in steady state with close-to-plug flow hydraulic properties. Total COD elimination ranged from 51% to 82% for reactors.[69]

S. H. Liu et al.[70] designed a tubular microbial fuel cell/electro-Fenton (TMFC/EF) combined system for field-scale applications. This TMFC/EF combined system demonstrated the treatment of sulfolane-contaminated groundwater by driving the Fenton reaction at the cathode with bioelectricity generated at the anode. It was observed that the tubular MFC/EF combined system produced 525 mV of output voltage and 691 mW/m^3 of power density. The sulfolane (50 mg/L) removal efficiency was 100%.

S. H. Liu et al.[71] created a tubular biotrickling filter microbial fuel cell (Tube-BTF-MFC) system, TMFCs were placed on top of a biotrickling filter-microbial fuel cell (BTF-MFC). It was observed that the addition of TMFCs significantly increased the power output of a Tube-BTF-MFC in the treatment of ethyl acetate (EA). Circulating water spray generates a progressive water flow pattern; the Tube-MFC can reduce electrolyte sharing between MFC anode chambers and enhance

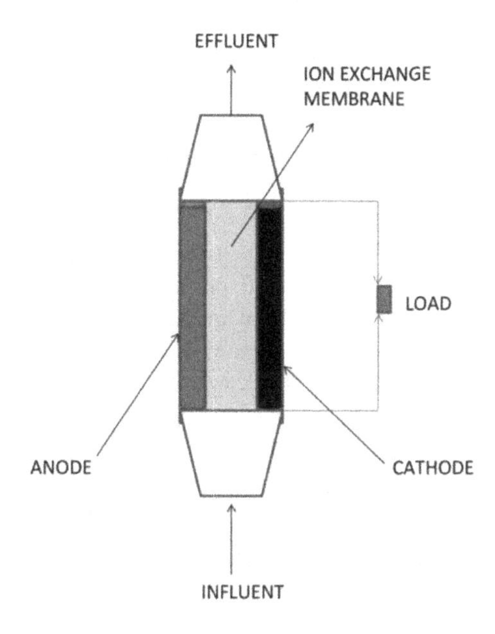

FIGURE 11.5 Tubular MFC.

TBTF-MFC power output. Subsequently, the TBTF-MFC had an EA removal efficiency of 93% and a closed circuit voltage of 186.5 mV with 129.8 g/m^3 of EA.

As the anode of a TMFC, a composite electrode composed of a graphite fibre brush and carbon granules (MFC-GFB/GG) was demonstrated by J. Li et al.[72] In comparison to an MFC with graphite granules (MFC-GG) and an MFC with a graphite fibre brush (MFC-GFB), MFC-GFB/GG had a longer start-up lag time while reaching the highest operating voltage at 50V. Furthermore, it was observed that the MFC-GFB/GG achieved the highest power density of 66.9 1.6 Wm^3, which was approximately 5.3 and 1.2 times that of the MFC-GG and MFC-GFB, respectively, during stable operation. The highest performance of the MFC-GFB/GG can be attributed to the electrode's highest active biomass content and the MFC's lowest internal resistance. The optimum COD concentration for MFC-GFB/GG was determined to be 500 mg COD L^{-1}.

Marassi et al.[73] demonstrated the development of a tubular air-cathode microbial fuel cell (TMFC) bioreactor with carbon cloth electrodes separated by nafion membrane inoculated with a consortium of *Shewanella oneidensis* and *Clostridium butyricum* and fed with dairy wastewater. Total chemical and biochemical oxygen demand (TCOD and TBOD) removals were 94% and 96%, respectively. Clearly, the MFC bioreactor achieved a maximum power density of 2.4–3.5 W m^3 and a current density of 1.1–2.4 A m^3, owing to other metabolic pathways other than the electrogenic one.

To support its production, the beef packing industry consumes a large amount of water and energy. To make this industry more sustainable, treatment of wastewater from a Midwestern beef packing plant was demonstrated in continuous fed mode, a bench-scale TMFC by J. Li et al.[74] When the MFC was fed 1 g L^{-1} beef extract solution, the maximum current density was 8.8±0.2 A m^{-3} and the organics removal was 28.2±5.9%. Changing the feeding solution to real beef packing wastewater had no discernible effect on system performance. Clearly, the achieved current density was 8.4±0.2 A m^{-3} and the organics removal rate was 35.9±9.7%.

11.5.7 Biochemical Fuel Cell

Biochemical fuel cells are appealing and promising. Biochemical fuel cells, as opposed to chemical fuel cells, operate at moderate reaction conditions, such as desirable operational temperature and pressure. They also use a neutral electrolyte and cheap catalysts such as platinum. The catalyst in biological fuel cells is either a simple microbe like Baker's yeast or an enzyme. Biochemical fuel cells immediately transform the chemical energy of carbohydrates like sugars and alcohols into electric energy. Because most organic substrates burn with the evolution of energy, biocatalyzed oxidation of organic compounds by oxygen at the two electrode interfaces provides a method for converting chemical energy to electrical energy. A substrate, such as carbohydrate, is oxidized without the participation of oxygen in typical microbial catabolism, while its electrons are picked up by an enzyme-active site, which acts as a reduced intermediate.[75]

The active metal is oxidized to metal ions in the anodic process, while hydrogen is evolved in the cathodic reaction. If the hydrogen accumulated, the cathodic reaction would come to a halt due to a sort of polarization known as concentration polarization. Because widespread anaerobic corrosion was known to occur, sulphate-reducing bacteria, which thrive in these conditions, could use the hydrogen to reduce sulphate. Thus, the organisms would remove the accumulating hydrogen, a process known as depolarization, and allow the corrosive reaction to continue.[76]

It is evident that in a biochemical fuel cell, the biosystem can only induce a difference in electrode potential in a medium by altering the electrochemical activity of the medium when compared to sterile medium. This may only result from a change in the relative concentrations of the medium's electrochemically active ingredients and must be accomplished through the addition or removal of an electrochemically active component or a combination of both of these processes. As a result, there are three fundamental types of biochemical fuel cells: (1) Depolarization cell, in which an electrochemical product is eliminated by the biological system, such as the depolarization of the electrode process in an oxygen concentration cell. (2) Product cell, in which the biological system generates an electrochemically active product, such as hydrogen, which is subsequently employed in a normal hydrogen-oxygen cell. (3) Redox cell, in which the biological system turns electrochemical products back into electrochemical reactants, such as a cyclic regenerative system including biological ferricyanide reduction and electrochemical ferrocyanide oxidation.[77] The bioelectrochemical dehalogenation system is a prospective green energy technology that uses electroactive bacteria as catalysts to breakdown halogenated organic molecules. The foundation of the system is the dehalogenation metabolism of OHRB enriched at the cathode. Furthermore, reactor configurations have been shown to affect system stability and feasibility, as well as final dehalogenation efficiency.[78]

11.6 CONCLUSION

Bioelectrochemical oxidation systems are highly efficient and operate in mild conditions with low capital investment, making them preferential wastewater treatment techniques in this century. In this chapter, wastewater treatment and energy recovery by bioelectrochemical oxidation systems were discussed with suitable illustrations, making it a reader's guide. BES is of two types, viz., MFC and MEC. In MFC, microbiota are used as catalysts, while in MEC, chemical substrates are employed. Major factors affecting MFC are pH, temperature, electrode material used, and substrate employed.

Treatment of wastewater has become the need of the hour as it may result in ground table water contamination, reducing the water availability for human consumption.

REFERENCES

1. Kornboonraksa, T., Lee, H. S., Lee, S. H., & Chiemchaisri, C. (2009). Application of chemical precipitation and membrane bioreactor hybrid process for piggery wastewater treatment. *Bioresource Technology, 100*(6), 1963–1968.
2. Galambos, I., Molina, J. M., Járay, P., Vatai, G., & Bekássy-Molnár, E. (2004). High organic content industrial wastewater treatment by membrane filtration. *Desalination, 162*(1–3), 117–120.
3. Rajeshwari, K. V., Balakrishnan, M., Kansal, A., Lata, K., & Kishore, V. V. N. (2000). State-of-the-art of anaerobic digestion technology for industrial wastewater treatment. *Renewable & Sustainable Energy Reviews, 4*(2), 135–156.
4. Aghalari, Z., Dahms, H. U., Sillanpää, M., Sosa-Hernandez, J. E., & Parra-Saldívar, R. (2020). Effectiveness of wastewater treatment systems in removing microbial agents: A systematic review. *Globalization and Health, 16*(1), 1–11.
5. Dhote, J., Pradeep Ingole, S., & Chavhan, A. (2012). PDF review on waste water treatment technologies. *Review on Wastewater Treatment Technologies, 1*(5), 1–18.
6. Mohan, S. V., Raghavulu, S. V., Peri, D., & Sarma, P. N. (2009). Integrated function of microbial fuel cell (MFC) as bio-electrochemical treatment system associated with bioelectricity generation under higher substrate load. *Biosensors and Bioelectronics, 24*(7), 2021–2027.
7. Mohanakrishna, G., Al-Raoush, R. I., & Abu-Reesh, I. M. (2021). Integrating electrochemical and bioelectrochemical systems for energetically sustainable treatment of produced water. *Fuel, 285*, 119104.
8. De Andrade, J. R., Oliveira, M. F., Da Silva, M. G. C., & Vieira, M. G. A. (2018). Adsorption of pharmaceuticals from water and wastewater using nonconventional low-cost materials: A review. *Industrial and Engineering Chemistry Research, 57*(9), 3103–3127.
9. Sonune, A., & Ghate, R. (2004). Developments in wastewater treatment methods. *Desalination, 167*(1–3), 55–63.
10. Lekang, O. I., Marie Bomo, A., & Svendsen, I. (2001). Biological lamella sedimentation used for wastewater treatment. *Aquacultural Engineering, 24*(2), 115–127.
11. Loures, C. C. A., Alcântara, M. A. K., Filho, H. J. I., Teixeira, A. C. S. C., Silva, F. T., Paiva, T. C. B., & Samanamud, G. R. L. (2013). Advanced oxidative degradation processes: Fundamentals and applications. *International Review of Chemical Engineering (IRECHE), 5*(2), 102.

12. Poyatos, J. M., Muñio, M. M., Almecija, M. C., Torres, J. C., Hontoria, E., & Osorio, F. (2010). Advanced oxidation processes for wastewater treatment: State of the art. *Water, Air, and Soil Pollution*, *205*(1–4), 187–204.
13. Buckner, C. A., Lafrenie, R. M., Dénommée, J. A., Caswell, J. M., Want, D. A., Gan, G. G., Leong, Y. C., Bee, P. C., Chin, E., Teh, A. K. H., Picco, S., Villegas, L., Tonelli, F., Merlo, M., Rigau, J., Diaz, D., Masuelli, M., Korrapati, S., Kurra, P., & Mathijssen, R. H. J. (2016). We are IntechOpen, the world's leading publisher of open access books built by scientists, for scientists TOP 1%. *Intech*, *11*, 13.
14. Borole, A. P. (2015). Microbial fuel cells and microbial electrolyzers. *Electrochemical Society Interface*, *24*(3), 55–59.
15. Logan, B. E., Wallack, M. J., Kim, K. Y., He, W., Feng, Y., & Saikaly, P. E. (2015). Assessment of microbial fuel cell configurations and power densities. *Environmental Science and Technology Letters*, *2*(8), 206–214.
16. Pant, D., Singh, A., Van Bogaert, G., Irving Olsen, S., Singh Nigam, P., Diels, L., & Vanbroekhoven, K. (2012). Bioelectrochemical systems (BES) for sustainable energy production and product recovery from organic wastes and industrial wastewaters. *RSC Advances*, *2*(4), 1248–1263.
17. Ahanchi, M., Jafary, T., Yeneneh, A. M., Rupani, P. F., Shafizadeh, A., Shahbeik, H., Pan, J., Tabatabaei, M., & Aghbashlo, M. (2022). Review on waste biomass valorization and power management systems for microbial fuel cell application. *Journal of Cleaner Production*, *380*(P1), 134994.
18. Du, Z., Li, H., & Gu, T. (2007). A state of the art review on microbial fuel cells: A promising technology for wastewater treatment and bioenergy. *Biotechnology Advances*, *25*(5), 464–482.
19. Kondaveeti, S., Govindarajan, D., Mohanakrishna, G., Thatikayala, D., Abu-Reesh, I. M., Min, B., Nambi, I. M., Al-Raoush, R. I., & Aminabhavi, T. M. (2023). Sustainable bioelectrochemical systems for bioenergy generation via waste treatment from petroleum industries. *Fuel*, *331*(P1), 125632.
20. Ling, T., Huang, B., Zhao, M., Yan, Q., & Shen, W. (2016). Repeated oxidative degradation of methyl orange through bio-electro-Fenton in bioelectrochemical system (BES). *Bioresource Technology*, *203*, 89–95.
21. Chen, P., Guo, X., Li, S., & Li, F. (2021). A review of the bioelectrochemical system as an emerging versatile technology for reduction of antibiotic resistance genes. *Environment International*, *156*, 106689.
22. Zhang, P., Liu, J., Qu, Y., & Feng, Y. (2017). Enhanced *Shewanella oneidensis* MR-1 anode performance by adding fumarate in microbial fuel cell. *Chemical Engineering Journal*, *328*, 697–702.
23. Al-Sahari, M., Al-Gheethi, A., Radin Mohamed, R. M. S., Noman, E., Naushad, M., Rizuan, M. B., Vo, D. V. N., & Ismail, N. (2021). Green approach and strategies for wastewater treatment using bioelectrochemical systems: A critical review of fundamental concepts, applications, mechanism, and future trends. *Chemosphere*, *285*, 131373.
24. Verma, M., & Mishra, V. (2022). Bioelectricity generation by microbial degradation of banana peel waste biomass in a dual-chamber S. Cerevisiae-based microbial fuel cell. *SSRN Electronic Journal*, *168*, 106677.
25. Liu, S. Y., Charles, W., Ho, G., Cord-Ruwisch, R., & Cheng, K. Y. (2017). Bioelectrochemical enhancement of anaerobic digestion: Comparing single- and two-chamber reactor configurations at thermophilic conditions. *Bioresource Technology*, *245*(Pt A), 1168–1175.
26. Samsudeen, N., Radhakrishnan, T. K., & Matheswaran, M. (2015). Bioelectricity production from microbial fuel cell using mixed bacterial culture isolated from distillery wastewater. *Bioresource Technology*, *195*, 242–247.

27. Raghavulu, S. V., Mohan, S. V., Goud, R. K., & Sarma, P. N. (2009). Effect of anodic pH microenvironment on microbial fuel cell (MFC) performance in concurrence with aerated and ferricyanide catholytes. *Electrochemistry Communications*, *11*(2), 371–375.
28. Behera, M., & Ghangrekar, M. M. (2009). Performance of microbial fuel cell in response to change in sludge loading rate at different anodic feed pH. *Bioresource Technology*, *100*(21), 5114–5121.
29. Li, S., & Chen, G. (2018). Factors affecting the effectiveness of bioelectrochemical system applications: Data synthesis and meta-analysis. *Batteries*, *4*(3), 34.
30. More, A. G., & Gupta, S. K. (2018a). Evaluation of chromium removal efficiency at varying operating conditions of a novel bioelectrochemical system. *Bioprocess and Biosystems Engineering*, *41*(10), 1547–1554.
31. Behera, M., Murthy, S. S. R., & Ghangrekar, M. M. (2011). Effect of operating temperature on performance of microbial fuel cell. *Water Science and Technology*, *64*(4), 917–922.
32. Liu, Y., Climent, V., Berná, A., & Feliu, J. M. (2011). Effect of temperature on the catalytic ability of electrochemically active biofilm as anode catalyst in microbial fuel cells. *Electroanalysis*, *23*(2), 387–394.
33. Feng, Y., Wang, X., Logan, B. E., & Lee, H. (2008). Brewery wastewater treatment using air-cathode microbial fuel cells. *Applied Microbiology and Biotechnology*, *78*(5), 873–880.
34. Wen, Q., Wu, Y., Cao, D., Zhao, L., & Sun, Q. (2009). Electricity generation and modeling of microbial fuel cell from continuous beer brewery wastewater. *Bioresource Technology*, *100*(18), 4171–4175.
35. Elakkiya, E., & Niju, S. (2021). Bioelectrochemical treatment of real-field bagasse-based paper mill wastewater in dual-chambered microbial fuel cell. *3 Biotech*, *11*(2), 42.
36. Elakkiya, E., & Matheswaran, M. (2013). Comparison of anodic metabolisms in bioelectricity production during treatment of dairy wastewater in microbial fuel cell. *Bioresource Technology*, *136*, 407–412.
37. Venkata Mohan, S., Mohanakrishna, G., Velvizhi, G., Babu, V. L., & Sarma, P. N. (2010). Bio-catalyzed electrochemical treatment of real field dairy wastewater with simultaneous power generation. *Biochemical Engineering Journal*, *51*(1–2), 32–39.
38. Krishna, K. V., Sarkar, O., & Venkata Mohan, S. (2014). Bioelectrochemical treatment of paper and pulp wastewater in comparison with anaerobic process: Integrating chemical coagulation with simultaneous power production. *Bioresource Technology*, *174*, 142–151.
39. Mohanakrishna, G., Venkata Mohan, S., & Sarma, P. N. (2010). Bio-electrochemical treatment of distillery wastewater in microbial fuel cell facilitating decolorization and desalination along with power generation. *Journal of Hazardous Materials*, *177*(1–3), 487–494.
40. Lu, N., Zhou, S., Zhuang, L., Zhang, J., & Ni, J. (2009). Electricity generation from starch processing wastewater using microbial fuel cell technology. *Biochemical Engineering Journal*, *43*(3), 246–251.
41. Karuppiah, T., Pugazhendi, A., Subramanian, S., Jamal, M. T., & Jeyakumar, R. B. (2018). Deriving electricity from dye processing wastewater using single chamber microbial fuel cell with carbon brush anode and platinum nano coated air cathode. *3 Biotech*, *8*(10), 1–9.
42. Naveenkumar, M., Senthilkumar, K., Sampathkumar, V., Anandakumar, S., & Thazeem, B. (2022). Bio-energy generation and treatment of tannery effluent using microbial fuel cell. *Chemosphere*, *287*(P1), 132090.
43. Heilmann, J., & Logan, B. E. (2006). Production of electricity from proteins using a microbial fuel cell. *Water Environ. Res*, *78*, 531–537.

44. Perazzoli, S., De Santana Neto, J. P., & Soares, H. M. (2018). Prospects in bioelectrochemical technologies for wastewater treatment. *Water Science and Technology*, *78*(6), 1237–1248.
45. Sonawane, J. M., Yadav, A., Ghosh, P. C., & Adeloju, S. B. (2017). Recent advances in the development and utilization of modern anode materials for high performance microbial fuel cells. *Biosensors and Bioelectronics*, *90*, 558–576.
46. Kumar, G. G., Sarathi, V. G. S., & Nahm, K. S. (2013). Recent advances and challenges in the anode architecture and their modifications for the applications of microbial fuel cells. *Biosensors and Bioelectronics*, *43*(1), 461–475.
47. Mustakeem (2015). Electrode materials for microbial fuel cells: Nanomaterial approach. *Materials for Renewable and Sustainable Energy*, *4*(4), 1–11.
48. Wei, J., Liang, P., & Huang, X. (2011). Recent progress in electrodes for microbial fuel cells. *Bioresource Technology*, *102*(20), 9335–9344.
49. Ngoc-Dan Cao, T., Mukhtar, H., Yu, C. P., Bui, X. T., & Pan, S. Y. (2022). Agricultural waste-derived biochar in microbial fuel cells towards a carbon-negative circular economy. *Renewable and Sustainable Energy Reviews*, *170*, 112965.
50. Obileke, K. C., Onyeaka, H., Meyer, E. L., & Nwokolo, N. (2021). Microbial fuel cells, a renewable energy technology for bio-electricity generation: A mini-review. *Electrochemistry Communications*, *125*, 107003.
51. Choudhury, P., Uday, U. S. P., Mahata, N., Nath Tiwari, O., Narayan Ray, R., Kanti Bandyopadhyay, T., & Bhunia, B. (2017). Performance improvement of microbial fuel cells for waste water treatment along with value addition: A review on past achievements and recent perspectives. *Renewable and Sustainable Energy Reviews*, *79*, 372–389.
52. Park, D. H., & Zeikus, J. G. (2003). Improved fuel cell and electrode designs for producing electricity from microbial degradation. *Biotechnology and Bioengineering*, *81*(3), 348–355.
53. Logroño, W., Pérez, M., Urquizo, G., Kadier, A., Echeverría, M., Recalde, C., & Rákhely, G. (2017). Single chamber microbial fuel cell (SCMFC) with a cathodic microalgal biofilm: A preliminary assessment of the generation of bioelectricity and biodegradation of real dye textile wastewater. *Chemosphere*, *176*, 378–388.
54. Ahn, Y., & Logan, B. E. (2010). Effectiveness of domestic wastewater treatment using microbial fuel cells at ambient and mesophilic temperatures. *Bioresource Technology*, *101*(2), 469–475.
55. Ullah, Z., & Zeshan, S. (2020). Effect of substrate type and concentration on the performance of a double chamber microbial fuel cell. *Water Science and Technology*, *81*(7), 1336–1344.
56. Lee, J. H., Yun, E. T., Kim, H. S., Ham, S. Y., Sun, P. F., Jang, Y. S., Park, J. H., Kim, N. P., & Park, H. D. (2023). Chromium (VI) reduction by two-chamber bioelectrochemical system with electrically conductive wall. *Electrochimica Acta*, *440*, 141738.
57. Ali, A., Arshiq, M., Abu, B., Kim, H., Ahmad, A., Alshammari, M. B., & Suriaty, A. (2022). Oxidation of food waste as an organic substrate in a single chamber microbial fuel cell to remove the pollutant with energy generation. *Sustainable Energy Technologies and Assessments*, *52*(8), 102282.
58. Subha, C., Kavitha, S., Abisheka, S., Tamilarasan, K., Arulazhagan, P., & Rajesh Banu, J. (2019). Bioelectricity generation and effect studies from organic rich chocolaterie wastewater using continuous upflow anaerobic microbial fuel cell. *Fuel*, *251*(April), 224–232.
59. Tamilarasan, K., Banu, J. R., Jayashree, C., Yogalakshmi, K. N., & Gokulakrishnan, K. (2017). Effect of organic loading rate on electricity generating potential of upflow anaerobic microbial fuel cell treating surgical cotton industry wastewater. *Journal of Environmental Chemical Engineering*, *5*(1), 1021–1026.

60. Ghorab, R. E. A., Pugazhendi, A., Jamal, M. T., Jeyakumar, R. B., Godon, J. J., & Mathew, D. K. (2022). Tannery wastewater treatment coupled with bioenergy production in upflow microbial fuel cell under saline condition. *Environmental Research, 212*(Pt B), 113304.
61. J ayashree, C., Tamilarasan, K., Rajkumar, M., Arulazhagan, P., Yogalakshmi, K. N., Srikanth, M., & Banu, J. R. (2016). Treatment of seafood processing wastewater using upflow microbial fuel cell for power generation and identification of bacterial community in anodic biofilm. *Journal of Environmental Management, 180*, 351–358.
62. Wang, Y., Pan, Y., Zhu, T., Wang, A., Lu, Y., Lv, L., Zhang, K., & Li, Z. (2018). Enhanced performance and microbial community analysis of bioelectrochemical system integrated with bio-contact oxidation reactor for treatment of wastewater containing azo dye. *Science of the Total Environment, 634*, 616–627.
63. Srivastava, R. K., Sarangi, P. K., Vivekanand, V., Pareek, N., Shaik, K. B., & Subudhi, S. (2022). Microbial fuel cells for waste nutrients minimization: Recent process technologies and inputs of electrochemical active microbial system. *Microbiological Research, 265*, 127216.
64. Wang, X., Zhang, G., Jiao, Y., Zhang, Q., Chang, J. S., & Lee, D. J. (2022). Ferrous iron oxidation microflora from rust deposits improve the performance of bioelectrochemical system. *Bioresource Technology, 364*, 128048.
65. Aelterman, P., Rabaey, K., Pham, H. T., Boon, N., & Verstraete, W. (2006). Continuous electricity generation at high voltages and currents using stacked microbial fuel cells. *Environ Sci Technol, 40*, 3388–3394.
66. Yang, W., Li, J., Ye, D., Zhang, L., Zhu, X., & Liao, Q. (2016). A hybrid microbial fuel cell stack based on single and double chamber microbial fuel cells for self-sustaining pH control. *Journal of Power Sources, 306*, 685–691.
67. Fujimura, S., Kamitori, K., Kamei, I., Nagamine, M., Miyoshi, K., & Inoue, K. (2022). Performance of stacked microbial fuel cells with barley–shochu waste. *Journal of Bioscience and Bioengineering, 133*(5), 467–473.
68. Yazdi, H., Alzate-Gaviria, L., & Ren, Z. J. (2015). Pluggable microbial fuel cell stacks for septic wastewater treatment and electricity production. *Bioresource Technology, 180*, 258–263.
69. Kim, J. R., Premier, G. C., Hawkes, F. R., Rodríguez, J., Dinsdale, R. M., & Guwy, A. J. (2010). Modular tubular microbial fuel cells for energy recovery during sucrose wastewater treatment at low organic loading rate. *Bioresource Technology, 101*(4), 1190–1198.
70. Liu, S. H., Tsai, Y. N., Chen, C., Lin, C. W., & Zhu, T. J. (2022). Enhanced sulfolane-contaminated groundwater degradation and power generation by a mini tubular microbial fuel cell/electro-fenton combined system. *Process Safety and Environmental Protection, 170*, 1261–1268.
71. Liu, S. H., Fu, S. H., Chen, C. Y., & Lin, C. W. (2020). Enhanced processing of exhaust gas and power generation by connecting mini-tubular microbial fuel cells in series with a biotrickling filter. *Renewable Energy, 156*, 342–348.
72. Li, J., Liu, C., Liao, Q., Zhu, X., & Ye, D. (2013). Improved performance of a tubular microbial fuel cell with a composite anode of graphite fiber brush and graphite granules. *International Journal of Hydrogen Energy, 38*(35), 15723–15729.
73. Marassi, R. J., Queiroz, L. G., Silva, D. C. V. R., Silva, F. T., da, Silva, G. C., & Paiva, T. C. B. (2020). Performance and toxicity assessment of an up-flow tubular microbial fuel cell during long-term operation with high-strength dairy wastewater. *Journal of Cleaner Production, 259*, 120882.
74. Li, J., Ziara, R. M. M., Li, S., Subbiah, J., & Dvorak, B. I. (2020). Understanding the sustainability niche of continuous flow tubular microbial fuel cells on beef packing wastewater treatment. *Journal of Cleaner Production, 257*, 120555.

75. Shukla, A. K., Suresh, P., Berchmans, S., & Rajendran, A. (2004). Biological fuel cells and their applications. *Current Science*, *87*(4), 455–468.
76. Lewis, K. (1966). Symposium on bioelectrochemistry of microorganisms. IV. Biochemical fuel cells. *Bacteriological Reviews*, *30*(1), 101–113.
77. Young, T. G., Hadjipetrou, L., & Lilly, M. D. (1966). The theoretical aspects of biochemical fuel cells. *Biotechnology and Bioengineering*, *8*(4), 581–593.
78. Zhu, X., Wang, X., Li, N., Wang, Q., & Liao, C. (2022). Bioelectrochemical system for dehalogenation: A review. *Environmental Pollution*, *293*, 118519.

12 Bioelectrochemical Oxidation Systems
Basic Concepts and Types

Jemina JohnRajan, Anwesha Anurupa, Vidya Sriraman, Shobana Sugumar

12.1 INTRODUCTION

The most significant threats to human health and the ecosystems we depend on are rising energy and water demands, which have increased by 1300 times in the previous 200 years. The world's energy needs are anticipated to grow by 28% by 2040, while water demand is anticipated to grow by 30% by 2050 due to the continued expansion of the global population. Modern wastewater treatment techniques, created in the 1900s, use a lot of energy and can make up as much as 3% of major economies' power consumption. Moreover, 80% of the wastewater produced worldwide enters receiving waters untreated. With rising energy prices and a growing global population, this scenario is neither feasible nor cheap.

Biologically based electrochemical cells are one such neoteric energy-generating device that converts energy from a substrate or sunlight into chemical and electrical energy. Microbiology, chemistry, and physics came together to form the foundation of biological fuel cell technology (or, more specifically, electrochemistry). The reaction of microbial oxidation is the fundamental tenet of BES. However, how electrons are used on the cathode demonstrates how appealing this technology is. Since any reduction reaction may be carried out in the cathode chamber, there is a wide range of possible applications. Therefore, BES has been specified into many categories, such as MxC, where M means microbial, C represents a cell, and x stands for various uses, such as fuel (microbial fuel cells (MFCs)) or desalination (MDC) (Perazzoli et al., 2018).

Wastewater treatment is regarded as both a fundamental task and a critical step in BES development and, thus, as crucial to the potential application of this technology. In BESs, the removal effectiveness of organics is highly varied and depends on several variables. Organic waste removal is one of the main goals of wastewater treatment. BESs do a decent job of removing organic matter from low-strength wastewater or simple organic substrates. Still, removal efficiency would be substantially lower when dealing with high-strength wastewater or complex organic compounds.

Therefore, BES systems, with more research on optimization for energy production, are a treasured and crucial addition to industries for more eco-friendly and profitable waste management. This chapter details the BES system, its history, the principle behind its operation, and the different types of BES systems developed to date. The types of BES systems are also summarized in Figure 12.1.

DOI: 10.1201/9781003368472-12

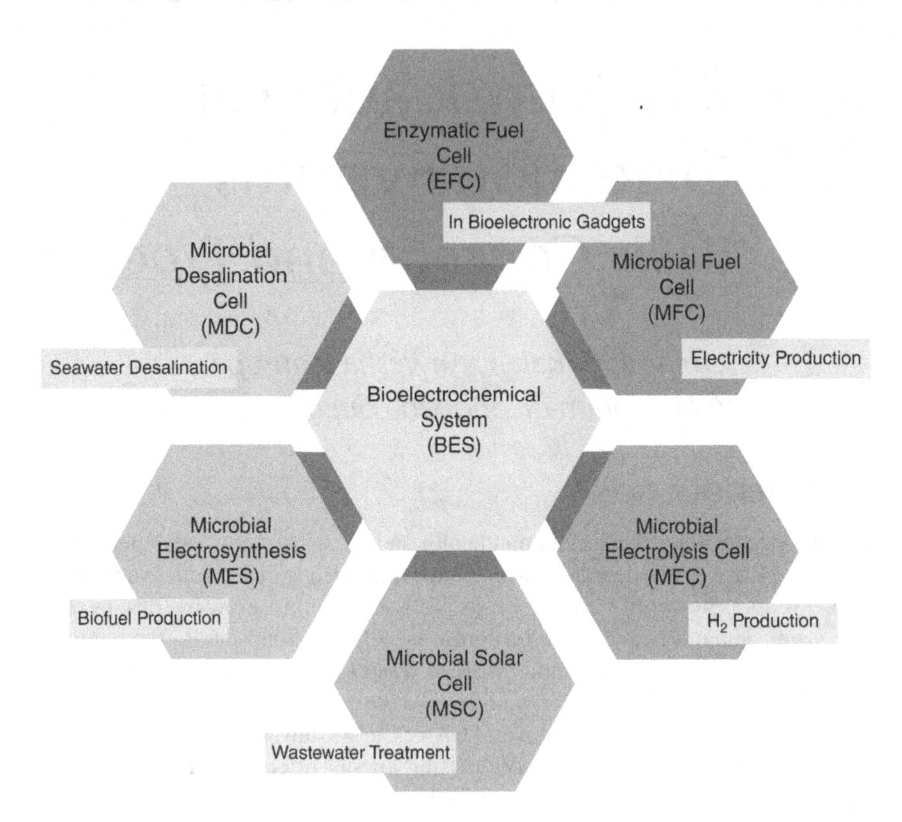

FIGURE 12.1 In general, the BES system can be divided into six. They are Enzymatic Fuel Cell (EFC), Microbial Fuel Cell (MFC), Microbial Electrolysis Cell (MEC), Microbial Solar Cell (MSC), Microbial Electrosynthesis (MES), and Microbial Desalination Cell (MDC). With more research going on today, there are new additions to this list, especially with the application of microbes. Therefore, collectively, they are also termed as MxCs. The figure gives a flow chart of different types of BES.

12.2 HISTORY

It all started when research in electrophysiology revealed that any physiological process accompanied by chemical change involves an associated electrical change. Waller recognized that the functions of assimilation and respiration may be mutually adversarial in terms of observable electric effects and approached the existence of two opposing forces in the presence of analytical and MES processes in a very provocative manner. It is also interesting because he believed that "the product of dissociation.... gives current from the focus of dissociation, whereas a product of association, during the formation, gives rise to a current in the opposite direction" (Potter, 1911). With this understanding, M.C. Potter devised experimentation in which a maximum voltage of 0.3–0.5 was measured using *Saccharomyces cerevisiae* as the test organism, platinum as the electrode, and glucose as the substrate (Potter, 1911).

Lavoisier and Laplace discovered that the hydrogen released when hydrochloric acid reacts with iron is charged positively. This created the question of whether

CO_2 escaping from the fermentation of a saccharine solution might carry an electric charge and be ionized. A gold leaf electroscope and a Dolezalek electrometer were employed in many tests to ascertain this fact. The technique involved suspending a metal plate with a rolled edge a few centimetres above the surface of glucose-fermenting due to the action of yeast, connecting the metal plate to an electroscope or electrometer, and appropriately screening the entire apparatus in a box lined with tinfoil. It was concluded from this experiment that the gases released during the putrefaction of organic matter are also ionized since the CO_2 liberated during the fermentation of glucose by the action of yeast carries both positive and negative ions (Potter, 1915).

Barnet Cohen brought actual attention to the field of BES in 1931. A healthy culture in the mainstream media will have an advantage of 0.5 and 1 V over the control. When $K_3[Fe(CN)_6]$ or benzoquinone-type compounds are added to a culture medium, they partially decrease the medium and maintain a readily reversible oxidation-reduction system. He exploited this phenomenon to build several microbial half-fuel cells, which could generate over 35 volts when connected in series but only with a current of 2 milliamps (Cohen, 1931). Following these initial initiatives, interest in biofuel cells was rekindled in the early 1960s with the advent of manned space travel because of these cells' potential to turn biowaste into energy in spacecraft.

Two worldwide and US patent applications made between 1967 and 1987 were the precursors to the recent decade's intense interest in developing MFCs. The Mobil Corporation granted John Davis the first patent detailing MFCs in 1967. This patent describes an externally mediated (addition of methylene blue) MFC that converts hydrocarbons (ethane, n-propane, and n-butane) to alcohols, aldehydes, and carboxylic acids while producing current from a primary alkaline fuel cell using *Nocardia salmonicolor* or environmental bacteria isolated from sludge. These alkaline MFC tests produced a current of 0.05 mA with the addition of a mediator and linear alkanes of different lengths (CH_4-C_{20}) as the only electron source. The concentration of the methylene blue redox mediator determined the fuel cell's open circuit potentials (the difference in potential between the cathode and anode electrodes when no current is collected). Even though *Nocardia sp.* has not been employed in MFCs since the 1967 experiment, they are still used for oxidizing alkanes (Biffinger & Ringeisen, 2008).

John Davis published a similar invention that same year, using the same overall MFC concept but with the addition of aerobic pre-treatment of the microbial component for the anodic reaction. The bacteria would oxidize alkanes of various lengths while they were oxygenated. After being nitrogen-purged, the oxidized alkanes and the microbial culture were injected into the MFC design to produce electricity. This invention was the first to use *Escherichia coli* in anaerobic circumstances with glucose, as opposed to the original patent, which solely used *Nocardia sp.* The *E. coli*-containing fuel cell produced an open-circuit voltage of roughly 600 mV, or 0.5 mA, using methylene blue as the redox mediator. Considering the fundamental nature of the MFC architecture and the stage of MFC research at that time, 0.5 mA is exceptional even by today's standards (Biffinger & Ringeisen, 2008).

However, research in this field did not start in earnest until the late 1990s and the 2000s, and since then, it has produced notable advancements and several possible applications. When the predicted surface area of the anode was measured, the power density of MFCs went from 0.001 to 0.01 milliwatts per square metre (mW m^{-2}) in 1999 to 787 mW m^{-2} in 2003, and finally to levels of 2770 mW m^{-2} in 2008. Recent advancements in microelectronics have considerably reduced the power requirements for electrical devices, which has led to a resurgence of interest in microbe-catalyzed tiny fuel cells as an alternative to fuel cells using inorganic catalysts (Pant et al., 2012).

In the late 2000s, hydrogen gas started to gain popularity as a renewable energy source. But it was generated only from fossil fuels, such as natural gas. Therefore, Cheng and Logan showed that by using electrohydrogenesis, hydrogen might be produced effectively and sustainably from any biodegradable organic material. In this procedure, a small voltage is added to the circuit to catalyze the formation of hydrogen gas from protons and electrons generated by exoelectrogenic bacteria (Cheng & Logan, 2007). These reactors were known as Microbial Electrolytic Cells. With the aid of microorganisms or enzymes, MECs were used to convert electrical energy to chemical energy to produce valuable chemicals like formate, methanol, ethanol, or hydrocarbons. A voltage is applied to the bacteria's output at the cathode, which is entirely anoxic, to promote hydrogen evolution. Using the hydrogen evolution reaction (HER), protons and electrons mix at the cathode to create hydrogen: $2H^+ + 2e^- \rightarrow H_2$. Bacteria near the anode break down organic materials and generate a voltage of about 0.3 V, whereas the HER requires a voltage of about 0.41 V and only needs a theoretical input of 0.11 V. For the HER to occur in an MEC, a more significant voltage input of 0.25–0.8 V is necessary for practice (Logan, 2008).

The very recent and on-brand trend in the field of BES is Plant-Microbial Fuel Cell (PMFC). The PMFC concept was developed using a bioenergy source in situ. Microorganisms and plants were present in the PMFC to turn solar energy into environmentally friendly power. The basic concept is that plants create rhizodeposits, primarily in the form of carbohydrates, and bacteria use fuel cells to turn these rhizodeposits into electrical energy. In 2008, Strik and his team demonstrated this concept using Reed mannagrass and achieved a maximal electrical power production of 67 mW m^{-2} anode surface (Strik et al., 2008).

Researchers have concentrated on selecting and optimizing electrode materials, designing electrochemical devices, and screening electrochemically active or inert model microorganisms to increase the use of bioelectrochemical systems (BESs). BESs have recently been widely used in desalination, materials science, solid waste processing, nitrate removal, and other fields. Notably, every one of these applications and research relates to energy transformation and electron transfer, or the exchange of chemical and electrical energy.

12.3 PRINCIPLE OF BIOELECTROCHEMICAL SYSTEMS

The term "bioelectrochemical systems" refers to systems that use microorganisms as biocatalysts in electrode reactions. They combine microbial and electrochemical processes to create bioelectricity, hydrogen, or other valuable compounds. An external conductive wire connects anode and cathode electrodes in a traditional BES system to complete an electrical circuit.

By fermenting (oxidizing) the substrate, the bacteria found in the anode section theoretically produce proton (H^+) and electron (e^-) in the compartment. Additionally, it makes it easier for bioanion (electron) to migrate to the anode. Therefore, this part of the BES is often sustained in anaerobic conditions since oxygen restricts electricity production while oxygen from the air is allowed in the cathode chamber. The H^+ migrates through the solution to the cathode electrode to create an interface across the ion-selective membrane. As a result, a potential difference (PD) between the two electrodes is created. This PD transmits the election to the external load via the external electrical circuit. This procedure often produces an electric current (Pant et al., 2012; Shah, 2020).

Extracellular Electron Transfer (EET) is the crucial mechanism in BESs that connects bacteria with insoluble electron donors/acceptors and is the foundation for many applications. The two modes of EET are direct extracellular electron transfer (DET) and indirect extracellular electron transfer (IDET). Microorganisms attached to electrodes directly transfer extracellular electrons in DET, while microbes not attached to electrodes indirectly transfer extracellular electrons in IDET using electron shuttles.

12.3.1 Direct Extracellular Electron Transfer

DET occurs when a redox-active membrane organelle, such as cytochromes, makes physical contact with the BES anode during the transfer of electrons from a microbe to a solid acceptor. These membrane-bound cellular redox centres enable the electron transfer between the microbial cells and a solid terminal electron acceptor without the involvement of any diffusive species. But only bacteria inside the first monolayer at the anode surface may be electrochemically active because the DET via outer membrane redox proteins requires physical contact (adherence) of the bacterial cell to the BES anode. The overall performance of BES is significantly constrained by this limitation (Harnisch et al., 2011; Shah, 2021).

To overcome this issue, some *Geobacter* strains can develop electronic conducting molecular pili, also known as nanowires (Reguera et al., 2005), which enable the microbe to utilize a BES anode and more distant solid electron acceptors without making direct physical contact with them. The nanowires connected to a membrane-bound protein and redox-active protein that enable transmembrane electron transport need to be attached to the electrode. Such nanowires might promote the growth of thicker electroactive biofilms, leading to improved anode performances.

12.3.2 Indirect Extracellular Electron Transfer

In the IDET process, electrons are often carried onto the electrode surface by mediator shuttles or microbial carriers, which then promote EET. These mediators create a favourable environment for the bacteria to be electrochemically active and to reduce various substances to produce different products. The prosthetic groups, which often conceal the electrochemically active groups in charge of redox reactions, impact how the cell interacts with the electrode surface. As a result, these cells can interact with the electrode surface with a mediator or an electron shuttle. But this is restricted to MFCs only when the host bacteria are incapable of DET (Ivase et al., 2020).

Some of the properties of mediators for higher efficiency are as follows (Wilkinson, 2000):

1. Enter the microorganism cells when oxidized,
2. Easily interact with the electron source,
3. Quickly leave the cell when reduced,
4. Be electrochemically active at the anode surface,
5. Be stable for a long time,
6. Be soluble in the anolyte medium,
7. Not be toxic to the biocatalyst,
8. Not be metabolized by the biocatalyst, and
9. Have a matching formal redox potential with that of the redox couple providing the reducing action.

12.3.3 Outward Electron Transfer

Electroactive bacteria breathe through a distinctive process in which the electrons are exported from the intracellular electron transport chain to extracellular electron acceptors (A. Kumar et al., 2017). An electrode in the anode compartment of a BES system promotes bacterial growth and biofilm formation. Utilizing the DET mentioned above and IDET mechanisms, the electrons generated from bacterial metabolism travel to the electrodes. Essential redox proteins in DET for iron reduction include the multi-copper proteins OmpB and OmpC (Holmes et al., 2008) and the outer-membrane multiheme cytochromes (OMCs). *Geobacter sulfurreducens* and *Shewanella oneidensis* MR-1 are two significant microorganisms connected to DET.

Self-secreted and synthetic mediators are used in indirect ET for anodes. The most prevalent examples of self-secreted mediators are flavins, specifically riboflavin (vitamin B2; RF) and flavin mononucleotide (FMN). It's crucial to note that some researchers have also suggested that secreted flavins may be linked to OMCs, acting as cofactors to enable EET in *Shewanella* (Okamoto et al., 2014). Methyl viologen, methylene blue, resazurin, anthraquinone 2,6-disulfonate (AQDS), and other well-known mediators involved in Mediated Electron Transfer are examples of manufactured EET mediators (MET).

New mechanistic foundations for EETs have been discovered in recent research on EET mechanisms. In the case of *Listeria monocytogenes*, a recent study reveals a new ET chain that aids microbial development, including insoluble electron acceptors. The well-known multi-heme assembly is absent from this recently identified EET strategy (as seen in the case of Gram-negative EET bacteria *Shewanella* and *Geobacter*). In this bacteria, EET is shown to be regulated by an eight-gene locus. A unique NADH dehydrogenase that codes for the discovered locus is in charge of transporting the electrons to a quinone pool in the membrane. Additionally, the assembly of an extracellular flavoprotein, which involves free flavins and aids in mediating electron transfer, is made more accessible by various proteins (Light et al., 2018; Shah, 2021).

More studies show promising microorganisms across diverse environments capable of EET other than the environmental resources. Therefore, more research on the mechanisms involved in EET in human pathogens and microbiomes can open up an exciting new field of study and an exciting venture for medicine.

12.3.4 Inward Electron Transfer

This phenomenon of transfer of Electrons from a cathode to a microbe involves microorganisms that can take up electrons directly from the cathode, and these kinds of microorganisms are called Electrotrophs. These organisms can take up electrons in numerous ways, like using OMCs for direct uptake, soluble redox shuttles for indirect uptake, or oxidizing hydrogen by several other pathways. The method of choice depends on the operating condition, bacterial inoculum, and target reactions.

It was proposed that Gram-positive bacteria utilize a cascade of membrane-bound complexes, such as membrane-bound Fe-S proteins, oxidoreductase, and periplasmic enzymes, to receive electrons in the absence of cytochrome C. In addition to *Shewanella* and *Geobacter*, the highly studied bacteriae for MFCs and MECs, DET from a cathode to microbes has been seen in a biocathode for microbial communities, including *beta proteobacteria* and *firmicutes* (Croese et al., 2011). Without the need for a mediator, *Clostridium pasteurianum* boosted butanol synthesis by cathode electron transfer (Choi et al., 2014). Based on the idea that iron-reducing bacteria can deliver electrons to anodes, a concept of iron-oxidizing bacteria (FeOB) being able to absorb electrons from cathodes was proposed and observed in two FeOB, *Mariprofundus ferrooxydans*, and *Rhodopseudomonas palustris*, in recent studies (Bose et al., 2014; Doud & Angenent, 2014).

Increased current consumption was seen when the cathode was altered by adding positively charged species such as chitosan or by expanding the interfacial area utilizing a porous three-dimensional scaffold electrode. Nano-nickel, carbon nanotubes (CNTs), conjugated oligo electrolytes (COEs), and CNTs on reticulated vitreous carbon (NanoWeb-RVC) were used to try and connect nanoparticles to a cathode. Additionally, the bioelectrochemical generation of hydrogen in an MES system was improved using a graphene-modified biocathode. The positively charged anode produced an enriched biofilm on the anode. In contrast, the negatively charged cathode repels microorganisms since most bacterial cell walls have a net negative overall charge. A charge barrier prevents microorganisms from adhering to a cathode (Choi et al., 2014). Therefore, modifications should be tested following changes in cell surface properties on a cathode.

Numerous bioelectrochemical reductions could be catalyzed by the EET from the cathode to the microorganism. The metabolic process was changed by electro-fermentation, which employed electrons from the cathode as reducing power to generate more reduced molecules, such as alcohols, than acids. Artificial photosynthesis might produce electro fuel by fixing carbon using electrical energy rather than solar energy.

12.4 CONSTRUCTION AND COMPONENTS OF THE BES SYSTEM

BES, in its fuel cell form, comprises Proton Exchange Membrane (PEM), at least one anode, and a cathode. A PEM acts as a barrier between the anodic and cathodic chambers in fuel cells to prevent oxygen migration from the cathode to the anode. Only H^+ or other cations can travel across the membrane from the anode to the cathode. A conductive substance simultaneously connects the two electrodes to complete the external circuit. Organic and inorganic substrates are utilized as electron donors in the anode chamber and then oxidized by the natural element coupled to the anode. Reduction processes happen when the electrons reach the cathode (generally O_2 reduction). Equations illustrate the primary cathodic and anodic processes and the overall response.

1. $C_6H_{12}O_6 + 6H_2O \rightarrow 6CO_2 + 24H^+ + 24e^-$ (Substrate Oxidation)
2. $4e^- + 4H^+ + O_2 \rightarrow 2H_2O$ (O_2 Reduction)
3. $C_6H_{12}O_6 + 6H_2O + 6O_2 \rightarrow 6CO_2 + 12H_2O$ (Overall process)

In its Electrolysis cell form, BES can be of a single chamber or double chamber type. The double chamber is similar to fuel cells in construction, and similarly, the exoelectrogenic bacteria transform the organic matter (oxidize it) at the anode to CO_2, H^+, and electrons. With the help of an external PD, these electrons are moved from the anode to the cathode through the external circuit. The cathode is where hydrogen ions move to be reduced. The equations corresponding to an electrolysis cell are:

1. $CH_3COO^- + 4H_2O \rightarrow 2HCO_3^- + H^+ + 4H_2$
2. $CH_3COO^- + 4H_2O \rightarrow 2HCO_3^- + 9H^+ + 8e^-$
3. $2H^+ + 2e^- \rightarrow H_2$

Generalized structure of a BES cell is visually represented in Figure 12.2. The essential components that makeup BES are electrode, membrane, and substrate (fuel).

12.4.1 Electrode

Generally, an electrode can be defined as a small piece of metal, or any other material or substance, utilized in contact with a non-metallic component of a circuit to carry electricity to or from a power source. It could also be a piece of machinery or a living thing that can create an electric circuit. High electrical conductivity, a broad surface area with accessible pores, improved mass transfer properties, mechanical strength, chemical stability, biocompatibility, low cost, and scalability are the fundamental requirements for electrode materials.

12.4.1.1 Anode

Carbon felt, carbon mesh, carbon cloth, graphite felt, and graphite fibre brush are common anode materials. Platinum-coated electrodes are frequently utilized because of how well they generate power and how effectively they do so. Catalytic

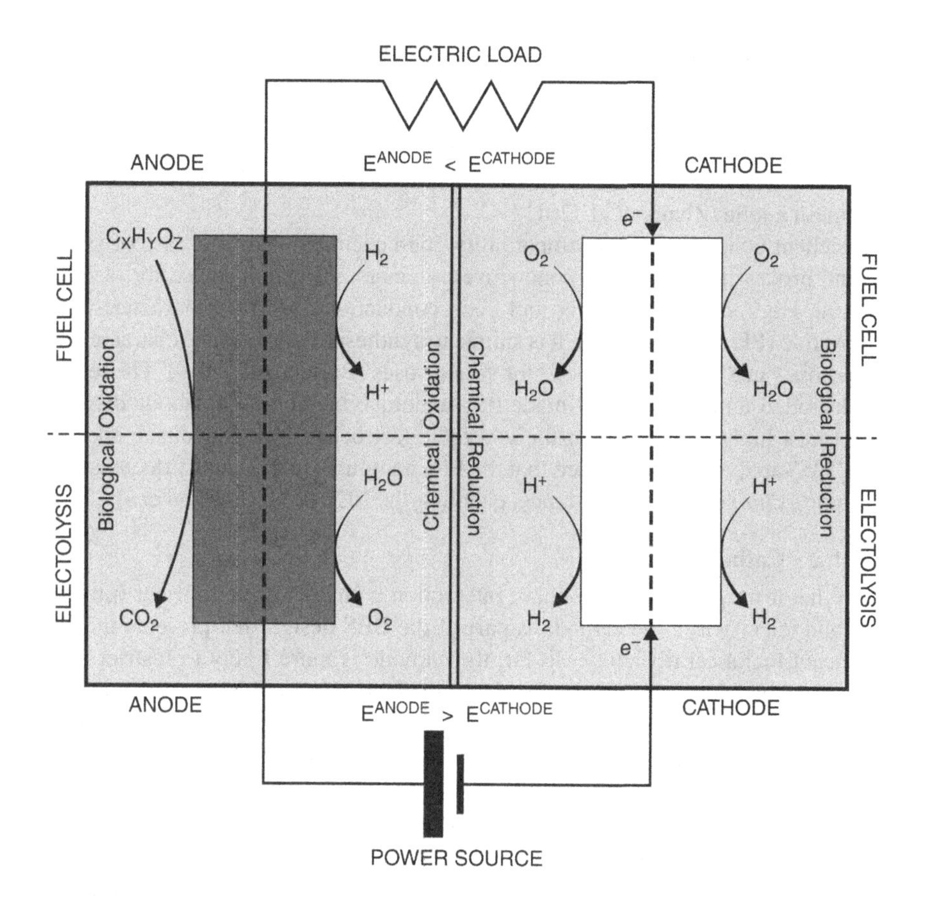

FIGURE 12.2 This figure is an overview diagram of potential microbial and chemical catalysis pairings in BESs, with the top half representing a fuel cell and the bottom half an electrolysis cell. The blue rectangle corresponds to the anode, where the biological oxidation of substrate occurs, and the white rectangle corresponds to the cathode, where biological reduction occurs. In fuel cells, energy is produced ($E^{anode} < E^{cathode}$), and in Electrolysis Cells, Energy is used ($E^{anode} > E^{cathode}$). In both cases, electricity is generated and can be used to power the current load. Still, in the electrolysis cell, additional current is supplied through an external power source, and value-added products like biofuels are produced.

activity and efficiency, when compared to other electrodes, are substantial in the presence of oxygen.

Due to its exceptional conductivity and mechanical stability, CNT is a desirable option for electrode materials (Yazdi et al., 2016). The performance of bioanodes is significantly impacted by the geometry of CNTs. According to the study's findings, CNTs that are longer and loosely packed are ideal for the effective EET process. Less amorphous carbon in the CNTs, especially in MFCs, makes it easier for microbes to connect with electrodes (Erbay et al., 2015).

Two-dimensional (2D) nanomaterial graphene has excellent biocompatibility, outstanding mechanical stability, and high electrical conductivity. Graphene also has a larger specific surface area than other nanostructures made of carbon. Due to these characteristics, graphene has been widely used as an effective electrode material in numerous energy devices (Filip & Tkac, 2014). In the first investigation of graphene-based bioanodes, graphene was used as a bioanode and was secured to a stainless steel mesh. The improved anode produced more current than the standard stainless steel mesh anode (Zhang et al., 2011).

Excellent conductivity, biocompatibility, high chemical stability, ease of synthesis, and processing ease make conductive polymers a popular choice for electrode material. High stability in water and good conductivity are two characteristics of polypyrrole (PPy). Additionally, it is simple to synthesize, and surface characteristics like porosity can be easily adjusted for various uses (Balint et al., 2014). The reactive self-degraded template method-made PPy nanotube membrane bioanode displayed a six times higher current than the control. Electrochemical impedance spectroscopy (EIS) analysis demonstrated that the PPy nanotubes significantly decreased the electrode's charge transfer resistance, enabling the EET process (Zhou et al., 2016).

12.4.1.2 Cathode

Given that it must have a three-phase interaction with air (oxygen), water (protons), and solid (electricity), the cathode is part of the BES design that presents the most significant technical difficulty. So far, the cathode is more likely to restrict power production than the anode. Platinum, platinum black, and graphite are potential cathode materials. There is significant interest in cathodes that increase current densities over those of the primary material by utilizing only microorganisms (biocathodes).

Due to their low cost, high stability, good biocompatibility, and vast surface area, carbon-based materials are frequently used as fuel cell cathodes. To create well-performing fuel cells, they must be considerably upgraded, as they have weak reduction activities. The best cathodic performance is demonstrated by Pt catalysts deposited on carbon-based materials. However, the high price of Pt catalyst makes it impractical for practical use; as a result, low-cost materials should be researched for cost-effective electrodes. A viable strategy for creating low-cost cathodes is anchoring metal oxide nanoparticles on carbon supports. For instance, comparable to Pt cathodes, Mn_2O_3 and Fe_2O_3 cathodes showed good ORR performances (Martin et al., 2011).

Compared to Pt and activated carbon (AC) cathodes, the Fe-AAPyr cathode demonstrated superior performance. After the MFC was run for a prolonged period, a biofilm formation was discovered on the Fe-AAPyr cathode. Biofilm growth impairs the cathode's performance by obstructing active catalytic sites. It's interesting to note that despite the creation of a biofilm from prolonged exposure to wastewater, the Fe-AAPyr cathode didn't show any performance degradation. On the other hand, the ORR activity significantly decreased during biofilm growth on the Pt cathode, perhaps due to the inactive catalyst. These findings unmistakably demonstrate that the Fe-AAPyr is highly stable for MFC operation over an extended period without altering its ORR activity, even in the worst wastewater conditions. Comparable to Pt cathodes, with strong ORR performances (Santoro et al., 2015).

The use of mediators, cathode modification with catalyst, increasing temperature, increasing the presence of the oxidant, increasing the active sites at the cathode, increasing the number of active sites at the cathode, and optimizing the operational conditions within the cathodic compartment are some of the methods that have been suggested to lower the activation barrier for the ORR (Palanisamy et al., 2019). Due to this, cathodes rather than anodes are preferred for these composite materials. Utilizing photosensitive enzymes and their combinations with nanocatalysts to boost the kinetics of relevant electrochemical reactions, such as CO_2 reduction, is one of the more recent methods of cathode enhancement research (Sahoo et al., 2020).

By using microbes as catalysts to speed up electrochemical reduction on the cathode surface, biocathodes provide a cutting-edge method for creating sustainable cathodes. The biocathodes reduce construction and operating costs and offer flexibility in manufacturing essential commodities by doing away with the requirement for pricey chemical catalysts.

The ideal physiological conditions that encourage microbial growth on the cathode surface are necessary for the biocathode. With the help of unique methods like electrically inverting electrodes, biofilms can form on the cathode. For instance, by electrically inverting an organics-oxidizing bioanode, the following process can be used to produce hydrogen-evolving biocathodes:

i. Obtain a conventional MFC with an organics-oxidizing bioanode and a ferricyanide-reducing chemical cathode.
ii. Change the bioanode to a hydrogen-producing biocathode and the ferricyanide-reducing chemical cathode to a ferrocyanide-anode simultaneously (Rozendal et al., 2008).

Similarly, Pisciotta et al. (2012) obtained a biocathode that could fix CO_2 and produce hydrogen while using an electrode inversion technique. The ability to train the biofilm-covered cathode in an oxic environment before switching it to an anode that generates current has only been partially demonstrated in scientific investigations. This suggests that when the cathode is shifted from oxic to anoxic conditions, both exoelectrogenic and electrotrophic microorganisms can be maintained in the electrode biofilms. This method produced biocathodes for oxygen reduction successfully (aerobic biocathodes). It was unclear, however, whether bioanodes would inevitably include anoxic-adapted electrotrophic bacteria (Pisciotta et al., 2012).

Graphite felt, carbon fabric, and other carbon-based materials are frequently utilized as biocathode materials. It has also been demonstrated that stainless steel is an excellent biocathode candidate. Both pure cultures and microbial consortia of mixed cultures are used to create biocathodes. Using cathodes made of graphite felt, carbon paper, and stainless steel mesh, Zhang and He (2012) have methodically investigated the role of cathode materials in developing MFC biocathodes. An anaerobic sludge was utilized to inoculate the cathodes. Over carbon paper and stainless steel mesh biocathodes, the graphite-felt-biocathode displayed the highest ORR activity. Accelerating the performance of biocathodes can be accomplished by increasing the surface area for microbial adhesion. Granules of graphite activated carbon (AC) and AC powder added to the cathode chambers of fuel cells boosted fuel cell performance by increasing the surface area available for forming biofilms, which improved ORR activities of the biocathodes. The enhanced biocathodes significantly lowered the fuel cell systems' internal resistance (Tursun et al., 2016).

12.4.2 Membranes or Separators

Substrate crossover and oxygen diffusion can be significant problems in fuel cells. Therefore, membranes are used to separate the cathode and anode. But this membrane can increase the cost of these fuel cell designs and internal resistance, which can create problems with performance. They are also not long-lasting due to membrane biofouling and affect the long-term application of fuel cells. Electrolysis cells also require membranes to separate hydrogen production from the anode region (Mohanakrishna et al., 2017).

The accumulation of bacteria, their by-products, extracellular polymeric substances (EPSs), and inorganic salts is known as biofouling. Biofilm growth on the membranes, in contrast to that on the anode, has no positive effects on the functionality of BESs. Such a biofilm layer increases the system's internal resistance by reducing the membrane's cation permeability. Therefore, lower power densities are produced due to the decrease in ion conductivity (Ji et al., 2011).

Cell membranes are classified as organic, inorganic, or mixed, depending on the type of structural material. The first kind comprises polymers, such as sulfonated polymers or Nafion. Investigations have also been done on several different polymer varieties. Although natural polymer-based membranes have certain distinctive qualities, they may also be vulnerable to degradation and biofouling (Pasternak et al., 2019). Conversely, synthetic polymers like expanded polystyrene may offer long-term durability but also require a longer time to start up (Mathuriya & Pant, 2019). Ceramic separators dominate the category of inorganic separators, whereas composite membranes make up the third category (Dharmalingam et al., 2019). Ceramic separators have outstanding power performance but also high porosity, which can cause substrate crossover and high oxygen back-diffusion, both of which can cause biofouling.

Some physical methods used for cleaning the membrane include physical cleaning, using pressure generated by magnets to peel off the biofilm layer, and using ultrasonic waves (Rossi et al., 2018). Out of these three, ultrasonic waves have shown performance regain similar to a fresh new membrane (Xue et al., 2021). Still, when using ultrasound as a long-term strategy for membrane maintenance, the membrane lifetime after much cleaning needs to be carefully examined. The efficiency of these methods also highly depends on the type of BES used (Pasternak et al., 2022).

Chemicals commonly used in cleaning are hydrochloric acid, Sulphuric acid, Sodium hydroxide, Sodium hypochlorite (Xue et al., 2021), and some other combinations that act as lysis solutions. Although it was discovered that chemical cleaning of the membranes was efficient, it should be emphasized that sulphuric acid and hydrogen peroxide are quite harsh chemicals. Nafion membranes can be treated using this chemical combination, but other polymer membranes will be harmed and deteriorate (Pasternak et al., 2022).

Based on the charge they allow to pass through, membranes can be classified into cationic and anionic exchange membranes. Due to their strong proton conductivity, cation exchange membranes (CEMs) are typically used as membrane separators for fuel cells. The most popular membrane is Nafion because of its strong proton conductivity and negatively charged hydrophilic sulphonate group. It has been demonstrated that thinner Nafion membranes are more efficient than thicker ones because the former contribute less ohmic resistance (Jung et al.,

1998). However, the higher oxygen permeability and substrate crossover caused by the thinner membranes reduce the system's Coulombic efficiency. The membrane thickness should be optimized for optimum performance to eliminate this problem. For the Nafion-based systems, pH drop in the anode chamber is another crucial worry. The oxidation of the substrate at the anode generates protons, which may result in a pH drop. This pH drop is significant for Nafion membranes. Low pH impairs performance and is known to limit bacterial respiration at the anode (Kim et al., 2007). Additionally, the Nafion membrane permits cations (such as Na^+) other than protons to be transported, which results in a charge imbalance between the anode and cathode (Mohanakrishna et al., 2017).

Anion exchange membranes (AEMs) often generate more current than CEMs. The OH ions from the cathode through the AEM devour the protons created at the anode, preventing the anodic pH drop. This characteristic allows the AEM to reduce both the cathode resistance brought on by the precipitation of transported cations and the ion transport resistance (Piao et al., 2013).

12.4.3 Substrate

Substrates are organic compounds the microbe consumes during catalytic or enzymatic reactions in bioelectrochemistry. It is regarded as one of the crucial biofactors influencing electricity production. The substrate acts as a carbon energy source by giving the bacterial cells the energy they need to thrive. It also has an impact on fuel cell performance and economic viability. Cheng & Logan (2011) claimed that the fuel cell's microbial population and power production are influenced by the substrate's concentration. As a result, a wide range of substrates can be employed as feedstock in BES. These include wastewater, cellulose, and acetate, as well as simple sugars like glucose, pure sugar, low molecular weight sugar, and complex sugars.

Acetate is simple and highly utilized organic substrate. This is due to its resistance to other microbial processes, including fermentation and room-temperature methanogenesis (Pham et al., 2009). In terms of power production and columbic efficiency, the electrical energy generated from an acetate substrate is often higher than that from other substrates (Pant et al., 2012). The by-product of higher-order carbon source metabolism, like glucose, is acetate. As a result, glucose is another substrate widely utilized as a fuel in a fuel cell. Other substrates include cellulose, glucose, lactate, and sodium acetate (Pant et al., 2012), petroleum refinery wastes, and residues from the agro-process wastes such as distillery, dairy, and vegetable wastes (Ivase et al., 2020).

12.4.4 Cell Design

There are standard three-cell designs. They are Dual chamber, Single chamber, and Stack design.

12.4.4.1 Dual Chamber

The original and most popular design for BESs is a dual chamber configuration. Anode and cathode chambers are taken into account in the design. A CEM divides the two chambers. An external resistor connects the anode from the anode chamber and the cathode from the cathode chamber. Batch mode is frequently used when operating the double-chamber BES. Because the design is straightforward and the BES functions effectively, early-stage researchers can use it as a model.

By moving each chamber's location and employing various chamber shapes, numerous BES designs can be produced. Even though the anode chamber is the only place where the biocatalyst and electron generation can occur, the cathode reaction is just as significant. As a result, the same volumes of anode and cathode chambers were used in most investigations (Mohanakrishna et al., 2017).

12.4.4.2 Single Chamber

After a double-chamber setup, a single-chamber design was explored. It only has an anode chamber, and the air can contact the cathode electrode. Here, the catholyte is atmospheric air or oxygen applied to the cathode surface, which causes an oxygen reduction reaction (ORR). This design proved less effective at producing power than the two-chamber type (Cheng & Logan, 2011). But the design was considered advantageous to scale up the process because it needs less room for setup and building. When wastewater treatment is a primary goal of the process, this design can provide comparable treatment performance with less bioelectricity generation. The design can transform the current anaerobic digesters into BESs and is similar to the traditional anaerobic digesters for methane production (Pawar et al., 2022).

12.4.4.3 Stack Design

To scale up fuel cell technology, it was necessary to integrate multiple MFCs into a single operating system. Except for low-power electronic modules and actuators, which have allowed the employment of MFCs in real applications as exemplified by Gastrobot (Wilkinson, 2000) and EcoBots-I and -II (Melhuish et al., 2006), the maximum potential of a single MFC (regardless of anolyte volume) unit that can be achieved is practically around 1.0 V, which is very little. No proper application can be run with this potential. Multiple MFCs with tiny volumes were created and integrated into a single operating system since power generation benefits from vast volumes of MFCs are limited.

Ieropoulos et al. (2008) reported to have conducted the first investigation using MFC stacking, in which a current generation rate roughly 50 times higher than that of the large MFC was seen. The findings of this study imply that rather than enlarging a single unit, it may be preferable to connect several small-sized units to accomplish MFC scale-up. It was also recognized that the hydraulic/fluidic and electrochemical connections affect the MFC scaling.

12.5 ENZYMATIC FUEL CELLS

A typical fuel cell comprises two electrodes continually fed with fuel and oxidant: a fuel-oxidizing anode and an oxidant-reducing cathode linked to one another by an external load and placed in an electrolyte. For fuel oxidation and oxidant reduction, chemical fuel cells rely on non-selective metal catalysts like platinum and its alloys. The present fuel cell study focuses on identifying less susceptible catalysts platinum to break down through surface poisoning, reducing catalyst load to lower costs, and studying electrolyte characteristics to avoid fuel-oxidant crossover across half-cells.

The Enzyme-based Biofuel Cell (EFC) is a well-known model of a bioelectronic gadget that employs enzymes to oxidize fuel and reduce oxidants (such as oxygen or peroxide) to produce electricity. The practice of EFCs has only developed in recent decades and now finds applications in implanted medical equipment. Theoretically, these devices are more eco-sustainable than traditional fuel cells, which frequently rely on high-priced rare metal catalysts and run on processed fossil fuels. Further, they function at room temperature and in an ambience with a neutral pH, which substantially impacts system activity. However, relative to transition metal catalysts, biocatalysts have some limitations. Redox proteins typically only display their better catalytic abilities in their natural environments, meaning that nature did not develop enzymes specifically for bioelectric-catalytical systems. The inability of the protein to electrically communicate with the electrode surface, the biocatalyst's assembly, and the electrodes' low stability are typical manifestations of this. The reduced volumetric catalyst density of enzymes is another disadvantage. Since enzymes are giant molecules, they often have fewer active sites than traditional metal electrodes (Ivanov et al., 2010).

An enzymatic fuel cell may employ cascades and combinations of enzymes to promote the metabolism of a microorganism. When the specific type of enzyme is used, the substrate can undergo deep oxidation, eventually producing CO_2. The cathode and anode electrodes of the fuel cell may be assembled without the need for membranes or noble metals, significantly lowering the volume of the biofuel cell (Barelli et al., 2019). The oxidoreductase family of enzymes is those utilized in fuel cells. They can be divided into three classes depending on the type of electrical communication they utilize or the redox cofactors that are associated with them. Pyrroloquinoline quinone-dependent dehydrogenases, such as alcohol dehydrogenase (AlcDH), glucose dehydrogenase, and glycerol dehydrogenase, make up the first category. The second category consists of enzymes like glucose dehydrogenase and AlcDH that have nicotinamide adenine dinucleotide ($NADH/NAD^+$) or nicotinamide adenine dinucleotide phosphate ($NADPH/NADP^+$) cofactors. Extraction of the electrons is challenging for enzymes in the third group because the FAD redox cofactor is firmly bound and buried deep inside the protein structure. GOx comes under this category and is the most widely utilized enzyme.

Enzymatic biofuel cells operate according to the same fundamental principles as other fuel cells: they create electricity by forcing electrons to go around an electrolyte barrier through the wire after being separated from the parent molecule by a catalyst. An anode in an enzymatic biofuel cell is catalyzed by oxidases appropriate for converting preferred fuels or combining such enzymes to fully oxidize biofuels. The cathode often contains an oxidoreductase that catalyzes reduction to water in mildly acidic or neutral environments, using molecular oxygen as the final electron acceptor. The fuels that an enzymatic fuel cell accepts and the catalysts they employ differ from those of normal fuel cells. An enzymatic biofuel cell employs enzymes produced from living cells instead of the metal catalysts used in most fuel cells, such as nickel and platinum. This benefits enzymatic biofuel cells in a variety of ways. In contrast to precious metals, which must be mined and have a non-elastic supply, enzymes are immensely numerous and inexpensive to mass produce. Additionally, enzymes do not have the contamination issues that more conventional metallic catalysts have, rendering them unusable.

Traditional fuel cells cannot meet the operational needs of enzymatic biofuel cells. The enzymes must be immobilized close to the anode and cathode for the fuel cell to function correctly; otherwise, they would disperse into the cell's fuel, and most freed electrons would not reach the electrodes, decreasing the efficiency. It offers a way for electrons to be moved from and to the electrodes even while they are immobilized. It is possible to transport electrons from an enzyme to an electrode directly (DET) or indirectly (with the aid of other substances that can do this) (mediated electron transfer). The former method is only feasible with specific enzymes where activation sites are close to the surface of an enzyme. Still, it has fewer toxicity hazards for the fuel cells designed for usage within the human body. Finally, the complex fuels employed in enzymatic fuel cells demand various enzymes at each stage of the metabolic process. Producing the necessary enzymes and retaining them at the necessary levels might present challenges (Atanassov et al., 2007).

Enzymatic fuel cells have gotten close to power densities of 1 mW cm^2 in recent years. High energy-density fuels like ethanol and glycerol offer considerable promise for portable electronic applications. More effective systems have also resulted from changes in the overall system architecture. Examples include the development of a single chamber, air-breathing systems employing tiny MEAs, and the removal of the separator membrane without noticeably reducing power output. But many planned applications need more significant performance enhancements (higher power densities and energy efficiencies).

EFCs need electrode materials that are more catalytic while still performing well, especially in light of issues with the cluttering of the active surfaces and a decline in enzyme activity. Additionally, it is crucial to investigate time-dependent performance over realistic timescales, especially with an emphasis on long-term modifications in the enzyme activity.

12.6 MICROBIAL FUEL CELL

MFC is a sustainable method of energy generation using simple and accessible materials such as wastewater, which contain high amounts of organic content and serve as a substrate to produce electricity. By conversion of organic compounds in the medium, microorganisms convert the chemical energy obtained from the breakdown of carbon sources by serial biochemical reactions into electrical energy. With increased production of industrial effluents and domestic waste, this method provides an eco-friendly way of generating electricity and a solution to the disposal of growing masses of waste. The ability of bacteria to digest varied materials is utilized to generate a reaction like that of a dry cell, where the bacteria catalyze the chemical reactions that lead to energy production. Much like a battery, an MFC also has an anode and a cathode where, in the anode, the bacteria catalyze the breakdown of the substrate, which leads to the production of electrons and protons. The electrons are then transferred to the cathode through a conductive material. The cathode is aerobic in nature; hence, oxygen often functions as an electron acceptor for the cathode in MFC. The protons generated from the anode move through a PEM, where the protons combine with electrons and oxygen once in the cathode to generate water, which is obtained as a by-product (R. Kumar et al., 2017). While theoretically, this method seems plausible, the energy output per fuel cell is limited and needs to be compounded to produce a significant power output.

On a design basis, MFCs can be divided into single-chamber MFCs, double-chamber MFCs, stacked MFCs, and upflow MFCs. Single-chambered cells are found to be most commonly used due to their higher efficiency (Gyaneshwar et al., 2022). Other types of MFCs are mediator and mediator-free cells. Mediator cells have chemicals that shuttle electrons from the cell to the electrode. These cells are generally less preferred due to the toxic nature of the mediator chemicals. Mediator-free cells use electrochemically active bacteria, which perform the electron transfer through inbuilt mechanisms that vary between species.

For electrode materials generally, graphene electrodes, carbon-based materials like paper, brushes, mesh, and felt, and graphite-based materials like rods, plates, granules, and fibre are used owing to their biocompatibility, electrical conductivity, and high pore stability. Metals like platinum, platinum black, copper, aluminium, and zinc have also been used (Shabangu et al., 2022).

12.6.1 Anode Materials

The anode component of an MFC is an essential aspect in determining the energy output of a fuel cell, as well as in providing a place for bacterial growth. The anode material should be judged based on its architecture, biocompatibility, surface area, and conductivity. Despite metals offering higher conductivity, metals are less preferred due to their corrosive nature, which fails to meet the criteria of electrode materials. Non-corrosive metals like stainless steel, when used, showed a decreased power density output as compared to traditional carbon-based electrodes. Some metals, such as gold, silver, and platinum, proved to be better electrode materials. However, this increased cost of materials and did not facilitate bacterial adhesion (Yaqoob et al., 2020).

Research studying the effects of metals like iron, aluminium, and zinc as anode materials and their dependency on the surface area available and the shape of the anode on *Rhodobacter capsulatus* for the photosynthetic MFC. It showed that zinc as an anode material had higher bioelectricity production than iron and aluminium. By area availability, for higher voltage generation, a 20 cm^2 area was most suitable, while a 16 cm^2 anode area was found to give higher current output. When changing the shape of the anode, a circular anode provided the highest voltage output as obtained against a square or rectangular anode (Mirza et al., 2022).

While using carbon materials as an anode, there is a recorded loss of potential over a long distance while moving towards the current collector. Traditionally, titanium-based collectors are used while connecting electrodes to an external circuit. By altering the configuration and surface area of the collector to the anode, its effects on power output were tabulated. Maintaining the anode surface area and increasing the contact area of the collector from 28 cm^2 to 70 cm^2 showed almost double the power output production. By decreasing the distance of the collector from the anode while keeping the area constant, the power output showed slight improvement (Paitier et al., 2022).

12.6.2 Cathode Materials

In an MFC, oxygen is reduced in three layers, air, electrolyte, and the electrode, to give water or hydrogen peroxide. Most materials considered for anode can be used in cathode as well. However, specific properties of cathode materials include mechanical strength, high conductivity, and catalytic properties (Mustakeem, 2015). Platinum is most commonly used as a cathode due to its high reductional capacity. The use of this metal limits the practical application of MFCs. Biocathodes are a cheaper alternative to expensive platinum cathodes. Biocathode entails using cheaper material alongside a microbial community in the electrode or the electrolyte to perform the cathode reactions. This utilizes the formation of biofilms by electrogenic bacteria, which function as electron acceptors.

MFCs can also function as toxicity sensors, where the presence of toxins slows bacterial metabolism. Slowing in metabolic activity can be measured and compared to standardized results. A study used MFC as an online toxicity sensor for heavy metals. Response of the sensor to heavy metal from mining effluents showed a shift in the microbial community of the MFC, as well as a deviation of t_{max} estimation with increasing metal concentrations (Adekunle et al., 2019).

12.6.3 Microbes Used in MFC

The microbes used are called exoelectrogens or electricigens. They are electrochemically active and transfer electrons. The electrons generated reach the anode in three possible ways:

1. Protein carriers of the microbial cell surface transport them.
2. They can be released by membranous projections such as nanowires, which are conductive appendages modified pili or flagella in microbes like *Geobacter* and *Shewanella*.
3. They can be secreted through mediators. These can be exogenous mediators like Quinine, phenoxazine, phenothiazine, neutral red, etc., or endogenous mediators like 2-amino-3-carboxy-1,4-naphthoquinone (Matsena & Chirwa, 2022).

The microbes' pure cultures are *Archaebacteria*, *Acidobacteria*, *Cyanobacteria*, etc. However, predominantly *Proteobacteria* is used, with *γ-Proteobacteria* being the most studied electricigen (Y. Cao et al., 2019). A study based on yeast as electricigens is being done. A recent study based on yeast *Pichia fermentans*, through the development of anodic biofilm with wheat straw hydrolysate as substrate, showed a maximum response of 20.13 ± 0.052 mW/m^2 with hydrolysate prepared with *Phlebia floridensis* and 20.42 ± 0.071 mW/m^2 with substrate preparation by *Phanerochaete chrysoporium* (Pal et al., 2023). Mixed cultures in MFC have shown higher treatment efficiency and a significant increase in electricity generation. A study showed that mixed inoculum of *Pseudomonas aeruginosa*, *Klebsiella variicola*, and *Saccharomyces cerevisiae* attained a maximum power density of 500 mW/m^2 at 0.657V as compared to 130 mW/m^2 at 0.158V, 98 mW/m^2 at 0.178V, and 50 mW/m^2 at 0.1V for individual pure culture, respectively. Mixed culture MFC also showed the highest removal of COD, BOD, nitrate nitrogen, grease, total suspended solids, and suspended solids as opposed to pure culture inoculum in the treatment of Palm Oil Mill Effluent (POME) (Sarmin et al., 2021).

Despite being marketed for various uses, MFCs are yet to be made mainstream due to their dependency on factors like temperature of operation, the pH conditions, growth and efficiency of microbes, the actual power output itself, etc., which limit their usage. Methods of upscaling and commercializing MFCs are being explored and show great promise in addressing the growing energy crisis.

12.7 MICROBIAL ELECTROSYNTHESIS

With its setup being like an MFC, the Microbial Electrosynthesis (MES) cell is a biochemical mechanism through which electricity is provided to microbes, which then catalyze the conversion of CO_2 or other organic compounds to form multi-carbon species whose harvesting is economically feasible. Typical carbon sources are CO_2 (from gaseous, liquid, or solid sources), methane, and acetate. CO_2 is preferred due to its abundant availability in the atmosphere. Carbon sequestration is of prime importance, with rising CO_2 emissions by industries, the military, and automobiles reaching as much as 37.5 billion tonnes in 2022. In MES, the electricity to drive the reaction is externally supplied to aid the conversion of CO_2 at the cathode. Microbial capture cell (MCC) is a similar setup that does not require an external energy supply and therefore has lower yields than MES. A significant disadvantage posed by using CO_2 as an electron acceptor is the large requirement for electrons to reduce into multi-carbon species. For example, the reduction of butyrate to butanol requires only four electrons.

In contrast, reducing CO_2 to butanol has a sixfold higher electron requirement (24 electrons) despite having the same reduction potentials for the reactions. Due to efficiency loss, additional issues arise from the multiple-step conversion when CO_2 is used in the cathode (Rabaey & Rozendal, 2010). Conditions of operation are similar to those of MFCs for biocompatibility, high electron transfer rate, surface area, structure, etc. of the electrode materials used. Microbes used for MES usually can DET, wherein an external electron shuttle is not required.

12.7.1 Electron Transfer Mechanism (EET)

The microbes have different methods of direct EET where intracellular electrons are exchanged with an extracellular electron donor/acceptor like an electrode or a metal compound through the cell membrane (Tanaka et al., 2018). This is commonly facilitated by either membrane-bound c-type cytochrome or soluble carriers like flavins (Gurumurthy et al., 2021). While the idea remains the same, each organism's mechanism may differ. Direct EET is a property of electroactive species, making them biocatalysts suitable for MES or MFC. Traditionally, the microbes used for MES are methanogenic bacteria, metal-reducing bacteria (*Shewanella* species and *Geobacter* species) and acetogenic bacteria (Shin et al., 2017). Indirect EET utilizes artificial mediators like Methyl viologen, Neural red, etc., which are redox mediators. These participate in the electron exchange, acting as electron donors or acceptors.

12.7.2 Electrode Materials

The most commonly used electrode materials for MES are carbon-based materials like carbon cloth, felt, rods, brushes, and graphene. Their high chemical stability and porous nature make them ideal as an electrode. A study showed that carbon could be extended beyond electrodes and can be used to increase compound synthesis. The experimental setup used titanium mesh (with iridium and ruthenium) as the anode and carbon felt as the cathode. Granular activated carbon (GAC) was added in various concentrations to support the cathode compartment. To produce acetate, a multi-microbial inoculum was harnessed, including Sulfurospirillum, Paludibacter, Acetobacterium, and Bacteroides. The results showed that the increase in the functional surface area provided by the GAC aided the CO_2 reduction and increased its efficiency, giving 2.8 times higher acetate productivity in the cell. While the cell with 16 gL^{-1} of added GAC produced nearly 3.9 ± 0.5 gL^{-1} of acetate, the control with no added GAC produced only 1.5 ± 0.5 gL^{-1} of acetate (Dong et al., 2018).

Traditional electrodes are made either of metal or carbon; however, metal-carbon hybrid electrodes are being explored. A composite cathode made of reduced graphene oxide-coated copper foam was used to generate acetate from CO_2. Two types of copper-graphene oxide hybrid electrodes were created: direct coating in the presence of nitrogen (rGO-CuF) and electroplating (epCu-rGO). Comparative analysis of rGO (with no metal component) and rGO-CuF revealed an 8.3 times lower acetate production in rGO. The higher performance of the metal-enriched cathode was attributed to the increased surface area (1.9 times) and to the copper's electrical conductivity. The performance of epCu-rGO, however, was less than that of the rGO-CuF. The decrease in performance was chalked up to the bactericidal property of copper with which the microbes were in direct contact. In the rGO-CuF, however, reduced graphene oxide (rGO) covered the Cu foam electrode's active areas to prevent Cu's antibacterial property from acting as a conducting buffer between the microorganisms and the electrode (Aryal et al., 2019).

Other novel approaches to hybrid cathodes include copper nanoparticles functionalized on carbon cloth electrodes inoculated with anaerobic granular sludge to convert CO_2 to methane (Georgiou et al., 2022) and copper ferrite supported reduced graphene oxide nanocomposites, which were used to enrich the cathode of the electrosynthesis to produce volatile fatty acids (isobutyric acid and acetate) from CO_2. The research showed 1.72- and 1.35-times higher production of isobutyric acid and acetate compared to MES with carbon cloth electrodes (Thatikayala & Min, 2021).

Several approaches to increase the yield of the desired product have been studied to make product recovery economically and quantitatively feasible. Genetically, engineered microbes show promise by allowing modifications to the microbe's extracellular electron transport efficiency by altering the c-type cytochrome expression or to the chemical pathway of product formation by either removing unwanted pathways or inserting desired metabolic enzymes (Surti et al., 2021).

Other structural modifications to MES designs to increase product extraction include that of the triple chamber circular MES (CMES). This novel reactor has three chambers—anode, cathode, and extraction chamber. The anode and cathode were separated by a PEM, and so were the anode and extraction chamber. The cathode and the extraction chamber were separated using an AEM. The design allows for the diffusion of the produced product (acetate) from the cathodic chamber to the extraction chamber via the AEM during the product biosynthesis. This increases the ease of product extraction for further processing, thereby significantly reducing the need for further downstream processing, as well as boasts a higher acetate production rate in the extraction chamber (104.26 ± 1.75 g/m^2 day) when compared to the cathodic chamber (81.35 ± 1.15 g/m^2 day) (Das et al., 2021).

Integrating sustainability into MES, photoanodes are used, which harness renewable solar energy to maintain the cathode potential and reduce the overpotential for oxygen evolution. The photoanode made of bismuth vanadate on fluorine-doped tin oxide (FTO) glass showed acetate production amounting to a theoretical conversion of 0.97 ± 0.19% of received input sunlight into acetate, which could be sustained for as long as 7 days (Bian et al., 2020).

The substrate carbon source can vary, allowing domestic, industrial, and agricultural wastewater to be converted into economically viable products. This broadens the scope for application and development of MES in industries to reduce carbon footprint as well as remain energy efficient. Products isolated include volatile fatty acids, hydrogen, methane, etc. While the technology is highly innovative and promising, MES has several limitations, including substrate pre-treatment, product yield, and long-term operation, that need to be addressed before its commercialization.

12.8 MICROBIAL DESALINATION CELL

Despite 70% of the earth's surface being covered with water, freshwater is only 3%, of which only 1/3rd is accessible in rivers, groundwater, lakes, etc. With global warming increasing the variability in rain patterns, the need for stable potable and domestic water sources is crucial. As an extension of MFC and BES systems, the desalination cell tackles the need for freshwater while being a green innovation. Average commercial desalination theoretically consumes about 3 kJ/kg to desalinate 1 m^3 of brine. However, the actual energy usage is estimated to be anywhere between 5 and 26 times the predicted value, depending on the process.

An MDC is typically a three-chambered setup, where the chambers are demarcated by membranes like AEM, CEM, ultrafiltration membrane, etc. (X. Cao et al., 2009). The anodic chamber contains electrically active microbes, which form biofilm attached to the anode. The microbe oxidizes the organic matter (substrate) to release protons, electrons, and CO_2. The electrons generated in the process are mobilized through an external circuit to the cathode, where they are accepted by oxygen and further reduced to form water as a by-product. To maintain the charge balance, the ions like Cl^{-1} and SO_4^{-2} move to the anode from the saline water in the 3rd chamber (middle chamber). Similarly, positive ions like Na^{+1} and Ca^{+2} migrate to the cathode. The PD generated by the movement of electrons from the anode to the cathode facilitates desalination self-sufficiently (Gujjala et al., 2022).

The reactions of the cell can be written as (Saeed et al., 2015):

$$\text{Anode: } (CH_2O)_n + nH_2O \rightarrow nCO_2 + 4ne^- + 4nH^+$$

$$\text{Cathode: } O_2 + 4nH^+ \rightarrow 2H_2O$$

Several configurational alterations have been made and studied to obtain a more efficient and economical model of MDC. One of the studied models is the stacked MDC, where multiple ion exchange membranes are placed between the electrodes, promoting higher ion movement resulting in higher energy recovery and freshwater. The increased membrane counts resulted in a pH imbalance and an increase in internal resistance. This problem was addressed using a tubular flow stacked MDC (USMDC) in which the tubular structure helps reduce ohmic resistance. The study was performed using a 1L capacity cell. The results, as obtained from 36 hours of functioning with sodium acetate as anode substrate, show that the desalination ratios were 90.3%, 60.6%, and 76.9% with USMDC, Upflow MDC, and Stacked MDC, respectively, with 50 ohms as applied external resistance. The desalination efficiency was also found to be between 87.2% and 96% over 120 days (Wang et al., 2020).

Air cathode MDC utilizes oxygen as a terminal electron acceptor by use of platinum/carbon as a catalyst. A study published combined wastewater treatment with desalination to create a dual-purpose air cathode microbial cell. The microbes in the anode were used to treat the petroleum refinery wastewater while simultaneously desalinating saltwater. The model was studied for 8 months. The maximum power generated from the cell was 570.86 $\mu W/m^2$ (with 1KΩ resistance), with a COD removal of 96%. A salinity removal rate of 150.39 ppm/hr was reached with 35,000 ppm saltwater (Jaroo et al., 2021).

In air cathode MDC generally, the cost of metal, slow redox activity, and high power input have proved to be a setback.

Other models include:

- Biocathode MDC – Biocathode MDC utilizes the biofilm formation of aerobic microorganisms like yeast, algae, and bacteria, where they act as a catalyst.
- Capacitative MDC – Adsorption of ions instead of their movement prevents pH imbalances and salt scaling.
- Osmotic MDC – Integrating osmotic MFC into MDC, the AEM separating the anode and middle desalination chamber is replaced with forward osmosis (FO) membrane. This change allows freshwater harvesting from the anode due to osmosis while the middle chamber is desalinated using a conventional MDC mechanism by electricity generation (B. Zhang & He, 2012).
- Photosynthetic MDC – A modification of biocathode MDC where photosynthetic algae are used for sustainable in-situ oxygen production and to aid COD removal in the cathode (Kokabian & Gude, 2013).

- Upflow MDC – Tubular structured cell comprised traditionally of two chambers (inner anode and outer desalination chamber). The conventional cathode is replaced with a carbon-based cloth to surround the desalination chamber (Tawalbeh et al., 2020).

Integrated systems combining electrolysis cells and desalination cells to remove metal (from industrial waste) simultaneously, nitrogen (from municipal waste), and salt from brine were created and studied. The experiment showed 95.1% nitrogen oxidation in the cathode and a total nitrogen removal rate of 4.07 mg/L hr by heterotrophic denitrification. A desalination rate of 63.7% was achieved, while, with the harvested power, 99.5% of lead(II) was removed within 48 hours (Li et al., 2017). Other such systems include the recirculatory MDC-MEC, which can simultaneously generate power, treat waste, and perform desalination. A comparative study of dual-chambered, single-chambered, and MDCs with ferricyanide as a cathode electrolyte was done on the MDC-MEC system as well (Koomson et al., 2021).

MDCs have gained steady interest. However, the model faces limitations in its durability, pH fluctuation, batch mode of operation, low initial salt concentration, high resistance, and low diversity in inoculum culture (*Microbial Desalination Cells: Progress and Impacts – Journal Of Biochemical Technology*, n.d.). Despite this, several studies have proved that MDC is an indispensable green technology.

12.9 MICROBIAL ELECTROLYSIS CELL

Fossil fuels are used to create most of the hydrogen gas, resulting in the uncontrolled emission of CO_2, a factor in environmental change. Water electrolysis may produce sustainable hydrogen using energy from renewable sources like wind, sun, or biomass; however, the process has high energy needs (5.6 kWh/m^3H_2), and standard electrolyzer energy efficiency is only 56–73% (Logan, 2008). MECs are a type of microbial bioelectrochemical technology that uses electricity and was first created for the high-efficiency synthesis of biological hydrogen from waste streams. Compared to conventional biological technologies, MECs can produce high-yield hydrogen from various organic materials under relatively favourable circumstances by overcoming thermodynamic constraints. The anode receives the electrons generated by MECs, which then go to the cathode through a power supply-containing electrical circuit. Hydrogen is created at the cathode by reducing protons or water. Cation movement from the anode to the cell cathode balances the flow of negative charge outside the cell. Theoretically, 0.14 V has to be provided to stimulate hydrogen generation in natural settings. However, in reality, it is necessary to provide more than 0.14 V, partly because of the cathode overpotential (Jeremiasse et al., 2010). Since MECs are a relatively new technology, many researchers might not know how these reactors are built or the variables that might impact performance. Since the anode design and the electrogenic processes are comparable, MECs and MFCs have many similarities.

An MEC integrates an MFC with electrolysis to produce biohydrogen, whereas MFCs employ microbes as catalysts to oxidize organic and inorganic compounds and generate energy. However, with an MEC, an external voltage is required to get over the thermodynamic barriers. The ability of the two systems to manufacture hydrogen in MECs and use the decrease of dissolved oxygen to generate an electric current in MFCs to produce energy, respectively, is the primary distinction between them from an electrochemical. The oxidation of organic materials (electron donors) and the electron transfer of electroactive microorganisms to the anode determine the current in MFC and MEC (Yasri et al., 2019). Different MEC combinations have been adjusted to improve efficiency. Prior designs included a primary H-type cell having gas-collecting components connected to a cathode chamber. Later, essential advancements were achieved to produce dual-compartmental MECs, which were easy to use.

Electrochemically active bacteria oxidize organic material in an MEC, producing CO_2, electrons, and protons. The protons are released into the solution as the bacteria transmit electrons towards the anode. The unbound protons in the solution are combined with the electrons after the electrons pass via a wire to a cathode. This, however, does not happen on its own. MEC reactors need an external voltage source to create hydrogen at the cathode from the fusion of these electrons and protons in a physiologically aided neutral pH state. However, compared to conventional water electrolysis, MECs only need a small amount of energy input (0.2–0.8 V) (Kadier et al., 2016). Since the electrons that go to the cathode eventually become hydrogen gas, current directly correlates to the rate at which hydrogen is produced. Improved reactor designs, compact electrode spacing, varied membrane materials, and large surface area anodes have rapidly increased MEC current densities and hydrogen recoveries.

Additionally, several experimental findings show that various operating conditions, such as inoculation, applied voltage, electrolyte resistance strength, electrode physicochemical properties, operating temperature, hydraulic retention time (HRT), and organic loading rate, have an impact on MFC performance. More research would be required to understand the impact of various operating parameters further and maximize the MFC's hydrogen generation. Various prospects are linked with microbial community analysis's future in MEC that need to be explored further.

12.10 MICROBIAL SOLAR CELLS

A recently created technology called microbial solar cells (MSCs) uses solar energy to generate energy or chemicals. MSCs gather solar energy using photoautotrophic microorganisms or higher plants and employ the BES's electrochemically active microorganisms to produce electricity. MSCs operate on the basic tenets of photosynthesis, organic matter transport to the anode compartment, anodic oxidation of organic matter by electrochemically active bacteria, and cathodic oxygen reduction. Plant-based MSCs are referred to as PMFCs. Plant roots excrete rhizodeposits that directly feed the electrochemically active microbes at the anode of PMFCs. The excretion of organic molecules by plant roots into the soil is known as rhizodeposition and includes components such as sugars, organic acids, polymeric polysaccharides, enzymes, and dead cell matter. About 20–40% of a plant's photosynthetic output comes from the rhizodeposits, and various bacteria can degrade these substances. Electricity is produced on-site continually while the plant grows with its roots in the MFC (Strik et al., 2008). MSCs operate on the basic tenets of photosynthesis, organic matter transport to the anode compartment, anodic oxidation of organic matter by electrochemically active bacteria, and cathodic oxygen reduction. The simplicity of the method is its significant advantage. As a result, there are fewer requirements for selecting the right bacteria, and the same applies to selecting the electrode material. The biggest drawback is the requirement for regular exogenous chemical addition, as, in the case of a fuel cell, it is technically and energetically unfeasible, ecologically controversial, and has eventually resulted in the widespread abandonment of the strategy. The use of artificial redox mediators may be energy-reasonable for MSCs where there isn't a continual supply and removal of substrate solutions; nevertheless, this requires a longer retention duration of the mediator in the system and maximal energetic output from the cell (Rosenbaum & Schröder, 2010).

In cases when organic substrates are accessible, MFCs are an up-and-coming option for producing energy in an eco-friendly manner. However, creating self-sustaining systems becomes challenging due to oxygen requirements in the cathode and the anode's substrates that can be changed. Numerous researchers have used photosynthetic systems to satisfy the oxygen demand of MFCs. Such systems operate on microalgae that emit oxygen while absorbing CO_2 from the environment during the light phase. The use of solar energy technologies employing MFCs has drawn much interest recently because of the connection between oxygen production and CO_2 collection (Gouveia et al., 2014). A phototrophic biofilm develops on the anode of a fuel cell, converting solar energy into electrical energy. To function for prolonged periods of more than 20 days, MSCs with phototrophic biofilms have a self-organizing biofilm, including Chlorophyta and/or Cyanophyta. So far, all investigations have employed mixed populations of microbes, which likely contain electrochemically active bacteria.

MSCs can also be included in a marine environment along the coast. Ecosystems that use solar energy create phytoplankton that float in the water, including macroalgae and zooplankton. If these substrates are gathered, an MFC can produce energy. Pumps might feed raw saltwater to a 40-km-long tubular MFC to create power in a real-world application. According to estimates, MSCs in coastal zones, which make up 10% of the ocean, may produce electrical power of 2.4–16 TWh per year or 0.01–0.05 mW/m^2 when split by surface area. This PD is significantly lower than MSCs using higher plants or phototrophic biofilms by more than three orders of magnitude. Estimates of the energy intake are 18 times higher than the possible electrical production, according to state of the art (Girguis et al., 2010).

12.10.1 Construction of MSC

In an MSC, the anode absorbs electrons from microorganisms either free-floating in solution or growing on the anode surface (biofilm). The cathode and anode materials must be specific for a device to function well. The optimal anode promotes direct electron transmission from the microorganisms without needing an intermediary electron. The anode cannot obstruct light from passing through the photosynthetic cells, which is an additional criterion for MSCs. To avoid causing the device's rate limiter, the cathode should speed up the process by which protons and oxygen combine to generate water (Schneider et al., 2016). MSCs may have a single or dual chamber, and they can also include either planktonic cells or a biofilm-forming on (or near) the electrode surface. The difference between the redox potential of the fuel oxidation redox reaction at the anode and the redox reaction at the cathode determines the maximum device voltage (usually oxygen reduction). Because there is more energy in the fuel than is needed to power the metabolic activities of the microbes, some extra electrical energy can be produced. In contrast to the protons, which diffuse to the cathode and combine with oxygen to generate water, the electrons are gathered at the anode. While protons diffuse towards the cathode and combine with oxygen to generate water, the electrons are gathered at the anode.

An MSC requires a long-lasting anode material that is also biocompatible. It must be moderately conductive and inexpensive to manufacture. It must also have a surface suitable for cell adhesion and be electrochemically robust in the potential window of interest. Letting light into the anode chamber, illuminating the phototroph-containing biofilm, is the last prerequisite for MSC electrodes. As the most prevalent anode material in MSCs, carbon strikes a fair balance between these needs (Wei et al., 2011). The fact that carbon-specific electrodes might prevent light from reaching the microbes in MSC applications is a disadvantage of these electrodes. Therefore, transparent electrodes composed of indium tin oxide (ITO) or FTO coated on glass are often used in MSCs. Typically, carbon sheets, carbon cloths, or composites of carbon fibres with platinum coating serve as the cathodes in an MSC. There have also been uses for stainless steel, gold, platinum, and raw graphite or carbon cloth. Due to its high redox potential, oxygen is a suitable electron acceptor at the cathodes. Platinum is the preferred catalyst to speed up water formation from the interaction between protons and oxygen. Low oxygen solubility in the catholyte can be a concern since an effective cathode requires a high oxygen concentration to be accessible at all times. One option is to include cultures that produce oxygen under the light in the cathode chamber. The decrease in oxygen content in the dark is a disadvantage of this strategy (McCormick et al., 2011). A different approach to the poor oxygen availability in the catholyte is to use an air cathode in place of the liquid compartment.

According to numerous studies, phototrophic biofilms and PMFCs have the highest power generation (50 and 7 mW/m^2, respectively) and the highest estimated net power potential (67 and 61 mW/m, respectively). According to these studies, these two technologies also have the highest theoretical power generation expectations and performance results. The most effective MSC systems are, hence, phototrophic biofilms and PMFCs. Overall, MSCs are strong, with operational durations ranging from 5 to 175 days (Schamphelaire et al., 2008). Other MSCs, in contrast, employ chemical catalysts, which deplete the system in hours and prevent it from self-sustaining. MSCs with catalysts produce fuels in-situ, such as hydrogen, which is subsequently oxidized in traditional fuel cells. Since none of the MSCs created so far is intended for scale-up, they are all laboratory-scale systems.

Furthermore, no significant processes have complete or enough data accessible to make correct estimations. For instance, there are no measurements of MSCs' coulombic efficiency (CE – the percentage of entirely oxidized electron donors' electrons that are delivered to the anode). Currently, it is difficult experimentally to determine the precise carbon and electron fluxes. MSCs must compete with other renewable energy systems for "real-life" applications. Several variables, including energy output and associated environmental benefits, will affect how this market turns out. According to the best long-term power output of 50 mW/m^2, there are exciting applications for MSCs. For instance, weather sensors for temperature, pressure, and humidity that needed 24 mW were put on a buoy and powered by a sediment MFC. We anticipate that these sensors and other low-power applications, such as LED lights, can be run on MSCs.

12.11 UPSCALING

Despite the concept and working of BES-based systems being proven and the technology itself being almost three decades old, BES-based systems are yet to be implemented in the real world on a large scale. The scale of functioning of these systems has remained at a lab level. For example, MDCs have yet to experiment on a scale more significant than a couple of litres. With the world becoming increasingly aware of sustainability and green technology, microbial systems are critical in reversing the effects caused by modern industrialization and excessive CO_2 release. Upscaling of BES systems can be done by integrating modern technology with established models of BES to achieve a sustainable multipurpose green machine. The lack of commercialization can be attributed to a lack of standardized parameters of productivity, focused funded research, and lack of awareness. Governmental bodies and other research organizations should address the promise of green energy and sustainable development. The availability of subsidies will accelerate the adoption and upscaling of BESs.

12.12 ADVANTAGES AND APPLICATIONS

Table 12.1 provides us the advantages and application of different types of BESs.

TABLE 12.1
Advantages and Application of Different Types of BESs

S.no	Name	Advantages	Applications
1.	Enzymatic Fuel Cell	• Cost efficient • Operating condition – Mild • pH – Near neutral • Ambient temperature operation • Catalytic activity – High • Overvoltage for substrate conversion – Low (Barelli et al., 2019)	Pressure and Sphincter sensors, Neurostimulators, drug carriers, cochlear, and retinal implants

(Continued)

TABLE 12.1 *(Continued)*
Advantages and Application of Different Types of BESs

S.no	Name	Advantages	Applications
2.	Microbial Fuel Cell	• Energy benefits – High • Environmental impact – Low • Operating stability – Good • Economic Efficiency – High • Sludge – Low • Operational flexibility in extreme conditions like low substrate and concentrations and extreme temperatures is high	Dye decolourization, Bioenergy production, wastewater treatment (Industrial, municipal, and domestic), Biosensors
3.	Microbial Electrolysis Cell	• The oxidation of organic molecules, which takes the role of oxidizing water, uses much less energy than oxidizing the potent water molecule • Naturally occurring biofilms form on the anode surface • Inexpensive electrode materials, including some carbon-based compounds, can be used • Self-assembly and self-maintenance property enable maintenance-free long-term MEC operation • Less oxygen is produced, so safer • Chlorine production is not seen due to low potential	Production of Bioelectricity, Biofuel (Hydrogen), and Integrated systems to produce methane, formic acid, and hydrogen peroxide
4.	Microbial Solar Cell	• Generate a variety of fuels and chemicals in addition to electricity • Urban areas or landscapes may readily include PMFCs, which help to "green" the city • Closed MSC systems can protect the organisms' nutrients, resulting in long-term, low-maintenance energy generation • No demand for expensive or hazardous special catalysts	Biodiesel, Bioelectricity generation, Wastewater treatment, Carbon sequestration, Brewery industries, etc.
5.	Microbial Electrosynthesis	• Energy cost – low • Easier recovery • It is designed to be modular and doesn't need much space to set up • Can incorporate renewable energy sources into their activities • Value-added chemicals can be produced and employed as alternative energy sources	Green biorefineries, Industrial production of multicarbon species like butanol, acetic acid, ethanol, etc.

(Continued)

TABLE 12.1 *(Continued)*
Advantages and Application of Different Types of BESs

S.no	Name	Advantages	Applications
6.	Microbial Desalination Cell	• Sustainable and self-sustaining • Low energy input required • Enhanced desalination of water • Shorter startup period • It can be combined with a multiple-stage effect • Anaerobic ammonium oxidation bacterium addition increases the efficiency of nitrogen and carbon removal (Saeed et al., 2015)	Sustainable source of freshwater, Integrated systems with MEC to produce bioelectricity with simultaneous waste treatment and water softening.

Note: The following table lists some of the advantages and applications of the types of fuel cells that have been dealt with in this chapter.

12.13 CONCLUSION

Energy exchange is the foundation of all microbial electrochemical technologies, including those that reduce oxygen and convert it to electricity (EFCs), convert chemical energy into electricity (MFCs), convert electrical energy into chemical energy (MECs), and convert solar energy into electricity (MSCs). Energy conversion efficiency is the primary determinant of BES performance for all systems, particularly for the energy conversion phase using electrical energy. This phase involves a significant amount of electron transport, which suggests a potential future area of METs study. In the context of precise management of fermentation and degradation, the behaviour of the electron, intracellular reducing power, and energy metabolism in BESs are subjects of growing attention.

REFERENCES

Adekunle, A., Raghavan, V., & Tartakovsky, B. (2019). Online monitoring of heavy metals-related toxicity with a microbial fuel cell biosensor. Biosensors and bioelectronics, 132, 382–390.

Aryal, N., Wan, L., Overgaard, M. H., Stoot, A. C., Chen, Y., Tremblay, P. L., & Zhang, T. (2019). Increased carbon dioxide reduction to acetate in a microbial electrosynthesis reactor with a reduced graphene oxide-coated copper foam composite cathode. Bioelectrochemistry, 128, 83–93.

Atanassov, P., Banta, S., Barton, S. C., Cooney, M., Liaw, B. Y., Mukerjee, S., & Apblett, C. (2007). Enzymatic biofuel cells. The electrochemical society interface, 16(2), 28.

Balint, R., Cassidy, N. J., & Cartmell, S. H. (2014). Conductive polymers: towards a smart biomaterial for tissue engineering. Acta biomaterialia, 10(6), 2341–2353.

Barelli, L., Bidini, G., Calzoni, E., Cesaretti, A., Di Michele, A., Emiliani, C., & Sisani, E. (2019, December). Enzymatic fuel cell technology for energy production from bio-sources. In AIP Conference Proceedings (Vol. 2191, No. 1, p. 020014). AIP Publishing LLC.

Bian, B., Shi, L., Katuri, K. P., Xu, J., Wang, P., & Saikaly, P. E. (2020). Efficient solar-to-acetate conversion from CO_2 through microbial electrosynthesis coupled with stable photoanode. Applied energy, 278, 115684.

Biffinger, J. C., & Ringeisen, B. R. (2008). Engineering microbial fuels cells: recent patents and new directions. Recent patents on biotechnology, 2(3), 150–155.

Bose, A., Gardel, E. J., Vidoudez, C., Parra, E. A., & Girguis, P. R. (2014). Electron uptake by iron-oxidizing phototrophic bacteria. Nature communications, 5(1), 1–7.

Cao, X., Huang, X., Liang, P., Xiao, K., Zhou, Y., Zhang, X., & Logan, B. E. (2009). A new method for water desalination using microbial desalination cells. Environmental science & technology, 43(18), 7148–7152.

Cao, Y., Mu, H., Liu, W., Zhang, R., Guo, J., Xian, M., & Liu, H. (2019). Electricigens in the anode of microbial fuel cells: pure cultures versus mixed communities. Microbial cell factories, 18(1), 1–14.

Cheng, S., & Logan, B. E. (2007). Sustainable and efficient biohydrogen production via electrohydrogenesis. Proceedings of the national academy of sciences, 104(47), 18871–18873.

Cheng, S., & Logan, B. E. (2011). Increasing power generation for scaling up single-chamber air cathode microbial fuel cells. Bioresource technology, 102(6), 4468–4473.

Choi, O., Kim, T., Woo, H. M., & Um, Y. (2014). Electricity-driven metabolic shift through direct electron uptake by electroactive heterotroph *Clostridium pasteurianum*. Scientific reports, 4(1), 1–10.

Cohen, B. (1931). The bacterial culture as an electrical half-cell. Journal of bacteriology, 21(1), 18–19.

Croese, E., Pereira, M. A., Euverink, G. J. W., Stams, A. J., & Geelhoed, J. S. (2011). Analysis of the microbial community of the biocathode of a hydrogen-producing microbial electrolysis cell. Applied microbiology and biotechnology, 92(5), 1083–1093.

Das, S., Chakraborty, I., Das, S., & Ghangrekar, M. M. (2021). Application of novel modular reactor for microbial electrosynthesis employing imposed potential with concomitant separation of acetic acid. Sustainable energy technologies and assessments, 43, 100902.

Dharmalingam, S., Kugarajah, V., & Sugumar, M. (2019). Membranes for microbial fuel cells. In Microbial electrochemical technology (pp. 143–194). Elsevier.

Dong, Z., Wang, H., Tian, S., Yang, Y., Yuan, H., Huang, Q., & Xie, J. (2018). Fluidized granular activated carbon electrode for efficient microbial electrosynthesis of acetate from carbon dioxide. Bioresource technology, 269, 203–209.

Doud, D. F., & Angenent, L. T. (2014). Toward electrosynthesis with uncoupled extracellular electron uptake and metabolic growth: enhancing current uptake with rhodopseudomonas palustris. Environmental science & technology letters, 1(9), 351–355.

Erbay, C., Yang, G., de Figueiredo, P., Sadr, R., Yu, C., & Han, A. (2015). Three-dimensional porous carbon nanotube sponges for high-performance anodes of microbial fuel cells. Journal of power sources, 298, 177–183.

Filip, J., & Tkac, J. (2014). Is graphene worth using in biofuel cells? Electrochimica acta, 136, 340–354.

Georgiou, S., Koutsokeras, L., Constantinou, M., Majzer, R., Markiewicz, J., Siedlecki, M., & Constantinides, G. (2022). Microbial electrosynthesis inoculated with anaerobic granular sludge and carbon cloth electrodes functionalized with copper nanoparticles for conversion of CO_2 to CH_4. Nanomaterials, 12(14), 2472.

Girguis, P. R., Nielsen, M. E., & Figueroa, I. (2010). Harnessing energy from marine productivity using bioelectrochemical systems. Current opinion in biotechnology, 21(3), 252–258.

Gouveia, L., Neves, C., Sebastião, D., Nobre, B. P., & Matos, C. T. (2014). Effect of light on the production of bioelectricity and added-value microalgae biomass in a photosynthetic alga microbial fuel cell. Bioresource technology, 154, 171–177.

Gujjala, L. K. S., Dutta, D., Sharma, P., Kundu, D., Vo, D. V. N., & Kumar, S. (2022). A state-of-the-art review on microbial desalination cells. Chemosphere, 288, 132386.

Gurumurthy, D. M., Bilal, M., Nadda, A. K., Reddy, V. D., Saratale, G. D., Guzik, U., & Mulla, S. I. (2021). Evaluation of cell wall-associated direct extracellular electron transfer in thermophilic Geobacillus sp. 3 biotech, 11, 1–12.
Gyaneshwar, A., Selvaraj, S. K., Ghimire, T., Mishra, S. J., Gupta, S., Chadha, U., & Paramasivam, V. (2022). A survey of applications of MFC and recent progress of artificial intelligence and machine learning techniques and applications, with competing fuel cells. Engineering research express, 4(2), 022001.
Harnisch, F., Aulenta, F., & Schröder, U. (2011). 6.49-Microbial fuel cells and bioelectrochemical systems: industrial and environmental biotechnologies based on extracellular electron transfer. Comprehensive biotechnology (Second Edition) (pp. 643–659). Academic Press, Burlington.
Holmes, D. E., Mester, T., O'Neil, R. A., Perpetua, L. A., Larrahondo, M. J., Glaven, R., & Lovley, D. R. (2008). Genes for two multicopper proteins required for Fe (III) oxide reduction in Geobacter sulfurreducens have different expression patterns both in the subsurface and on energy-harvesting electrodes. Microbiology, 154(5), 1422–1435.
Ieropoulos, I., Greenman, J., & Melhuish, C. (2008). Microbial fuel cells based on carbon veil electrodes: stack configuration and scalability. International journal of energy research, 32(13), 1228–1240.
Ivanov, I., Vidaković-Koch, T., & Sundmacher, K. (2010). Recent advances in enzymatic fuel cells: experiments and modeling. Energies, 3(4), 803–846.
Ivase, T. J. P., Nyakuma, B. B., Oladokun, O., Abu, P. T., & Hassan, M. N. (2020). Review of the principal mechanisms, prospects, and challenges of bioelectrochemical systems. Environmental progress & sustainable energy, 39(1), 13298.
Jaroo, S. S., Jumaah, G. F., & Abbas, T. R. (2021, November). The operation characteristics of air cathode microbial desalination cell to treat oil refinery wastewater. In IOP conference series: earth and environmental science (Vol. 877, No. 1, p. 012002). IOP Publishing.
Jeremiasse, A. W., Hamelers, H. V., & Buisman, C. J. (2010). Microbial electrolysis cell with a microbial biocathode. Bioelectrochemistry, 78(1), 39–43.
Ji, E., Moon, H., Piao, J., Ha, P. T., An, J., Kim, D., & Chang, I. S. (2011). Interface resistances of anion exchange membranes in microbial fuel cells with low ionic strength. Biosensors and bioelectronics, 26(7), 3266–3271.
Jung, D. H., Lee, C. H., Kim, C. S., & Shin, D. R. (1998). Performance of a direct methanol polymer electrolyte fuel cell. Journal of power sources, 71(1–2), 169–173.
Kadier, A., Simayi, Y., Abdeshahian, P., Azman, N. F., Chandrasekhar, K., & Kalil, M. S. (2016). A comprehensive review of microbial electrolysis cells (MEC) reactor designs and configurations for sustainable hydrogen gas production. Alexandria engineering journal, 55(1), 427–443.
Kim, J. R., Cheng, S., Oh, S. E., & Logan, B. E. (2007). Power generation using different cation, anion, and ultrafiltration membranes in microbial fuel cells. Environmental science & technology, 41(3), 1004–1009.
Kokabian, B., & Gude, V. G. (2013). Photosynthetic microbial desalination cells (PMDCs) for clean energy, water and biomass production. Environmental science: processes & impacts, 15(12), 2178–2185.
Koomson, D. A., Huang, J., Li, G., Miwornunyuie, N., Ewusi-Mensah, D., Darkwah, W. K., & Opoku, P. A. (2021). Comparative studies of recirculatory microbial desalination cell–microbial electrolysis cell coupled systems. Membranes, 11(9), 661.
Kumar, A., Hsu, L. H. H., Kavanagh, P., Barrière, F., Lens, P. N., Lapinsonnière, L., & Leech, D. (2017). The ins and outs of microorganism–electrode electron transfer reactions. Nature reviews chemistry, 1(3), 1–13.
Kumar, R., Singh, L., & Zularisam, A. W. (2017). Microbial fuel cells: types and applications. Waste biomass management—A holistic approach, 367–384.

Li, Y., Styczynski, J., Huang, Y., Xu, Z., McCutcheon, J., & Li, B. (2017). Energy-positive wastewater treatment and desalination in an integrated microbial desalination cell (MDC)-microbial electrolysis cell (MEC). Journal of power sources, 356, 529–538.

Light, S. H., Su, L., Rivera-Lugo, R., Cornejo, J. A., Louie, A., Iavarone, A. T., & Portnoy, D. A. (2018). A flavin-based extracellular electron transfer mechanism in diverse Gram-positive bacteria. Nature, 562(7725), 140–144.

Logan, B. E. (2008). Microbial fuel cells. John Wiley & Sons.

Martin, E., Tartakovsky, B., & Savadogo, O. (2011). Cathode materials evaluation in microbial fuel cells: a comparison of carbon, Mn_2O_3, Fe_2O_3 and platinum materials. Electrochimica acta, 58, 58–66.

Mathuriya, A. S., & Pant, D. (2019). Assessment of expanded polystyrene as a separator in microbial fuel cell. Environmental technology, 40(16), 2052–2061.

Matsena, M. T., & Chirwa, E. M. N. (2022). Advances in microbial fuel cell technology for zero carbon emission energy generation from waste. In Biofuels and bioenergy (pp. 321–358). Elsevier.

McCormick, A. J., Bombelli, P., Scott, A. M., Philips, A. J., Smith, A. G., Fisher, A. C., & Howe, C. J. (2011). Photosynthetic biofilms in pure culture harness solar energy in a mediatorless bio-photovoltaic cell (BPV) system. Energy & Environmental Science, 4(16), 4699–4709.

Melhuish, C., Ieropoulos, I., Greenman, J., & Horsfield, I. (2006). Energetically autonomous robots: food for thought. Autonomous robots, 21(3), 187–198.

Mirza, S. S., Al-Ansari, M. M., Ali, M., Aslam, S., Akmal, M., Al-Humaid, L., & Hussain, A. (2022). Towards sustainable wastewater treatment: influence of iron, zinc and aluminum as anode in combination with salt bridge on microbial fuel cell performance. Environmental research, 209, 112781.

Mohanakrishna, G., Kalathil, S., & Pant, D. (2017). Reactor design for bioelectrochemical systems. In Microbial fuel cell (pp. 209–227). Springer, Cham.

Mustakeem, M. (2015). Electrode materials for microbial fuel cells: nanomaterial approach.

Okamoto, A., Kalathil, S., Deng, X., Hashimoto, K., Nakamura, R., & Nealson, K. H. (2014). Cell-secreted flavins bound to membrane cytochromes dictate electron transfer reactions to surfaces with diverse charge and pH. Scientific reports, 4(1), 1–8.

Paitier, A., Haddour, N., Gondran, C., & Vogel, T. M. (2022). Effect of contact area and shape of anode current collectors on bacterial community structure in microbial fuel cells. Molecules, 27(7), 2245.

Pal, M., Shrivastava, A., & Sharma, R. K. (2023). Electroactive biofilm development on carbon fiber anode by Pichia fermentans in a wheat straw hydrolysate based microbial fuel cell. Biomass and bioenergy, 168, 106682.

Palanisamy, G., Jung, H. Y., Sadhasivam, T., Kurkuri, M. D., Kim, S. C., & Roh, S. H. (2019). A comprehensive review on microbial fuel cell technologies: processes, utilization, and advanced developments in electrodes and membranes. Journal of cleaner production, 221, 598–621.

Pant, D., Singh, A., Van Bogaert, G., Olsen, S. I., Nigam, P. S., Diels, L., & Vanbroekhoven, K. (2012). Bioelectrochemical systems (BES) for sustainable energy production and product recovery from organic wastes and industrial wastewaters. RSC advances, 2(4), 1248–1263.

Pasternak, G., de Rosset, A., Tyszkiewicz, N., Widera, B., Greenman, J., & Ieropoulos, I. (2022). Prevention and removal of membrane and separator biofouling in bioelectrochemical systems-a comprehensive review. iScience, 25(7), 104510.

Pasternak, G., Yang, Y., Santos, B. B., Brunello, F., Hanczyc, M. M., & Motta, A. (2019). Regenerated silk fibroin membranes as separators for transparent microbial fuel cells. Bioelectrochemistry, 126, 146–155.

Pawar, A. A., Karthic, A., Lee, S., Pandit, S., & Jung, S. P. (2022). Microbial electrolysis cells for electromethanogenesis: materials, configurations and operations. Environmental engineering research, 27(1), 200484.

Perazzoli, S., de Santana Neto, J. P., & Soares, H. M. (2018). Prospects in bioelectrochemical technologies for wastewater treatment. Water science and technology, 78(6), 1237–1248.
Pham, T. H., Aelterman, P., & Verstraete, W. (2009). Bioanode performance in bioelectrochemical systems: recent improvements and prospects. Trends in biotechnology, 27(3), 168–178.
Piao, J., An, J., Ha, P. T., Kim, T., Jang, J. K., Moon, H., & Chang, I. S. (2013). Power density enhancement of anion-exchange membrane-installed microbial fuel cell under bicarbonate-buffered cathode condition. Journal of microbiology and biotechnology, 23(1), 36–39.
Pisciotta, J. M., Zaybak, Z., Call, D. F., Nam, J. Y., & Logan, B. E. (2012). Enrichment of microbial electrolysis cell biocathodes from sediment microbial fuel cell bioanodes. Applied and environmental microbiology, 78(15), 5212–5219.
Potter, M. C. (1911). Electrical effects accompanying the decomposition of organic compounds. Proceedings of the Royal Society of London. Series B, containing papers of a biological character, 84(571), 260–276.
Potter, M. C. (1915). Electrical effects accompanying the decomposition of organic compounds. II.-Ionisation of the gases produced during fermentation. Proceedings of the Royal Society of London. Series A, containing papers of a mathematical and physical character, 91(632), 465–480.
Rabaey, K., & Rozendal, R. A. (2010). Microbial electrosynthesis—revisiting the electrical route for microbial production. Nature reviews microbiology, 8(10), 706–716.
Reguera, G., McCarthy, K. D., Mehta, T., Nicoll, J. S., Tuominen, M. T., & Lovley, D. R. (2005). Extracellular electron transfer via microbial nanowires. Nature, 435(7045), 1098–1101.
Rosenbaum, M., & Schröder, U. (2010). Photomicrobial solar and fuel cells. electroanalysis: an international journal devoted to fundamental and practical aspects of electroanalysis, 22(7–8), 844–855.
Rossi, R., Yang, W., Zikmund, E., Pant, D., & Logan, B. E. (2018). In situ biofilm removal from air cathodes in microbial fuel cells treating domestic wastewater. Bioresource technology, 265, 200–206.
Rozendal, R. A., Jeremiasse, A. W., Hamelers, H. V., & Buisman, C. J. (2008). Hydrogen production with a microbial biocathode. Environmental science & technology, 42(2), 629–634.
Saeed, H. M., Husseini, G. A., Yousef, S., Saif, J., Al-Asheh, S., Fara, A. A., & Aidan, A. (2015). Microbial desalination cell technology: a review and a case study. Desalination, 359, 1–13.
Sahoo, P. C., Pant, D., Kumar, M., Puri, S. K., & Ramakumar, S. S. V. (2020). Material–microbe interfaces for solar-driven CO_2 bioelectrosynthesis. Trends in biotechnology, 38(11), 1245–1261.
Santoro, C., Serov, A., Villarrubia, C. W. N., Stariha, S., Babanova, S., Artyushkova, K., & Atanassov, P. (2015). High catalytic activity and pollutants resistivity using Fe-AAPyr cathode catalyst for microbial fuel cell application. Scientific reports, 5(1), 1–10.
Sarmin, S., Tarek, M., Roopan, S. M., Cheng, C. K., & Khan, M. M. R. (2021). Significant improvement of power generation through effective substrate-inoculum interaction mechanism in microbial fuel cell. Journal of power sources, 484, 229285.
Schamphelaire, L. D., Bossche, L. V. D., Dang, H. S., Höfte, M., Boon, N., Rabaey, K., & Verstraete, W. (2008). Microbial fuel cells generating electricity from rhizodeposits of rice plants. Environmental science & technology, 42(8), 3053–3058.
Schneider, K., Thorne, R. J., & Cameron, P. J. (2016). An investigation of anode and cathode materials in photomicrobial fuel cells. Philosophical transactions of the Royal Society A: mathematical, physical and engineering sciences, 374(2061), 20150080.
Shabangu, K. P., Bakare, B. F., & Bwapwa, J. K. (2022). Microbial fuel cells for electrical energy: outlook on scaling-up and application possibilities towards South African energy grid. Sustainability, 14(21), 14268.

Shah, M. P. (2020). Microbial bioremediation & biodegradation. Springer.

Shah, M. P. (2021). Removal of emerging contaminants through microbial processes. Springer.

Shah, M. P. (2021). Removal of refractory pollutants from wastewater treatment plants. CRC Press.

Shin, H. J., Jung, K. A., Nam, C. W., & Park, J. M. (2017). A genetic approach for microbial electrosynthesis system as biocommodities production platform. Bioresource technology, 245, 1421–1429.

Strik, D. P., Hamelers, H. V. M., Snel, J. F., & Buisman, C. J. (2008). Green electricity production with living plants and bacteria in a fuel cell. International journal of energy research, 32(9), 870–876.

Surti, P., Kailasa, S. K., & Mungray, A. K. (2021). Genetic engineering strategies for performance enhancement of bioelectrochemical systems: a review. Sustainable energy technologies and assessments, 47, 101332.

Tanaka, K., Yokoe, S., Igarashi, K., Takashino, M., Ishikawa, M., Hori, K., & Kato, S. (2018). Extracellular electron transfer via outer membrane cytochromes in a methanotrophic bacterium Methylococcus capsulatus (Bath). Frontiers in microbiology, 9, 2905.

Tawalbeh, M., Al-Othman, A., Singh, K., Douba, I., Kabakebji, D., & Alkasrawi, M. (2020). Microbial desalination cells for water purification and power generation: a critical review. Energy, 209, 118493.

Thatikayala, D., & Min, B. (2021). Copper ferrite supported reduced graphene oxide as cathode materials to enhance microbial electrosynthesis of volatile fatty acids from CO_2. Science of the total environment, 768, 144477.

Tursun, H., Liu, R., Li, J., Abro, R., Wang, X., Gao, Y., & Li, Y. (2016). Carbon material optimized biocathode for improving microbial fuel cell performance. Frontiers in microbiology, 7, 6.

Wang, Y., Xu, A., Cui, T., Zhang, J., Yu, H., Han, W., & Wang, L. (2020). Construction and application of a 1-liter upflow-stacked microbial desalination cell. Chemosphere, 248, 126028.

Wei, J., Liang, P., & Huang, X. (2011). Recent progress in electrodes for microbial fuel cells. Bioresource technology, 102(20), 9335–9344.

Wilkinson, S. (2000). “Gastrobots”—benefits and challenges of microbial fuel cells in food-powered robot applications. Autonomous robots, 9(2), 99–111.

Xue, W., He, Y., Yumunthama, S., Udomkittayachai, N., Hu, Y., Tabucanon, A. S., & Kurniawan, T. A. (2021). Membrane cleaning and performance insight of osmotic microbial fuel cell. Chemosphere, 285, 131549.

Yaqoob, A. A., Mohamad Ibrahim, M. N., Rafatullah, M., Chua, Y. S., Ahmad, A., & Umar, K. (2020). Recent advances in anodes for microbial fuel cells: an overview. Materials, 13(9), 2078.

Yasri, N., Roberts, E. P., & Gunasekaran, S. (2019). The electrochemical perspective of bioelectrocatalytic activities in microbial electrolysis and microbial fuel cells. Energy reports, 5, 1116–1136.

Yazdi, A. A., D’Angelo, L., Omer, N., Windiasti, G., Lu, X., & Xu, J. (2016). Carbon nanotube modification of microbial fuel cell electrodes. Biosensors and bioelectronics, 85, 536–552.

Zhang, B., & He, Z. (2012). Integrated salinity reduction and water recovery in an osmotic microbial desalination cell. Rsc advances, 2(8), 3265–3269.

Zhang, Y., Mo, G., Li, X., Zhang, W., Zhang, J., Ye, J., & Yu, C. (2011). A graphene modified anode to improve the performance of microbial fuel cells. Journal of power sources, 196(13), 5402–5407.

Zhou, X., Chen, X., Li, H., Xiong, J., Li, X., & Li, W. (2016). Surface oxygen-rich titanium as anode for high performance microbial fuel cell. Electrochimica acta, 209, 582–590.

Index

Note: Page numbers in **bold** and *italics* refer to tables and figures, respectively.

1-Amino-2-naphthol (1A2N), 67

A

Abiotic cathode, 69
Abiotic environment, 46
AC, *see* Activated carbon
Acetate, 54
Acetic acid, 17
Acetoanaerobium, 72
Acetoclastic group, 17
Acetoclastic methanogens, 18
Acetogenic microorganisms, 17
Acidithiobacillus ferrooxidans, 229
Acidogenesis, 34
Acidogenic microorganisms, 17
Acid orange 7 (AO7), 65
    decolorization, 67
Acinetobacter calcoaceticus, 107
Activated carbon (AC), 101, 170, 249
    adsorption, 3, 113
    cathodes, 248
    electrochemical regeneration process, 113
    filtration, 170–171
Activated sludge, 6, 160
Activated sludge process (ASP), 170
Activation polarization, 143
Active resistance, 118
AD, *see* Anaerobic digestion
Adenosine triphosphate (ATP), 86, 106
Adsorbable organic halogens (AOX), 15
Adsorption, 1, 6
Advanced oxidation processes (AOPs), 100, 108, 127, 171, 181, **219**
Aeromonas hydrophila, 14
Agricultural by-products, 6
Air cathode, 201–202
Alfalfa leaf, **116**
Algal biocathode, 90–94
Algal biomass, 89
Algal culture, 92
Algal microbial fuel cells (AMFCs), 86
Algal photosynthesis, 87
AMI7001, 20
Aminophenols (AP), 78
Ammonium nitrogen, 15
Anabaena sp., 88
Anaerobic bio-electrochemical reactor, 31
Anaerobic digestion (AD), 15–16, 101, 119, 218
    acetogenic microorganisms, 17
    acidogenic microorganisms, 17
    dry, 16
    hydrolytic microorganisms, 17
    methanogenic microorganisms, 17–18
    microbiology of, 16–18
    of waste sludge, 101
    wet, 16
Anaerobic ponds, 110
Anaerobic sludge, 65, 71, 74
    blanket reactor, 31
    digestion, 113
Anaerobic system (AS), 76
Anhydrous ethanol, 72
Anion exchange membrane (AEM), 19, 84
Anion-exchange membranes, 20
Anode, 246–248
    chamber, 71
    material, 27–28
Anode-respiring bacteria (ARB), 22
Anodic electroactive biofilms, 75
Anodic oxidation (AO), 127
Anolyte, 59, 67, 74–75, 79, 92, 195, 197, 204–206, 210, 226–228, 252; *see also* Catholyte
Anoxygenic photoautotrophic bacteria, 86
Anthraquinone 2,6-disulfonate (AQDS), 244
Anthropogenic sources, 218
AOP, *see* Advanced oxidation processes
Aromatic compounds, 80
Aromatic polyamide membranes, 46
Arsenic (As), 15
Atomic exchange, 4
Azo bonds, 69
Azotobacter sp., 53

B

Bacillus circulans, 79
Bacteria-encapsulating bioanode, 229
Bacterial electrolysis systems (BESs), 104
    MBfR system, *66*
Bacterial metabolism, 105
BESs, *see* Bioelectrochemical systems
BETs, *see* Bioelectrochemical technologies (BETs)
Bioanode, 24, 27, 68–69, 79–80, 142, 180, 229, 248–249
Bioaugmentation, 52

Biocathodes, 68, 199
MDC, 202–204, 260
Biochar-based materials, **116**
Biochemical fuel cells, 232–233
Biocides, 15
Biocompatibility of stainless steel, 28
Biodegradation, 47, 50, 52
Bioelectricity generation, 79, 86, 88–89
Bioelectrochemical oxidation system, sewage treatment and energy recovery, 100–105
electrochemical technology, 108–109
consumption of energy, 109
renewable energy generation, 109–111
energy generation, 111
obstacles, 112
microbial fuel cells (MFCs), 105
mechanisms, 105–106
microorganisms interact with the electrode surface, 106–108
municipal wastewater treatment plants, 112–115
wastewater treatment using MFC with energy recovery, 115–118
Bioelectrochemical oxidation systems, 166–168, 219–221, 239–240
advanced oxidation processes, 171
microbial fuel cell, 172–173
ozone (O_3), 171
ultraviolet (UV) irradiation, 171–172
advantages and applications, 265
biosensing, 180–181
cell design, 251
dual chamber, 251–252
single chamber, 252
stack design, 252
construction and components, 246
anode, 246–248
cathode, 248–249
electrode, 246
energy generation using bioelectrochemical oxidation system, 222
energy recovery technologies from wastewater, 175
biofuels, 177–179
hydroelectricity's power generation capacity, 177
hydropower, 177
methane, 175–176
nitrogenous fuels, 179
thermal energy, 176–177
enzymatic fuel cells, 252–254
factors affecting, 223
electrode material, 224–226
pH, 223
substrate concentration, 224
temperature of anode and cathode chambers, 223–224
fertilizer recovery technologies, 179–180
functional communities, 168–169
history, 240–242
hydrogen and methane generation, 168
membranes or separators, 250–251
microbial desalination cell, 259–261
microbial electrolysis cell, 261–262
microbial electrosynthesis, 257
EET, 257–258
electrode materials, 258–259
microbial fuel cell, 254–255
anode materials, 255–256
cathode materials, 256
microbes used in MFC, 256–257
microbial solar cells, 262–263
MSC, construction of, 263–265
processes, **220**
reactor configuration
biochemical fuel cell, 232–233
dual-chamber MFC (DCMFC), 227–228
oxygen exposed bioelectrode MFC, 229
single-chamber MFC (SCMFC), 226
stacked MFCs (SMFCs), 229–230
tubular MFC, 230–232
upflow MFC, 228–229
substrate, 251
system, *221*
upscaling, 265
wastewater treatment, 173–174, 221–222
sludge incineration, 175
system, 174–175
wastewater treatment energy demand, 181
water reclamation and reuse technologies, 169
activated carbon filtration, 170–171
membrane filtration, 169–170
Bioelectrochemical systems (BESs), 47, 63, 83, 104, 129, 190, 218–220, *240*
advantages and application, **265–267**
principle of, 242–243
direct extracellular electron transfer, 243
indirect extracellular electron transfer, 243–244
inward electron transfer, 245
outward electron transfer, 244–245
recovery of water, carbon, and nitrogen, *178*
removal of hazardous pollutants, 18, 45, 47, 63
aromatic compounds, 76–80
dyes, 65–71
heavy metals, 73–76
pharmaceutical wastes, 71–73
for removal of various pollutants, **64–65**
Bioelectrochemical technologies (BETs), 102, 111
Bio-electrohydrolysis system, 32
Biofilms, 160
forming microbes, 159
Bio flocs, 160
Biofouling, 250

Biofuels, 177–179
Biogas, 16, 31, 110, 115
Bio-H_2 processes, 35
Biohydrogen, 63
Biohydrogen production, 63
Biological oxygen demand (BOD), 14, 79
Biological removal of heavy metals from wastewater, 6
Biomass of bamboo, **116**
Biomass productivity, 92
Bio-photovoltaic systems (BPSs), 86
Biopiling, 52
Biopolymers, 6
Biorefinery configurations, 218
Bioremediation, 46–47, 49–51
  capabilities, 55
  effectiveness of, 50
  strategies, 58
Bioremediators, 47
Biosensing, 180–181
Biosorbents, 1
Biosorption, 1, 6
Biosparging, 52
Biotic environment, 46
Biotransformation, 51, 53–56, 58, 73, 229
Bioventing, 52
Bipolar membranes, 20
Brackish water desalination, 196–197
Bulk drug manufacturing industries, 15

C

Cadmium (Cd), 15
Capacitive deionization (CDI), 190
Capacitive MDC (cMDC), 206, 260
Carbamazepine (CBZ) degradation, 71
Carbohydrates, 16
Carbon, 50
Carbon adsorption, 1, 46
Carbon-based graphite, 94
Carbon brush, 65
Carbon capture and biofuel generation, 89
Carbon dioxide (CO_2), 16, 54, 101
  consumption rate, 91
  fixation, 84
Carbon fiber (CF) electrodes, 27
Carbon fixation, 91
Carbon fuel cells, **116**
Carbonisation of algae, 89
Carbon nanotubes (CNTs), 27, 197
Carbon reduction, 101
Catalytic nitrate reduction, 141
Cathode, 248–249
Cathode electrode, 47
Cathode fabrication, 29
Cathode material, 28–29
Cathode-on-top single-chamber MEC, 21
Cathode-stimulated treatment, 45, 48
Cathode-supported treatment, 45
Cathodic biofouling, 94
Cathodic oxygen reduction, 19
Catholyte, 4, 30, 49, 67, 74–75, 78–79, 92, 142, 201–202, 206, 209; *see also* Anolyte
Cation exchange membrane (CEM), 5, 29, 65, 94, 140, 193–197, 250
CE, *see* Coulombic efficiency (CE)
Cellular metabolism, 135
Cellulose acetate, 46
Cellulose triacetate (CTA) membrane, 74
Cementation, 3
Chaetoceros, 90
Charge-mosaic membranes, 20
Charge transfer efficiency (CTE), 204
Chemical mediators, 8
Chemical oxygen demand (COD), 14, 68, 170, 218
  removal, 89
Chemical precipitation, 1, 3–4
Chlorella sp., 88–90
Chlorination, 172
Chlorine, 170
Chloroplast, *87*
Chromium (Cr), 15, 49
Chryseobacterium sp., 76
Clostridium butyricum, 231
Co^{2+} separation, 75
Coagulation, 3–4
Coagulation-flocculation mechanism, 3–4
Cobalt oceteneoate, 69
Coconut shell, **116**
Colors, 15
Comamonadaceae sp., 76
Complexation, 3
Complexation-UF, 5
Concentration polarization, 232
Congo red decolorization (CR-DE), 67–68
Constructed wetland (CW)-bioelectrochemical oxidation systems, 126–133
  catalytic nitrate reduction, 141
  concerning other water treatment technology, **130–132**
  CW-MFC SYSTEMS, 144–146
  direct extracellular electron transfer (DIET), 134
  mediated extracellular electron transfer (MEET), 134–136
    electron transport from extracellular space to cell, 134
  nitrate controlling factors, 141
    cathodic substance, 141–142
    current density, 142
    nitrate loading rate, 143
    pH, 142
    TOC oxidation via bioelectrochemistry, 143
  non-spontaneous reaction systems, 137

spontaneous reaction systems
bioelectrochemical technology, 141
electricity, 138–140
nitrate reduction through bioelectrochemistry, 141
total organic carbon removal, 143
electrode substance, 143
pH, 143–144
simultaneous nitrate and organics removal, 144
wastewater treatment, 136
methods and cutting-edge facilities, 136
Contaminant
degradability, 56
magnitude of, 56
Cooler temperatures, 33
Copper (Cu), 15
Corn straw, **116**
Corynebacterium vitaeruminis, 79
Coulombic efficiency (CE), 69, 75, 90, 136, 199, 201–203, 211, 228, 251, 265
Coupling MEC
acidogenic bioreactors, 32–33
MEC-AD coupled system, 31
MEC with anaerobic membrane bioreactor (MBR), 32–33
thermoelectric microconverter-MEC coupled system, 33
Cropping waste, 225
CW-MFC SYSTEMS, 144–146
Cyanobacteria, 84
Cyclic voltammetry (CV), 72
Cytochrome-to-electrode contact, 106

D

Dairy wastewater, 14, 115
Dan Region Reclamation Scheme, 158
Dark fermentation (DF), 14, 63, 218
Decolourization, 221
Degradation of hydrocarbons, 57
Depolarization, 232
Depolarization cell, 232
Desalination, 190, 239
Desulfovibrio, 76
Dicarboxylic/carboxylic acids, 73
DIET, *see* Direct extracellular electron transfer
Diffusion resistance, 118
Direct electron transfer (DET), 106–107
Direct extracellular electron transfer (DEET), 134, 243
Direct interspecies electron transfer (DIET), 168
Direct oxidation of pollutants, 108
Discharge requirements, 101
Distillation, 197
Distillery wastewater, 115
DNA reactivation, 172
DNA sequencing, 58
Domestic wastewater, 14, 115
Dry AD, 16
Dry digestion, 16
Dual chamber cell, 251–252
Dual-chambered MECs, 20, *20*
Dual-chamber MFC (DCMFC), *227,* 227–228
Dyes, 65–71
acid orange 7 (AO7), 65
Congo red decolorization (CR-DE), 67–68
decolorization, 67–71
industry effluent, 117
reactive black 5 (RB-5), 69–70
reactive orange 16 (RO-16), 69–70

E

Efficiency of current, 109
Electrical energy per order (EEO), 181
Electricigens, 256
Electricity, 263
Electricmembrane bioreactor (EMBR), 76
Electroactive bacteria (EAB), 24, 27, 63
Electroactive biofilms, 105
Electroactive microorganisms (EAMs), 168
Electrochemical advanced oxidation processes (EAOPs), 127
Electrochemical coagulation (EC), 100
Electrochemical flotation (EF), 100
Electrochemical Heyrovsky reaction, 22
Electrochemical impedance spectroscopy (EIS), 72, 248
Electrochemically active bacteria (MFC), 75
Electrochemical oxidation (EO), 126
Electrochemical regeneration, 101
Electrochemical Volmer reaction, 22
Electrocoagulation, 4
Electroconductive pili, 134
Electrode, 246
Electrode/anode surface microbial interactions, 106
Electro-deionization (EDI), 190
Electrode materials, 105, 224–226
Electrode–microbe electron transfer, 24, 28
Electrodeposition, 3–4
Electrode stabilization of colloids, 4
Electrodialysis (ED), 5, 189
Electrodistillation (ED), 197
Electro-fermentation of wastewaters, 33
Electrofloculation, 76
Electro flotation, 4
Electrogens, 83
Electrohydrogenesis, 115
Electrolysis, 4
Electrolyte pH, 29–30
Electrolytic oxidation of coagulants, 4
Electromethanogenesis, 24, 115

Electronic commutation (EC) technology, 113
Electron transfer mechanism of photosynthetic microbes in MFC, 86–88
Electron transport chain (ETC), 106
in algal photosynthesis, 87, *87*
Electron transport proteins, 106
Electro-oxidation (EO), 4, 100
Electrostatic repulsion technique, 3
Elcctrotrophs, 9, 168, 203, 245
Emerging pollutants (EPs), 100
Endogenous redox mediators/electron shuttles, 106
Energy generation using bioelectrochemical oxidation system, 222
Energy recovery technologies from wastewater
biofuels, 177–179
hydroelectricity's power generation capacity, 177
hydropower, 177
methane, 175–176
nitrogenous fuels, 179
thermal energy, 176–177
Energy-rich biogas, 110
Engineered treatment wetland (ETW) CW, 127
Enterobacter cloacae, 14
Environmental management, 46
Environmental pollutants, 45
Enzymatic fuel cells, 252–254, **265**
Erythromycin (ERY), 72
Escherichia coli, 52, 79, 135
Ethanol, 54
Exoelectrogenic bacteria, 94
Exoelectrogens (electrotrophs), 21, 168, 256
Exoenzymes, 17

F

Faecal coliform (FC), 155
Fed-batch cycle, 104
Feedstock, 25–27
Fe(III)-reducing bacteria (FRB), 168
Fenton, 100
Fermentation, 54, 218
metabolism, 168
products, 53
Fermentative bacteria, 17
Fertilizer recovery technologies, 179–180
Flame spray oxidation, 28
Flavin adenine dinucleotide (FADH2), 106
Flavin mononucleotide (FMN), 244
Flocculation mechanism, 3–4
Foam flotation, 3
Food-processing wastewater, 115
Forward osmosis (FO) membrane, 195, 260
Fossil fuels, 13
fast depletion of, 102
Fresnel focal points, 92
Functional communities, 168–169

G

Gas chromatography-mass spectroscopy (GC-MS), 68
GC-thermal conductivity detector (GC-TCD), 68
General secondary treatment (CAS), 114
Genetically modified organisms (GMOs), 53
Genetic and metabolic burden, 57
Geobacter, 76, 106
Geobacter strains, 243
Geobacter sulfurreducens, 106
Geothrix fermentans, 14
Gluconobacter oxydans, 14
Glutamate, 53
Gram-negative rods, 55
Gram-positive bacteria, 245
Granular activated carbon (GAC), 116
Granules, 160
Graphene-modified graphite paper cathode BES, 74
Graphite, 27
activated carbon particles, 76
electrodes, 27
paper (GP), 74
Green algae *Chlorella vulgaris*, 84
Greenhouse gas (GHG) emissions, 101–102
Green sulphur bacteria, *87*
Green technology, 108

H

H_2-mediated methanogenesis, 24
Half-cell coupled MDCs, 205–206
Hazardous pollutants, bioelectrochemical system-based removal of, 63
aromatic compounds, 76–80
dyes, 65–71
heavy metals, 73–76
pharmaceutical wastes, 71–73
for removal of various pollutants, **64–65**
Heavy metals, 1, 49, 73–76
removal from industrial effluents, 1–3
atomic exchange, 4
biological approaches, 6
coagulation, 3–4
electrodialysis (ED), 5
electronic chemical procedures, 4
evaluation of processes, 7
flocculation, 3–4
future perspectives, 9
membrane filter, application of, 1, 5
microbial fuel cell (MFC), 7–9
physico-chemical techniques, 3, 7
precipitation, 1, 3
Hexadecatrienoic acid, 90
Hexavalent chromium (Cr(VI)), 49, 74
High molecular weight (HMW) PAHs, 77

High-performance liquid chromatography (HPLC), 72
Household wastewater, 104
HRT, *see* Hydraulic retention times (HRT)
Human and environmental receptor proximity, 56
Human health, 155
Hybrid system topologies, 166
Hydraulic retention times (HRT), 25, 67, 91, 104, 170, 195, 228, 262
Hydroelectricity's power generation capacity, 177
Hydrogels, 6
Hydrogen, 54, 105
  evolution reaction (HER), 22, 27–28
  and methane generation, 168
Hydrogenotrophic group, 17–18
Hydrogenotrophic methanogens, 18, 28–29, 34
Hydrogen peroxide (H_2O_2), 18, 70, 201
  production, 34
Hydrogen production, 33–34
Hydrogen production rate (HPR), 21
Hydrogen recovery, 119
Hydrogen-scavenging bacteria, 28
Hydrogen synthesis from fossil fuels, 18
Hydrolases, 50
Hydrolyses, 17, 32
  fermentation, 34
Hydrolytic microorganisms, 17
Hydrophobic organic pollutants, 170
Hydropower, 177

I

India, water pollution in, 153–155
  BOD of wastewater, 154
  class-1 cities, 153–154, 157
  Ministry of Environment and Forests (MoEF), 155, 157
  Ministry of Water Resources (MOWR), 155
  sewerage infrastructure, 157
  wastewater treatment facilities, 157
  water withdrawals, 156
Indirect electron transfer (IET), 106
Indirect extracellular electron transfer, 243–244
Indium tin oxide (ITO), 264
Industrial wastewaters, 14, 219
  composition, 15
Inward electron transfer, 245
Ion exchange, 1, 3–4
  capacities, 5
  resins, 189
Ion exchange membranes (IEMs), 105
Iron (Fe), 15
Isomerases, 50
Israel's agriculture, 158

J

JSP Enviro, 117

K

Kenya, 158
Kishon project, 158
Kocuria rosea, 79
K. pneumoniae, 14

L

Landfill wastewater, 14
Lead (Pb), 15
Leptospirillum ferrooxidans, 229
Ligases, 50
Light, 90
Lignin-derived compounds, 15
Lipids, 16
Listeria monocytogenes, 244
Livestock, 14
Low-cost ion-exchange membrane, 94
Low metal concentrations, 3
Low molecular weight (LMW) PAHs, 77
Lyases, 50

M

Mariprofundus ferrooxydans, 245
MCCs, *see* Microbial carbon cells
MDCs, *see* Microbial desalination cells
MEC-anaerobic digestion, 14, 24
MECs, *see* Microbial electrolysis cells
Mediated extracellular electron transfer (MEET), 134
  electron transport from extracellular space to cell, 134
Megasphaera, 76
Membrane biofilm reactor (MBfR), 65
Membrane bio-reactors (MBRs), 14, 32, 126
  acidogenic bioreactors, 32–33
Membrane filtration, 1, 5
Membrane free-BESs (MFBESs), 78
Membrane-free microbial electrolysis cell (MFMEC), 67
Membrane fuel cells (MFCs), 102
  applications of, *173*
Membrane operations, 3
Mercury (Hg), 15
Metabolism
  cellular, 135
  fermentation, 168
  microbial, 166
  syntrophic, 134
Methane (CH_4), 16, 31, 101, 175–176
Methane production, 34

Methanogenesis, 17–18, 34
Methanogenic activity, 90
Methanogenic archaea, 17–18
Methanogenic inhibition, 90
Methanogenic microbial electrolysis cells (MMECs), 24
Methanogenic microorganisms, 17–18
Methanogens, 18
Methylene blue, 244
Methyl ethyl ketone peroxide catalyst, 69
Methyl viologen, 244
MFC-Fenton system, 72
MFCs, *see* Microbial fuel cells
Microalgae, 86, 89
  and BES integrations, 86
  effectiveness of, 84
Microbacter, 76
Microbes and wastewater treatment, 45–47, 50
  adaption of, 55
  bacteria, 55–57
  bioelectrochemical process, 47–48
    cathode, treatment supported by, 49
    cathode, treatment through, 48–49
  bioremediation, 49–50
    factors, 50–51
    radical processes, 57–58
  characteristics of, 54–55
  genomic period, ahead of, 58
  microorganisms, role of, 51–53
    degradation of contaminants, 53–54
Microbial capacitive desalination cell (MCDC), 197
Microbial carbon capture cell, 83–84
Microbial carbon cells (MCCs), 83–84
  application of, 88
    bioelectricity generation, 88–89
    carbon capture and biofuel generation, 89
    methanogenic inhibition, 90
    wastewater treatment with value-added products recovery, 89–90
  biomass productivity, **93**
  COD removal efficiency achieved, **93**
  electrical power generation, **93**
  factors affecting performance
    algal biocathode, 90–94
    design considerations, 92–94
    nitrate concentration, 91
    operation conditions, 91–92
  photosynthetic microbes in, 88
  up-scaling, bottlenecks, 94–95
  wastewater treatment at anode and CO_2 capture at cathode, *85*
  working principle, 84–86
    electron transfer mechanism of photosynthetic microbes in MFC, 86–88
    microalgae and BES integrations, 86
Microbial desalination cells (MDCs), 166, 189–192, 196, 239–240, 259–261, **267**
  configurations, 200
  designs, 194–195
  functions
    air cathode, 201–202
    basic configuration, 200–201
    biocathode MDC, 202–204
    brackish water desalination, 196–197
    capacitive MDC (cMDC), 206
    half-cell coupled MDCs, 205–206
    microbial desalination cell configurations, 200
    microorganisms, 198–199
    osmotic MDC, 206–208
    oxidants used in cathodes of, 199–200
    recirculation MDC (rMDC), 205
    seawater desalination, 195–196
    stacked MDC, 204–205
    water Softening, 197–198
  historical progress, 193
  technology, 193–194
    COD removal from wastewater, 209–211
    desalination efficiencies, 208–209
Microbial electrochemical desalination (MED), 206
Microbial electrochemical systems, 13–14
  advantages and disadvantages, 35
  anaerobic digestion (AD) process, 15–16
    acetogenic microorganisms, 17
    acidogenic microorganisms, 17
    hydrolytic microorganisms, 17
    methanogenic microorganisms, 17–18
    microbiology of, 16–18
  cathode-on-top single-chamber MEC, 21
  coupling MEC
    acidogenic bioreactors, 32–33
    MEC-AD coupled system, 31
    MEC with anaerobic membrane bioreactor (MBR), 32–33
    thermoelectric microconverter-MEC coupled system, 33
  dual-chambered MECs, 20
  fundamentals of, 21–22
  industrial wastewater composition, 15
  obtaining value-added products
    hydrogen peroxide production, 34
    hydrogen production, 33–34
    methane production, 34
  parameters affecting design, 24–25
    anode material, 27–28
    applied potentials, 30–31
    cathode material, 28–29
    electrolyte pH, 29–30
    feedstock, 25–27
    temperature, 30
  proton exchange membranes, 20–21

single-chambered MECs, 19
wastewater treatment
mechanism of, 23–24
methods and limitations, 23
working principles, 22
Microbial electrochemical technologies (METs), 18, *139*
Microbial electrolysis cells (MECs), 14, 19, 63, 103, 111, 166, 218–220, 261–262, **266**
cathode-on-top single-chamber, 21
dual-chambered, 20, *20*
efficiency in wastewater treatment, *103*
fundamentals, 21–22
integration of with technologies for energy production, **26**
proton exchange membranes, 20–21
single-chambered, 19
wastewater treatment, 22–24
Microbial electrolytic cells, 242
Microbial electrosynthesis, 257, **266**
EET, 257–258
electrode materials, 258–259
Microbial fuel cells (MFCs), 7–9, *8*, 14, 18, 47, 63, 83, 102, **116**, 126, 166, 172–173, 218–220, 239, 254–255, **266**
anode materials, 255–256
cathode materials, 256
efficiency in wastewater treatment, *103*
electrical conductivity, 224
microbes used in MFC, 256–257
stability and durability, 224
surface area and porosity, 224
Microbial metabolism, 166
Microbial reverse-electrodialysis electrolysis cell systems, 14
Microbial solar cells, 262–263, **266**
Microelectrolysis, 76
Microorganism-associated anaerobic technology, 63
Microorganism fuel cells (MFCs), 106
Microorganisms, 6, 45, 47
accessibility to, 50
nutritional adaptability, 47
Milli-Q water, 75
Ministry of Water Resources (MOWR) of India, 155
Misplaced resource, 104
Mixed liquid suspended solid (MLSS), 78
Mixotrophic nitrification, 49
m-nitrophenol (MNP), 78
Modified microbial fuel cells (MFCs), 190
Monitoring ability, 56
MPEM-MFC I *(P. aeruginosa),* 70
MPEM-MFC II *(P. fluorescens),* 70
Multiple-effect evaporation, 189
Municipal solid waste (MSW), 16
Municipal wastewater treatment plants, 112–115

N

N^4-acetyl-metabolites, 73
N^4-acetyl-sulfadiazine, 73
N^4-acetyl-sulfamethoxazole, 73
Nafion, 20–21
Nanofiber-reinforced composite proton exchange membrane (NFR–PEM), 21
Nanofiltration (NF), 5, 168, 189, 197
Nanowire-forming bacteria, 106
Nanowires, 243
Naphthalene (NAP)-phenanthrene (PHE), 77
National Water Carrier, Israel, 158
Natural red, 8
Nernst equation, 194
Nickel (Ni), 15, 29
Nicotinamide adenine dinucleotide (NADH), 106
Nicotinamide adenine dinucleotide phosphate (NADP+), 87
Nitrate controlling factors, 141
cathodic substance, 141–142
current density, 142
nitrate loading rate, 143
pH, 142
TOC oxidation via bioelectrochemistry, 143
Nitric oxide (NO), 141
Nitrification, 126
Nitrobenzene, 48, 77, *78*
Nitrogen (N)
availability, 91
concentration, 91
Nitrogenous fuels, 179
Nitrogenous wastes, 15
Non-biodegradable organic materials, 15
Non-spontaneous reaction systems, 137
Nutrient recovery, 89

O

O_2 reduction reaction (ORR), 86
Ohmic resistance, 118
Oil refineries, 15
O-nitrophenol (ONP), 78
Open circuit voltage (OCV), 74
Operational circumstances, 133
Organic loading rates (OLRs), 25, 91, 228
Organic matter removal, 84
Organic removal of heavy metals, 6
Osmotic MDC, 206–208, 260
Osmotic microbial fuel cell (OsMFC), 74, 208
Outer-membrane multiheme cytochromes (OMCs), 244
Outward electron transfer, 244–245
Oxidation, 161
Oxidation-reduction potential (ORP), 74
Oxidation-reduction processes, 106
Oxidoreductases, 50

Oxygen, 199
diffusion, 94
exposed bioelectrode MFC, 229
Oxygenic cyanobacteria, 86
Ozonation, 100, 108
Ozone (O_3), 171

P

p-aminophenol, 73
Paracetamol (PAM), 72
PEMs, *see* Proton exchange membranes (PEMs)
Perchlorate, 48
Perfluoroalkyl acids (PFAAs), 15
Perfluorooctane sulfonate (PFOS), 15
Perfluorooctanoic acid (PFOA), 15
Petrochemical wastes, 15
pH, 142–143, 223
Pharmaceuticals, 170
Pharmaceutical wastes, 71–73
Phasin Pha-P, 53
Phenol, 15
Phenol-formaldehyde resin, 15
Phenolic compounds, 15
Phosphate buffer, 75
Phosphate buffer solution (PBS), 67
Phosphorus wastes, 15
Photo-assisted fuel cells (PFCs), 84
Photo-bioreactor (PBR), 86
Photosynthetic bacteria-assisted MFC (PSB-MFC), 86
Photosynthetic biocathode, 84
Photosynthetic MDC, 260
Photosynthetic microbial desalination cells (PMDCs), 199–200
Photosynthetic microbial fuel cells (PMFCs), 84–86
Photosynthetic oxygen, 88
Photosynthetic PMDCs, 86
Photosystem I (PS-I), 86
Photosystem II (PS-II), 86
Phthalocyanine (FePc), 202
Physical separation, 3
Physico-chemical techniques, 3, 7
Pichia fermentans, 106
Pinewood, **116**
Plant and cyanobacteria, *87*
Plant-microbial fuel cells, 84
Plastic degradation, 57
Platinum, 29
p-nitrophenol (PNP), 78
Pollutants, 13–15, 19, 55
biodegradable refractory, 15
heavy metal, 15
in industrial wastewater, 15
organic, 15
organic and inorganic, 14
Pollution sources
domestic and agricultural waste contamination, 155–156
pollution from industry, 156–157
Polycyclic aromatic hydrocarbons (PAHs), 76–77
Polyhydroxyalkanoate, 53
Polymerization, 69
Polymer matrix of RO membranes, 5
Polymer-supported ultrafiltration (PSU), 5
Polysaccharide-based compounds, 6
Polytetrafluoroethylene (PTFE), 76
Porous proton exchange device, 105
Potassium ferricyanide, 8, 199, 209
Potentiostatic mode, 104
Precipitation, 1, 3
Precipitation by hydroxide formation, 4
Primary substrates, 53
Product cell, 232
Propionate, 54
Propionibacterium, 76
Propionic acid, 17
Proteins, 16
Proteobacter, 9
Proton exchange membranes (PEMs), 19–21, *22*, 69, 83
sulfonated poly(arylene ether sulfone) (SPAES)/polyimide nanofiber (PIN), 21
Pseudomonas, 9
Pseudomonas aeruginosa, 106–107
Pseudomonas putida S12, 53
Pseudomonas sp., 57
Pulp/paper wastewater, 115
Pyrolyzed iron (II), 202

R

Reactive black 5 (RB-5), 69–70
Reactive orange 16 (RO-16), 69–70
Recirculation MDC (rMDC), 205
Redox cell, 232
Redox potentials, 29
Redox shuttles, 107
Refinery wastewater, 14
Remediation, 1
Resazurin, 244
Resin acids, 15
Reverse osmosis (RO), 5, 46, 168, 189, 197
Rhizodeposition, 262
Rhodoferax ferrireducens, 14
Rhodopseudomonas palustris, 245

S

Saccharomyces cerevisiae, 7
Sacrificial electrodes, 108
Salt removal from wastewater, 15

Scanning electron micrographs of SRB Biocathode, *78*
Scanning electron microscope (SEM), 72
Scenedesmus, 91
Seawater desalination, 195–196
Sediment microbial fuel cell (SMFC), 8
Semi-dry AD, 16
Sewage, 115
Shewanella, 106, 244
Shewanella oneidensis, 14, 231
Shewanella putrefaciens, 14, 84
Shewanella species, *222*
Single chamber cell, 252
Single-chamber MFC (SCMFC), 19, *19*, 226, *226*
Slaughterhouse wastewater, 115
Sludge digestion, 119
Sludge generation, 4
Sludge incineration, 175
Sludge processing plants, 115
Sludge treatment, 111
Sluggish metal precipitation, 4
Sodium ion (Na^+) batteries, 89
Soil fertilizer, 16
Solar-powered electrolysis, 18
Solvent extraction, 3
Sorption process, 6
South America, 158
South Asian wastewater, pollution level, 155
Spiral microbial electrochemical cell, 138, 140
Spontaneous reaction systems
  bioelectrochemical technology, 141
  electricity, 138–140
  nitrate reduction through bioelectrochemistry, 141
Stack design cell, 252
Stacked MDC, 204–205
Stacked MFCs (SMFCs), 229–230, *230*
Stainless-steel brush cathodes, 28
Sterols, 15
Substrate concentration, 224
Sulfadiazine, 73
Sulfadimidine, 73
Sulfanilic acid (SA), 67
Sulfate-reducer enriched biocathode, 77
Sulfathiazole, 73
Sulfonated polyether ether ketone (SPEEK), 21
Suspended solids, 15
Sustainable wastewater treatment, 101
Symbiosis, 92
Syntrophic metabolism, 134

T

Tafel reaction, 22
Tannery wastewater, 115
Tannins, 15
Temperature, 30
Temperature of anode and cathode chambers, 223–224
Thermal energy, 176–177
Thermincola sp, 14
Thermoelectric generators (TEG), 75
Thermoelectricity, 33
Thermoelectric microconverter, 14
  MEC coupled system, 33
Thonin, 8
Three-dimensional graphite foam (3DGF), 74
Tirupur (India), 117; *see also* India, water pollution in
Titanium wire, 27
Total chemical oxygen demand (TCOD) removal, 228
Total dissolved solids (TDS), 196
Total organic carbon (TOC), 170
Total organic carbon removal, 143
  electrode substance, 143
  pH, 143–144
  simultaneous nitrate and organics removal, 144
Toxicants, 50
Toxic contaminants, 15, 17, 27, 218
Toxic heavy metals, permissible limits on human health, **2**
Trace element solution, 75
Transferases, 50
Trickle filtration/air contact oxidation bed (TF-ACOB), 76
Triclosan (TCS), 70–71
  adsorption, 71
Tubular air-cathode microbial fuel cell (TMFC) bioreactor, 231
Tubular air-cathode system, 140
Tubular bio-contact reactor, 78
Tubular MDC, 194
Tubular MFCs, 138, 140, 230–232, *231*

U

Ultrafiltration (UF), 5
Ultraviolet (UV) irradiation, 171–172
Unsaturated polyester, 69
Upflow anaerobic microbial fuel cell (UAMFC), 228
Upflow anaerobic sludge beds (UASBs), 110
Upflow MDC, 260
Upflow MFC, 228–229
Upflow tubular microbial desalination cell (UMDC), 194
Urbanization and population growth, 159
Urban wastewater, 25
Urea, 15

V

Volatile fatty acid, 34

W

Waste-to-product conversion, 14
Wastewaters, 1, 13; *see also specific entries*
composition, 14
electro-fermentation of, 33
irrigation, 152, 158–159
irrigation in India, *154*
management, 159
nutrients in, 154
organic matter removal, 84, *85*
pollutants in, 218
production and treatment, 157
salt removal from, 15
used for treatment and the energy production, **225**
Wastewater treatment, 136, 173–174, 219; *see also specific entries*
at anode and CO_2 capture at the cathode of MCC, *85*
bio-electrochemical, 160–162
bio-electrochemical oxidation systems for, 221–222
energy demand, 181
mechanism of, 23–24
methods and cutting-edge facilities, 136
methods and limitations, 23
microbial fuel cell efficiency in, *103*
sludge incineration, 175
system, 174–175
with value-added products recovery, 89–90
working principles, 22
Wastewater treatment plants (WWTPs), 100
Water pollution in India, 153–154
Water quality, 101, 109
Water softening, 197–198
Water stress index (WSI), 159
Wet AD, 16
Wet digestion, 16
Wheat straw, **116**
Wolfe's vitamins, 75
World Water Assessment Programme, 189

X

Xenobiotics, 55

Z

Zeta potential measurement, 3
Zinc (Zn), 15

Printed in the United States
by Baker & Taylor Publisher Services